AF594968

Metal Machining—Recent Advances, Applications and Challenges

Metal Machining—Recent Advances, Applications and Challenges

Editor

Francisco J. G. Silva

MDPI • Basel • Beijing • Wuhan • Barcelona • Belgrade • Manchester • Tokyo • Cluj • Tianjin

Editor
Francisco J. G. Silva
ISEP—School of Engineering
Portugal

Editorial Office
MDPI
St. Alban-Anlage 66
4052 Basel, Switzerland

This is a reprint of articles from the Special Issue published online in the open access journal *Metals* (ISSN 2075-4701) (available at: https://www.mdpi.com/journal/metals/special_issues/metal_machining_advances_applications_challenges).

For citation purposes, cite each article independently as indicated on the article page online and as indicated below:

LastName, A.A.; LastName, B.B.; LastName, C.C. Article Title. *Journal Name* **Year**, *Volume Number*, Page Range.

ISBN 978-3-0365-2494-8 (Hbk)
ISBN 978-3-0365-2495-5 (PDF)

Contents

About the Editor

Francisco J. G. Silva is Associate Professor with Habilitation and has been the Director of the Master's Degree in Mechanical Engineering since November 2014. He is now member of the Technical and Scientific Council of ISEP (2020–2022), and a member of the Pedagogic Council of ISEP (2020–2022). He was Subdirector of the Mechanical Engineering Department from 2014 to 2016. He was Director of the Bachelor's in Mechanical Engineering from 2003 to 2006 at ESEIG, Vila do Conde, Portugal. He was a member of the General Council at the IPP—Polytechnic Institute of Porto— a member of the Scientific Council at ESEIG and member of the Pedagogic Council at ESEIG. He is author of the book "*Tecnologia da Soldadura - Uma Abordagem Técnico-Didáctica*", Publindústria, Porto (2016), coeditor of the book "*Lean Manufacturing - Implementation, Opportunities and Challenges*", Nova Science, NY, U.S.A. (2019), and coauthor of the book "*Cleaner Production—Toward a better future*", Springer Nature, Switzerland, 2020. He has supervised more than 120 MSc dissertations and cosupervised two PhD theses. He is coauthor of more than 150 scientific papers in journals such as *Wear, Surface and Coatings Technology, Vacuum, Thin Solid Films, Coatings, Metals, Materials, Composites Part B - Engineering, Robotics and Computer-Integrated Manufacturing, International Journal of Advanced Manufacturing Technology, Journal of the Brazilian Society of Mechanical Sciences and Engineering, Procedia Manufacturing*, etc. He is an Editorial Board Member of the scientific journals Coatings (MDPI), *Solids* (MDPI), *Encyclopedia* (MDPI) and *Machines* (MDPI). He has conducted more than 10 Special Issues in MDPI scientific journals. He founded and was Editor-in-Chief of the journal *Coating Science Technology*, and Co-Editor-in-Chief of the journal *Research Updates in Polymer Science* (LifeScience Global). He frequently conducts reviews for journal such as: *Journal of Cleaner Production* (Elsevier), *Robotics* (Elsevier), *Composites Part A* (Elsevier), *Metals* (MDPI), *Coatings* (MDPI), *Materials* (MDPI), *MicroMachines* (MDPI), *Journal of Materials: Design and Applications* (SAGE), *Production and Planning Control* (Taylor and Francis), *Theoretical and Applied Fracture Mechanics* (Elsevier), *Journal of the Brazilian Society of Mechanical Sciences and Engineering* (Springer), *Surface and Coatings Technology* (Elsevier), *Wear* (Elsevier), etc. He awarded the MDPI Top Reviewer Award in 2018, and the Metals Top Reviewer in 2019. One of his papers was awarded with the Best Paper Award in Coatings (MDPI, 2019).

Editorial

Metal Machining—Recent Advances, Applications, and Challenges

Francisco J. G. Silva

ISEP–School of Engineering, Polytechnic Institute of Porto, 4249-015 Porto, Portugal; fgs@isep.ipp.pt; Tel.: +351-228340500

Citation: Silva, F.J.G. Metal Machining—Recent Advances, Applications, and Challenges. *Metals* **2021**, *11*, 580. https://doi.org/10.3390/met11040580

Received: 29 March 2021
Accepted: 31 March 2021
Published: 2 April 2021

1. Introduction and Scope

Though new manufacturing processes that revolutionize the landscape regarding the rapid manufacture of parts have recently emerged, the machining process remains alive and up-to-date in this context, always presenting itself as a manufacturing process with several variants and allowing for high dimensional accuracy and high levels of surface finish [1–3]. Indeed, machining has numerous aspects that constantly need to be investigated due to the constant evolution of materials to be machined, the materials and geometry of tools, and the evolution of coatings normally applied to the tools' surfaces [4–6]. In view of this evolution, the parameters used in machining also need to be optimized, thus contributing to increased attention by researchers in this area of manufacturing [7–10]. In fact, metal alloys have significantly evolved in terms of properties, which poses additional challenges for research [11–13]. The market's demand for new alloys that need to meet increasingly demanding requirements is a constant, thus creating a greater diversity of alloys in the market and new challenges in their processing in order to achieve the characteristics required by customers. When the requirements are truly challenging, it becomes necessary to make polymeric or metallic matrix composite materials, creating even more demanding challenges in their processing that have further expanded the research field in the machining area [14–16]. Composite materials still have a huge margin of progression in terms of research, which will also allow the scientific community linked to the manufacturing processes to continue to have a lot of available topics to explore. For example, the chip that can be formed during the machining processes has been the subject of several studies because chip formation provides valuable and useful information about the way the machining process is being conducted and can provide information on the problems related to its removal from equipment and occupied space [17–19]. Dry machining has always been a great challenge [20] because lubrication causes environmental problems [21] and, in some cases, is not even allowed. Thus, aspects related to lubrication in machining have also been widely explored by using techniques that seek to minimize the use of lubrication (minimum quantity lubrication) [22,23]. On the other hand, the need to increase productivity levels has not only resorted to the science of materials and technological processes to offer the industry the necessary means to produce with the necessary quality at increasingly competitive costs but also captured the attention of the industrial engineering field [24,25]. This has led to numerous research projects aimed at the development of models and procedures that allow for the optimization of all operations involving machining processes, as well as some tools used in the process itself, such as more advanced jigs [26].

Recently, new research opportunities have opened up because machining operations are largely linked to the concepts of Industry 4.0. In fact, the operations traditionally developed between equipment can be integrated by using computer systems with greater decision power, making the whole production process much more agile [27,28]. The concepts of Industry 4.0 have also made it possible to develop other areas around machining, namely the concept of "machine learning," which allows for the creation of standard figures that can be recognized by equipment and thus allowing for greater integration between CAD (Computer-Aided Design) and CAM (Computer-Aided Manufacturing) [29,30]. Similarly, the measurement operations of tools and machined parts bring new challenges that are also being scientifically explored [31]. The vibration index and cutting forces developed in tools, which naturally evolve with wear, can also be properly monitored to bring added information to the process and allow for the automation of tool change decisions or maintenance interventions for equipment [32,33].

Bearing in mind all the previously mentioned factors, it is easy to realize that there is innumerable research to be continuously developed in this field of investigation. Thus, this Special Issue intended to gather contributions from different authors in the field of machining, allowing for its easy dissemination and thus contributing to an increase of knowledge of the scientific community that works hard in this area.

2. Contributions

The contributions received for this Special Issue are high-quality and show how active the research around machining processes is. Three of the studies contained in this Special Issue are related to the chip formation and cutting behavior that are registered during the machining process. In the work "Assessment of Chip Breakability during Turning of Stainless Steels Based on Weight Distributions of Chips" developed by Du et al. [34], the breakability of the stainless-steel chip is studied in the turning of these alloys by using a new methodology: the weight distribution of chips. This methodology was shown to present very consistent results in the evaluation of the way a part is trimmed, thus allowing one to perceive the machinability of a given alloy and allowing for a comparison with similar ones. The study was developed on an AISI 316L alloy, using one without treatment and another with treatment and showing that the treatment drastically modified the breakability of the chip. Even though the chip looked very similar for both cases, the developed method showed that the obtained results were significantly different, showing how this methodology can be useful in other analyses. On the other hand, the work entitled "Predicting Continuous Chip to Segmented Chip Transition in Orthogonal Cutting of C45E Steel through Damage Modeling" performed by Devotta et al. [35] integrated dynamic strain aging in the Johnson–Cook model, which is usually used to modelling machining processes, while also using the Voyiadjis–Abed–Rusinek approach. In this way, it became possible to predict the transition from a continuous chip to a discontinuous chip regarding the widely used C45E steel, depending on the rake angle and feed rate while keeping the cutting speed constant. The main outcome of the study was to discover that chip segmentation intensity and frequency are sensitive to fracture initiation strain models. Additionally, using the finite element method, but now based on an AA2024 T351 aluminum alloy, Muhammad Asad [36] studied the influence of the tool's geometry, namely hone and chamfer, on chip segmentation and burr formation. The study demonstrated an increasing trend in the degree of chip segmentation and end burr as hone edge tool radius or chamfer tool geometry macro parameters concerning chamfer length and angle increased. With the development of this work, a model that helps in the definition of the best tool geometry and the optimization of the cutting parameters was obtained, with the aim to increase productivity, minimize the formation of burr, and avoid the formation of a continuous chip. The quality of a machined surface is also present in this Special Issue. In order to minimize the problems reported in the quality of finishing of aluminum parts, Rubio-Mateos et al. [37] studied the introduction of elastomeric systems to support parts to be machined, with a view to dampening any vibrations during the finishing process of soft materials. For this, nitrile butadiene rubber (NBR) was used. A suitable flexible vacuum fixture was also developed, allowing for the easy implementation of the system in the manufacturing process. Different sets of parameters that varied the degree of compression imposed on the flexible system were tested, verifying that it perfectly accommodated these variations. Thus, the main outcome of this work was the establishment that the milling operations of the AA2024 alloy can benefit from more flexible fixations to the detriment of very rigid jigs. Del Sol and Rivero [38] also investigated the parameters that could give rise to skin panel and thin plate components obtained by machining, thus eliminating the need to use chemical milling in the manufacture of parts for the aeronautical industry because the rigidity presented for this type of parts is quite low. The study was essentially conducted by using the experimental pathway, measuring the cutting forces that developed during the process and keeping the surface roughness within the imposed limits. The correct selection of the cutting parameters led to a 40% reduction in the thickness variation of the components and a 20% decrease in the cutting forces, which makes the clamping process of parts easier. The study also resulted in the creation of a model capable of monitoring the quality of the process based on the measurement of equipment power consumption. The work of Berzosa et al. [39], entitled "Feasibility Study of Hole Repair and Maintenance Operations by Dry Drilling of Magnesium Alloy UNS M11917 for Aeronautical Components," also investigated the best set of parameters to be used for drilling holes in magnesium alloys, which are increasingly used in the aeronautical, aerospace, and even automotive industries. The study of the parameters allowed for an improvement of the surface quality obtained in holes made in that alloy, mainly in repair or maintenance operations. Additionally, based on the aeronautical industry, Martín Béjar et al. [40] investigated the macro-geometric deviations reported in the turning of a UNS A97075 alloy, verifying that the parts provided with a lower stiffness presented a greater sensitivity to macro dimensional deviations when adjusting parameters. It was once again verified that feed speed is the parameter with the greatest influence on the deviations recorded during the turning process. Based on the obtained results, models that allow for the prediction of macro dimensional deviations as a function of machining parameters were presented. Bañon et al. [41] carried out a quality study of the cut surface in structures composed of different materials. In that case, the abrasive waterjet cutting of a mixed structure of CFRTP (carbon fiber-reinforced thermo-plastic) with steel was studied, which presented quite different behaviors under the same cutting conditions. Two different stacking configurations were studied to investigate different sets of parameters that would lead to lower levels of roughness in waterjet cutting when using abrasives. The experimental work made it possible to draw several diagrams that enabled the correlation of the cutting parameters with the cut surfaces' quality. The sustainability related to the machining processes is also represented in this Special Issue. Indeed, sustainability can be explored in its most diverse aspects because productivity is fundamental but environmental impact—with important factors such as power consumption, the minimum use of lubricants/coolants, and social issues in which health conditions at work and ergonomics must be respected—cannot be ignored. Iqbal et al. [42] investigated the use of cryogenic coolants in the machining of the Ti6Al4V alloy, which is widely used in aeronautics. At the same time, they tried to optimize the parameters with a view to minimizing the consumption of tools by acting on the parameters of the milling process. As main outcomes, it was found that micro-lubrication was more effective than cryogenic cooling with CO_2 or liquid nitrogen; it could increase tool life while also improving the surface quality of machined parts, reducing energy consumption, and reducing the overall cost of process. These authors also verified that the high levels of cutter's helix angle and cutting speed clearly contributed to an increase in process sustainability. Diaz-Álvarez et al. [43] also investigated new cutting parameters in the turning of the Haynes 282 nickel alloy while avoiding the use of lubricants/coolants. The used coated tools allowed them to optimize the cutting parameters, making it possible to obtain roughness values in the machined parts as low as those obtained using lubricants/coolants. The proper selection of parameters also kept the cutting forces as low as those obtained with lubrication, as well as extending the tools' life. In this way, the process can become more environmentally sustainable without jeopardizing product quality or economic sustainability.

In addition to the aforementioned works, this Special Issue also presents two wide-revision works [5,8], one on the use of coated tools in turning and another that performs an in-depth study of the literature on TiAlN-based coatings for both the turning and milling processes, focusing on coatings developed around that same coating, providing information on recent uses of these coatings and what elements are used in the fabrication of these types of coatings, showing their mechanical properties, and providing information on their machining performance and application. Each of the reviews is based on more than one hundred references, thus allowing for the deepening and discussing of innumerable ideas taken from a wide range of works. These works constitute a great base of work for MSc and PhD students who are starting in the area by providing (in a concentrated way) a wide range of knowledge in these areas, from the cutting performance of various coated tools in machining processes to the study of the different wear patterns and mechanisms that these tools suffer during the machining process.

3. Conclusions and Outlook

Through the research collected in this Special Issue, it can be noted that there is much work regarding the machining process, in its most diverse aspects, to be continuously carried out because there is still a huge margin of progression in almost all aspects of machining. In this Special Issue, some excellent examples of the most recent developments in this area are shown, with special emphasis given to optimizing parameters, increasing the quality of machined surfaces, and improving the sustainability of the process. Very important information is also provided regarding tool coatings, with a view to extending cutting tools' working life, which will certainly be useful for those who conduct research in this area or for young students who want to start their studies in this field of knowledge.

The continuous search for greater productivity in machining and for increases of tools' working life (always based on the improvement of the global sustainability) will lead to more and more research in this area that will continue to be collected and disseminated, thus allowing this process to continue to be competitive and capable of producing high-quality parts. Thus, the research around machining processes will surely remain challenging.

Conflicts of Interest: The author declares no conflict of interests.

References

1. Rathod, V.; Doloi, B.; Bhattacharyya, B. Experimental investigations into machining accuracy and surface roughness of microgrooves fabricated by electrochemical micromachining. *Proc. Inst. Mech. Eng. Part B J. Eng. Manuf.* **2014**, *29*, 1781–1812. [CrossRef]
2. Kwon, S.B.; Nagaraj, A.; Yoon, H.S.; Min, S. Study of material removal behavior on R-plane of sapphire during ultra-precision machining based on modified slip-fracture model. *Nanotechnol. Precis. Eng.* **2020**, *3*, 141–155. [CrossRef]
3. Wang, J.; Fu, J.; Wang, J.; Du, F.; Liew, P.J.; Shimada, K. Processing capabilities of micro ultrasonic machining for hard and brittle materials: SPH analysis and experimental verification. *Precis. Eng.* **2020**, *63*, 159–169. [CrossRef]
4. Martinho, R.P.; Silva, F.J.; Baptista, A.P.M. Cutting forces and wear analysis of Si_3N_4 diamond coated tools in high speed machining. *Vacuum* **2008**, *82*, 1415–1420. [CrossRef]
5. Sousa, V.F.; Silva, F.J.G.; Alexandre, R.; Fecheira, J.S.; Silva, F.P.N. Study of the wear behaviour of TiAlSiN and TiAlN PVD coated tools on milling operations of pre-hardened tool steel. *Wear* **2021**, 203695. [CrossRef]
6. Kumar, R.; Singh, S.; Bilga, P.S.; Jatin, K.; Singh, J.; Singh, S.; Scutaru, M.-L.; Pruncu, C.I. Revealing the benefits of entropy weights method for multi-objective optimization in machining operations: A critical review. *J. Mater. Res. Technol.* **2021**, *10*, 1471–1492. [CrossRef]
7. Sousa, V.F.C.; Silva, F.J.G. Recent Advances on Coated Milling Tool Technology—A Comprehensive Review. *Metals* **2020**, *10*, 235. [CrossRef]
8. Sousa, V.F.C.; Silva, F.J.G. Recent Advances in Turning Processes Using Coated Tools—A Comprehensive Review. *Metals* **2020**, *10*, 170. [CrossRef]
9. Sousa, V.F.C.; Da Silva, F.J.G.; Pinto, G.F.; Baptista, A.; Alexandre, R. Characteristics and Wear Mechanisms of TiAlN-Based Coatings for Machining Applications: A Comprehensive Review. *Metals* **2021**, *11*, 260. [CrossRef]
10. Pangestu, P.; Pujiyanto, E.; Rosyidi, C.N. Multi-objective cutting parameter optimization model of multi-pass turning in CNC machines for sustainable manufacturing. *Heliyon* **2021**, *7*, e06043. [CrossRef]
11. Tanaka, H.; Kitamura, M.; Sai, T. An Evaluation of Cutting Edge and Machinability of Inclined Planetary Motion Milling for Difficult-to-cut Materials. *Procedia CIRP* **2015**, *35*, 96–100. [CrossRef]
12. Martinho, R.P.; Silva, F.J.G.; Martins, C.; Lopes, H. Comparative study of PVD and CVD cutting tools performance in milling of duplex stainless steel. *Int. J. Adv. Manuf. Technol.* **2019**, *102*, 2423–2439. [CrossRef]

13. Gouveia, R.M.; Silva, F.J.G.; Reis, P.; Baptista, A.P.M. Machining Duplex Stainless Steel: Comparative Study Regarding End Mill Coated Tools. *Coatings* **2016**, *6*, 51. [CrossRef]
14. Sun, W.; Duan, C.Z.; Yin, W.D. Chip formation mechanism in machining of Al/SiCp composites based on analysis of particle damage. *J. Manuf. Processes* **2021**, *64*, 861–877. [CrossRef]
15. Chen, Z.; Zhou, H.; Yan, Z.; Han, F.; Yan, H. Machining characteristics of 65 vol.% SiCp/Al composite in micro-WEDM. *Ceram. Int.* **2021**. in Press. [CrossRef]
16. Khan, M.A.; Jani, S.P.; Kumar, A.S.; Rajesh, S. Machining parameter optimization using Adam–Gene Algorithm while turning lightweight composite using ceramic cutting tools. *Int. J. Lightweight Mater. Manuf.* **2021**, *4*, 262–267. [CrossRef]
17. Chen, X.; Tang, J.; Ding, H.; Liu, A. A new geometric model of serrated chip formation in high-speed machining. *J. Manuf. Processes* **2021**, *62*, 632–645. [CrossRef]
18. Silva, F.J.G.; Martinho, R.P.; Martins, C.; Lopes, H.; Gouveia, R.M. Machining GX2CrNiMoN26-7-4 DSS Alloy: Wear Analysis of TiAlN and TiCN/Al_2O_3/TiN Coated Carbide Tools Behavior in Rough End Milling Operations. *Coatings* **2019**, *9*, 392. [CrossRef]
19. Arefin, S.; Zhang, X.; Kumar, A.S.; Neo, D.W.K.; Wang, Y. Study of chip formation mechanism in one-dimensional vibration-assisted machining. *J. Mater. Process. Technol.* **2021**, *291*, 117022. [CrossRef]
20. Cavaleiro, D.; Figueiredo, D.; Moura, C.W.; Cavaleiro, A.; Carvalho, S.; Fernandes, F. Machining performance of TiSiN(Ag) coated tools during dry turning of TiAl6V4 aerospace alloy. *Ceram. Int.* **2021**, *47*, 11799–11806. [CrossRef]
21. Silva, F.J.G.; Gouveia, R.M. *Cleaner Production—Toward a Better Future*; Springer Nature: Cham, Switzerland, 2020; ISBN 978-3-030-23164-4.
22. Hamran, N.N.N.; Ghani, J.A.; Ramli, R.; Haron, C.H.C. A review on recent development of minimum quantity lubrication for sustainable machining. *J. Clean. Prod.* **2020**, *268*, 122165. [CrossRef]
23. García-Martínez, E.; Miguel, V.; Manjabacas, M.C.; Martínez-Martínez, A.; Naranjo, J.A. Low initial lubrication procedure in the machining of copper-nickel 70/30 ASTM B122 alloy. *J. Manuf. Processes* **2021**, *62*, 623–631. [CrossRef]
24. Monteiro, C.; Ferreira, L.P.; Fernandes, N.O.; Sá, J.C.; Ribeiro, M.T.; Silva, F.J.G. Improving the machining process of the metalworking industry using the lean tool SMED. *Procedia Manuf.* **2019**, *41*, 555–562. [CrossRef]
25. Martins, M.; Godina, R.; Pimentel, C.; Silva, F.J.G.; Matias, J.C. A Practical Study of the Application of SMED to Electron-beam Machining in Automotive Industry. *Procedia Manuf.* **2018**, *17*, 647–654. [CrossRef]
26. Kumar, S.; Campilho, R.D.S.G.; Silva, F.J.G. Rethinking modular jigs' design regarding the optimization of machining times. *Procedia Manuf.* **2019**, *38*, 876–883. [CrossRef]
27. Mehdi, N.; Starly, B. Witness Box Protocol: Automatic machine identification and authentication in industry 4.0. *Comput. Ind.* **2020**, *123*, 103340. [CrossRef]
28. Barbosa, M.; Silva, F.J.G.; Pimentel, C.; Gouveia, R.M. A Novel Concept of CNC Machining Center Automatic Feeder. *Procedia Manuf.* **2018**, *17*, 95–959. [CrossRef]
29. Barton, D.; Gönnheimer, P.; Schade, F.; Ehrmann, C.; Becker, J.; Fleischer, J. Modular smart controller for Industry 4.0 functions in machine tools. *Procedia CIRP* **2019**, *81*, 1331–1336. [CrossRef]
30. Ulas, M.; Aydur, O.; Gurgenc, T.; Ozel, C. Surface roughness prediction of machined aluminum alloy with wire electrical discharge machining by different machine learning algorithms. *J. Mater. Res. Technol.* **2020**, *9*, 12512–12524. [CrossRef]
31. Alonso, V.; Dacal-Nieto, A.; Barreto, L.; Amaral, A.; Rivero, E. Industry 4.0 implications in machine vision metrology: An overview. *Procedia Manuf.* **2019**, *41*, 359–366. [CrossRef]
32. Sousa, V.F.C.; Silva, F.J.G.; Fecheira, J.S.; Lopes, H.M.; Martinho, R.P.; Casais, R.B.; Ferreira, L.P. Cutting Forces Assessment in CNC Machining Processes: A Critical Review. *Sensors* **2020**, *20*, 4536. [CrossRef]
33. Costa, S.; Silva, F.J.G.; Campilho, R.D.S.G.; Pereira, T. Guidelines for Machine Tool Sensing and Smart Manufacturing Integration. *Procedia Manuf.* **2020**, *51*, 251–257. [CrossRef]
34. Du, H.; Karasev, A.; Björk, T.; Lövquist, S.; Jönsson, P.G. Assessment of Chip Breakability during Turning of Stainless Steels Based on Weight Distributions of Chips. *Metals* **2020**, *10*, 675. [CrossRef]
35. Moris Devotta, A.; Sivaprasad, P.V.; Beno, T.; Eynian, M. Predicting Continuous Chip to Segmented Chip Transition in Orthogonal Cutting of C45E Steel through Damage Modeling. *Metals* **2020**, *10*, 519. [CrossRef]
36. Asad, M. Effects of Tool Edge Geometry on Chip Segmentation and Exit Burr: A Finite Element Approach. *Metals* **2019**, *9*, 1234. [CrossRef]
37. Rubio-Mateos, A.; Rivero, A.; Ukar, E.; Lamikiz, A. Influence of Elastomer Layers in the Quality of Aluminum Parts on Finishing Operations. *Metals* **2020**, *10*, 289. [CrossRef]
38. Del Sol, I.; Rivero, A.; Gamez, A.J. Effects of Machining Parameters on the Quality in Machining of Aluminium Alloys Thin Plates. *Metals* **2019**, *9*, 927. [CrossRef]
39. Berzosa, F.; de Agustina, B.; Rubio, E.M.; Davim, J.P. Feasibility Study of Hole Repair and Maintenance Operations by Dry Drilling of Magnesium Alloy UNS M11917 for Aeronautical Components. *Metals* **2019**, *9*, 740. [CrossRef]
40. Martín Béjar, S.; Trujillo Vilches, F.J.; Bermudo Gamboa, C.; Sevilla Hurtado, L. Parametric Analysis of Macro-Geometrical Deviations in Dry Turning of UNS A97075 (Al-Zn) Alloy. *Metals* **2019**, *9*, 1141. [CrossRef]
41. Bañon, F.; Simonet, B.; Sambruno, A.; Batista, M.; Salguero, J. On the Surface Quality of CFRTP/Steel Hybrid Structures Machined by AWJM. *Metals* **2020**, *10*, 983. [CrossRef]

42. Iqbal, A.; Suhaimi, H.; Zhao, W.; Jamil, M.; Nauman, M.M.; He, N.; Zaini, J. Sustainable Milling of Ti-6Al-4V: Investigating the Effects of Milling Orientation, Cutter's Helix Angle, and Type of Cryogenic Coolant. *Metals* **2020**, *10*, 258. [CrossRef]
43. Díaz-Álvarez, A.; Díaz-Álvarez, J.; Cantero, J.L.; Miguélez, H. Sustainable High-Speed Finishing Turning of Haynes 282 Using Carbide Tools in Dry Conditions. *Metals* **2019**, *9*, 989. [CrossRef]

 metals

MDPI

Review

Characteristics and Wear Mechanisms of TiAlN-Based Coatings for Machining Applications: A Comprehensive Review

Vitor F. C. Sousa [1], Francisco José Gomes Da Silva [1,2,*], Gustavo Filipe Pinto [1,2], Andresa Baptista [1,2] and Ricardo Alexandre [3]

1 ISEP—School of Engineering, Polytechnic of Porto, 4200-072 Porto, Portugal; vcris@isep.ipp.pt (V.F.C.S.); gflp@isep.ipp.pt (G.F.P.); absa@isep.ipp.pt (A.B.)
2 INEGI—Instituto de Ciência e Inovação em Engenharia Mecânica e Engenharia Industrial, 4200-465 Porto, Portugal
3 TeandM—Tecnologia, Engenharia e Materiais, S.A., 3045-508 Taveiro, Portugal; ricardo@teandm.pt
* Correspondence: fgs@isep.ipp.pt; Tel.: +351-228340500

Abstract: The machining process is still a very relevant process in today's industry, being used to produce high quality parts for multiple industry sectors. The machining processes are heavily researched, with the focus on the improvement of these processes. One of these process improvements was the creation and implementation of tool coatings in various machining operations. These coatings improved overall process productivity and tool-life, with new coatings being developed for various machining applications. TiAlN coatings are still very present in today's industry, being used due to its incredible wear behavior at high machining speeds, high mechanical properties, having a high-thermal stability and high corrosion resistance even at high machining temperatures. Novel TiAlN-based coatings doped with Ru, Mo and Ta are currently under investigation, as they show tremendous potential in terms of mechanical properties and wear behavior improvement. With the improvement of deposition technology, recent research seems to focus primarily on the study of nanolayered and nanocomposite TiAlN-based coatings, as the thinner layers improve drastically these coating's beneficial properties for machining applications. In this review, the recent developments of TiAlN-based coatings are going to be presented, analyzed and their mechanical properties and cutting behavior for the turning and milling processes are compared.

Keywords: machining; milling; turning; tool coating; TiAlN; TiAlN-based coatings; multilayer; nanolayer; wear mechanisms

Citation: Sousa, V.F.C.; Da Silva, F.J.G.; Pinto, G.F.; Baptista, A.; Alexandre, R. Characteristics and Wear Mechanisms of TiAlN-Based Coatings for Machining Applications: A Comprehensive Review. *Metals* **2021**, *11*, 260. https://doi.org/10.3390/met11020260

Academic Editor: Emin Bayraktar
Received: 31 December 2020
Accepted: 29 January 2021
Published: 4 February 2021

1. Introduction

Machining remains a very important process, with the machining industry in continuous growth in recent years, and still having a considerable expected growth in the following years. The turning and milling process are the most used machining processes; however, the drilling process is quite relevant for the machining industry as well. This importance of the machining processes is based on the high demand for high-quality and complex parts for various industry applications such as, the aeronautical and the aerospace industries [1,2]. These two industries benefit specially from the machining process, as it can produce highly complex parts accurately, by employing 5-axis machining methods or even 6-axis machining methods, enabling the complete production of complex parts, from the raw material to the final part, without the need to stop [3,4]. The need for these types of parts also creates the need for process improvement, by reducing machining times, improving tool-life or by applying new machining methods. These topics are still researched recently, with many being focused on the tool used for the machining process, as the tool directly influences the machining process's productivity, with studies being made on the creation of new and improved tool designs, such as the study performed by

Siddiqui et al. [5], where the development of a self-lubricated textured tool and its employment in the dry turning of aluminum alloy Al6061- T6 is described. The textured cutting tool allowed for the use of MoS_2 as a solid lubricant. It was found that this novel design significantly reduced the wear of the tool (by up to 35%), and also, the cutting temperature was reduced by up to 40% when compared to turning with the conventional tool under dry machining conditions. Still regarding the development of new tools and methods, there have also been improvements regarding the cemented carbide tools used, for example, gradient carbide tools [6] (with different layers with different values of hardness) have been developed. This enables one to tailor the base tool (uncoated substrate) to a certain machining application, conferring the tool with increased wear resistance. There have been some studies conducted on this topic, such as this study performed by Zhou et al. [7] where various gradient cemented carbides (coated and uncoated) are tested in machining tests of a titanium alloy. It was found that the gradient's layer thickness influences the cutting performance and that this thickness can be altered by changing the composition of the cemented carbide itself. Moreover, the tool with thicker layers exhibited the best cutting performance, suffering less wear damage. This is a very interesting study, as the base tool offers a better set of properties (compared to normal cemented carbide tools) that, when combined with a tool coating, will improve even further the tool's performance. Tool coatings have greatly contributed to the machining sector since they were first developed, but there is still room for improvement in this area and thus, recent studies made on machining tools also focus on the employment of coatings to machine certain materials, especially titanium alloys [8], aluminum alloys (these aluminum alloys are primarily applied to the aerospace and aeronautical industry, with some applications in the automotive industry) [9,10] and hard-to-machine materials such as Inconel [11], given that these materials are heavily used in the production of parts for the aeronautical industry. These studies are very important as they provide valuable information on what coatings are best suited for machining a certain material, or which coating to use when wanting to optimize the machining process [2,12].

Machining tools that are employed in today's machining processes are usually coated tools, either solid coated tools, machining tools with coated inserts, or just coated inserts in the case of the turning process. The use of coated tools has greatly improved the machining processes, enabling the machining of materials at higher speeds, when compared to regular uncoated tools and inserts (steel and solid carbide tools) [13–15]. These coatings have proved to be especially useful when it comes to improving tool-life and overall tool performance, when machining materials with low machinability [16,17]. This is due to the improvement of the mechanical properties of the tool by the coating, such as increased hardness, oxidation resistance, toughness, thermal stability (ability to retain microstructure at higher temperatures) and reduced friction coefficient. The employment of tool coatings also contributes for a better surface quality of the machined part [18] and reduction of the cutting forces developed during the machining process (especially due to reduction in friction coefficient), still a very important aspect in today's research of these cutting processes [19]. However, regarding tool-tip temperature and machining temperature, it has been reported that coated tools usually experience higher machining temperatures than uncoated tools [20]. Yet this fact does not seem to negatively impact the coated tool's life (in most cases), since there are other factors that occur at the tool-chip interface, such as the formation of a coating oxidation layer. Moreover, coatings can be tailored to fit a certain application, and the introduction of multiple layers influences the coating's properties [21]. These factors contribute to the tool's life, with coated tools exhibiting overall less wear when compared to uncoated tools, with numerous studies being conducted about this topic. For example, in the study performed by Thakur et al. [22], the performance of uncoated chemical vapor deposition (CVD) $TiCN/Al_2O_3$ bilayer and physical vapor deposition (PVD) TiAlN/TiN multilayer coated tools was evaluated. These tools were employed in the turning of Incoloy 825 at three different cutting speeds (51, 84 and 124 m/min). The coated tools outperformed the uncoated tools, however, this margin increased for the higher cutting speed values. Coated tools produced a better surface finish on the part when compared to the uncoated tools, furthermore, the PVD coatings suffered overall less wear than the CVD coating, exhibiting the lowest value for friction coefficient of all the tools (coated and uncoated). Regarding tool life, coatings improved greatly in this aspect, with the uncoated tools lasting only 90 s, and the CVD and PVD tool lasting for 28 and 40 min, respectively. Studies such as these are very important for the optimization of machining processes, providing valuable information regarding coating application and material machinability. Moreover, studies such as these highlight the value that tool coatings have when employed in machining processes, especially in improving tool-life and part production quality.

Coatings are usually obtained by two different processes, either by CVD or by PVD, with some differences between the two processes [23]. The CVD deposition process was the first to be invented, being used then for the deposition of TiN and TiC coatings in the 1960s, as a response to the tool life problem, then present in the machining industry. CVD produces coatings by having a precursor pumped inside of a reactor (the flux of this precursor is regulated by valves). The precursor molecules pass by the substrate (placed inside the reactor) and are deposited on its surface, giving origin to a thin hard film that has a relatively uniform thickness throughout the substrate's surface. The working temperature of this process is quite high, reaching temperatures of up to 900 °C. The PVD process was developed after the CVD, having some advantages when compared to it, such as a lower deposition temperature and the ability to create different types of coating (such as the TiAlN coating, created as an improvement over the TiN coating). The first coating deposited by this technique was TiN coating, achieved in the 1970s [24]. PVD consists of various methods, such as evaporation, sputtering and molecular beam epitaxy (MBE). Regarding sputtering, the coating is achieved by placing a magnetron near the target (containing the elements that are going to be part of the coating), in a vacuum reactor chamber. An inert gas is then introduced in the chamber, then a high voltage is applied between the target and the substrate also placed inside the reactor chamber, causing the release of atomic size particles from the target. These particles are projected onto the substrate, causing the formation of a thin solid film. In the evaporation technique, however, the target itself acts as an evaporation source, while the sample's material works as a cathode, the target material is heated at a high vapor pressure, which causes particle to release and be dispersed inside de reactor. The gas that is being pumped inside the chamber clashes with these particles, causing their acceleration, which in turn creates a plasma that will be deposited onto the substrate's surface. This process, contrary to the CVD process, runs at a much lower temperature, under 500 °C. Thus, PVD obtained coatings can be deposited onto steel substrate and cemented carbide tools without negatively impacting the properties of these types of substrate. Furthermore, the PVD process does not involve the use of any toxic precursors, unlike the CVD process, and is more energy efficient, having a considerably lower energy consumption than the CVD process [25,26].

Choosing the right coating deposition method is very important, as seen in the previous paragraph. Different techniques confer the coatings with different properties, being increased mechanical properties, adhesion properties and even residual stresses. Both CVD and PVD methods have certain advantages and disadvantages, for example, CVD coatings are very difficult to deposit onto steel substrates, due to high deposition temperature. However, there have been studies that seek to solve this problem, by implementing an interlayer, between the coating and the substrate, that will protect the substrate during the deposition of the outer coating [27,28]. Regarding the PVD process, due to its deposition temperatures, usually, good adhesive strength of coatings can be achieved when these are deposited onto steel substrates [29]; however, this process is a line-of-sight process, which means that coating deposition on complex geometries is considerably harder when compared to CVD. Moreover, the control of the thickness throughout the substrates surface is also harder. These problems can be attenuated by using a different PVD process, such as pulsed high-power sputtering [30]. However, this can come at a cost, such as inducing excessive residual stresses in the coating or even sacrificing adhesive strength. These two deposition processes (CVD and PVD) also influence machining performance, for example, PVD coatings are usually thinner than CVD coatings, however, there are some studies that report coatings with thicknesses up to 15 μm [31]. This coupled with the fact that PVD coatings exhibit compressive stresses, makes the cutting edge of the coated tool a very strong and resistant edge, making these types of coatings ideal for finishing operations, whereas in the case of CVD coatings, these exhibit tensile residual stresses and are usually thicker, making them more suited for roughing operations where, for example, a high material removal rate is preferred [32–34]. The control over these coatings properties makes them very versatile, moreover, they can be specifically made for a certain application, experiencing various combinations of coating's structures and compositions. Coatings are not only used on tools for metal cutting operations [35,36], their mechanical properties, high wear resistance, high temperature resistance and high corrosion resistance makes them very appealing for a wide range of applications. For example, they have seen some recent use in wood cutting processes [37], medical applications [38], mold industry [39], automotive (especially for brake pads) [40] and even being deposited in alloys used for nuclear fuel cladding [41,42].

As previously mentioned, the PVD process involves various methods for the deposition of coatings. These methods influence not only coating composition, but their properties as well. These methods are primarily divided into two groups: sputtering and evaporation. All these methods can be observed in Figure 1.

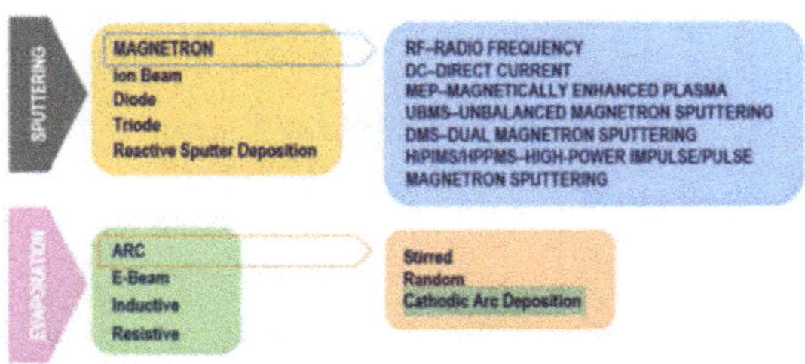

Figure 1. Physical vapor deposition (PVD) techniques currently being used in the production and deposition of coatings [27].

Currently, the most used method to produce PVD coatings is the direct current method (DC) for magnetron sputtering; however, there seems to be a shift in the use of these techniques to ones such as, unbalanced magnetron sputtering (UBMS) [25,26] and the high-power impulse/pulse magnetron sputtering (HiPIMS/HPPMS), the latter rising in popularity in recent years. There are several studies conducted regarding these deposition techniques, as they confer the coating with different properties, more suited to certain applications. For example, in the study carried out by Romero et al. [43], an evaluation of microstructure and tribological performance of TiAlTaN-(TiAlN/TaN) coatings has been observed. Various coatings of this type were deposited onto AISI M2 steel, the deposition consisted of a first layer of TiAlN/TaN followed by a second layer of TiAlTaN. In total, four combinations with different volume fractions for each layer were tested. Furthermore, the deposition was achieved by DC sputtering using two magnetrons and two targets, and by controlling the substrate rotation speed. The authors were able to control the coating's architecture, and thus their mechanical properties, by varying the rotation speed of the substrate during deposition. They were able to find the best values for the volume fractions of each layer, concluding that the combination of 48% TiAlTaN and 52% of TiAlN/TaN exhibited the best balance between adhesion properties, hardness (29 GPa) and friction coefficient (0.68). Recently, there have been some studies highlighting the benefit of some deposition techniques such as HiPIMS, linking this technique with an increase in mechanical properties and a betterment of adhesion properties. Zauner et al. [44] studied the influence of the HiPIMS parameter choice in the properties of TiAlN films, by using a TiAl composite target in mixed Ar/N_2 atmospheres. The parameters in question were the pulse frequency and duration, the N_2 flow ratio, target composition and substrate bias voltage. The authors found that changing these parameters has a great influence on the final coating's properties, even enabling the control of the coating's structure. The authors claim to obtain hardness of the TiAlN coating of up to 36 GPa. Still regarding the influence of deposition parameters in the coating's properties, in the work performed by Zhao et al. [45] the influence of the bias voltage chosen during deposition of TiAlN coatings is evaluated. Coatings have been fabricated using a multi-arc ion plating device. Various values for this parameter were tested, the lowest value of bias voltage being −40 V and the highest value being −120 V. The lowest value produced the coating with superior toughness, being the most suited for cutting applications. Increasing the bias voltage resulted in a loss of toughness, however, there was an increase in hardness and plasticity. Choosing the right deposition technique is of great importance when fabricating a tool coating. Thus, there have been some studies whose compare some of these techniques in the deposition of tool coatings, as seen in the study presented by Tillman et al. [46] where a comparison between DCMS (Direct Current Magnetron Sputtering) and HiPIMS in the deposition of TiAlN and TiAlN/TiAlCN coatings is made. The films were deposited in heat-treated AISI H11 steel, and the samples were evaluated regarding wear resistance and residual stresses. The authors found that the coatings obtained by HiPIMS had significantly higher residual stresses than the DCMS coatings. Furthermore, the adhesion of the coatings obtained by DCMS was higher. However, the TiAlN coatings deposited by HiPIMS displayed higher wear resistance than the other coatings obtained by DCMS. The problems presented in the last study have been researched as well, with some solutions for the adhesion problems and the higher compressive stresses of HiPIMS coatings being presented, such as the use of substrate surface texturing methods using etching process, which can help increasing the coating's adhesion and relieve excessive compressive stresses [47]. There are also some very recent studies on a novel deposition technique that can produce high ionization rates like the HiPIMS method. This method is the Continuous High-Power Magnetron Sputtering (C-HPMS). Liu et al. [48] studied TiAlN coatings obtained by this technique. The authors were able to obtain a coating with a very high hardness value (34.4 GPa) and a good adhesive strength (75 N). Moreover, the deposited coating presented very few particles on its surface. The method described in this paper paves the way to obtain droplet-free coatings and good mechanical properties by employing this method, presenting benefits such as fast deposition rate and efficient ionization.

Evaporation methods are seeing some recent research as well, with DC arc evaporation being the most common technique among them. However, some attention is being given to the cathodic arc deposition method, with studies being made relating deposition method and parameters to the coating's overall properties [49], even relating rotation speed during deposition and substrate orientation to the mechanical properties of deposited coatings [50], such as the study previously presented [43]. However, this deposition technique is being recently used mostly for the deposition and synthetization of borides and borides-related coatings, which are unable to be obtained by DC arc evaporation [49,51].

Regarding coating characterization as it was previously mentioned, coatings can be designed in order to fit a certain application, by controlling its architecture and composition, and thus, its microstructure and mechanical properties. There are various different coating designs applied to substrates for a wide range of applications. These can be observed in Figure 2.

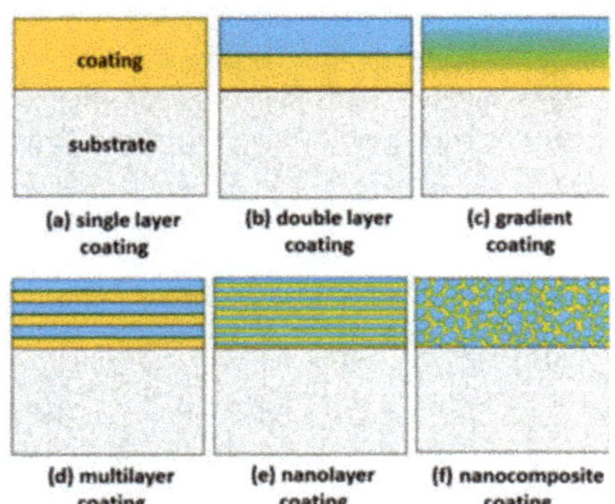

Figure 2. Different types of coating structure commonly applied to substrates [24].

The coatings can be identified as follows:

a. Monolayer (single layer) coating;
b. Bilayer (double layer) coating;
c. Gradient coating;
d. Multilayer coating;
e. Nanolayer coating;
f. Nanocomposite coating.

Different types of coating structure are chosen to deal with different kinds of problems, for example, the use of a multilayer coating can improve significantly upon the properties of the single layer coated tool. For example, a multilayered coating has significantly more crack propagation resistance than a single layered coating. The number of layers contributes to this, furthermore, an increase in the number of layers will also increase properties such as hardness [24,52,53]. A scheme of how crack propagation usually behaves depending on coating structure can be observed in Figure 3.

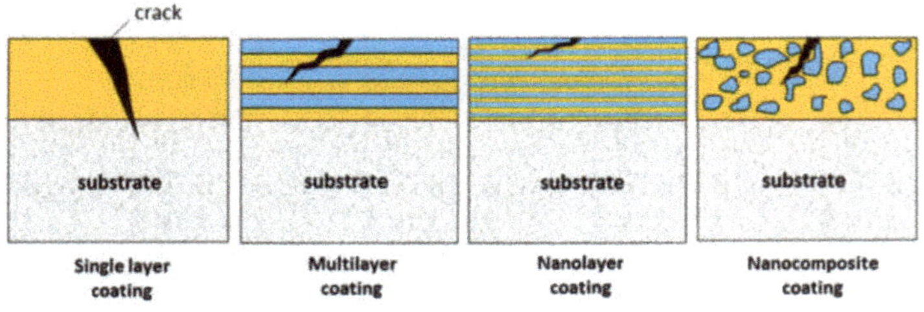

Figure 3. Crack propagation behavior for each of the common coating structure [24].

Layers can also be added to improve adhesion properties, enabling that the deposition of the outer layer, usually the "work" layer, with high adhesive strength. Usually this is done when the outer layer has problems with adhesion to the substrate. Coatings are also characterized by their chemical composition. The elements that constitute the coating confer it with different properties, as certain elements can improve properties such as, corrosion resistance, wear resistance and even thermal conductivity. For example, the first coatings (TiN) were improved by the addition of aluminum, creating the TiAlN coating. This coating proved itself to be very useful in high-speed machining application, being widely employed in many machining applications to this day, because they present an oxide layer created between the tool–workpiece interface, thus conferring this coating with high oxidation resistance [54–57]. Still regarding coating characterization, these are also characterized by their microstructure. Different coatings have different types of microstructure, depending on the deposition method, composition and their architecture, as described in the paper developed by Du et al. [58], where the effect of interlayers of Cr and Ti on the structure of TiAlN based coatings is studied. The authors deposited four types of coatings onto cemented carbide substrates, these being Cr/(Ti,Si,Al)N; Ti/(Ti,Si,Al)N; Cr/TiAlN; and Ti/TiAlN. It was found that the presence of these interlayers influenced the microstructure of the coatings. The Cr interlayer affects the growth of TiAlN based coatings, with the structure of the coatings containing this interlayer exhibiting a mix between columnar crystal morphology and equiaxed crystal morphology. In the case of Ti interlayer, the morphology was columnar. The Cr interlayer also promoted a better adhesion of the TiAlN based coatings onto it.

In this review paper the properties of TiAlN based coatings are going to be evaluated and presented, based on the information collected from recent articles conducted on this topic. The various types of coatings are going to be presented in the subsequent sections, mentioning in more detail these coating's properties such as, structure, microstructure and composition and its influence on the coating's mechanical properties, especially hardness values and young's modulus values. Moreover, the recent applications of these TiAlN based coatings in machining are going to be analyzed, highlighting the coating's performance on the various machining process (primarily turning and milling). Still regarding coating performance in machining, the wear mechanisms that these coatings suffer are also going to be analyzed and compared between each-other, as the analysis of these wear mechanisms gives very valuable information regarding the optimization and improvement of these machining processes [39,43]. This review intends to fill an existing gap about structured information regarding TiAlN-based coatings utilized in machining tools, mainly based on the most recent developments published in this field. This information, regarding wear behavior and the mechanical properties of the coatings, is going to be presented under the form of tables in order to convey a clear and easy-to-read message. Furthermore, the various types of structure of TiAlN-based coatings were divided into sections, with a section for monolayered, multilayered and nanolayered TiAlN-based coatings being created.

2. TiAlN-Based Coatings

Since its development in the 1970s, the TiAlN coating offered a great opportunity for enhancing tool life and performance for high-speed machining applications. Due to its properties, TiAlN and TiAlN-based coatings are still among the most used coatings for machining applications today, making them a very appealing research matter, with many new coatings being tested and evaluated for a wide variety of machining applications. Furthermore, recent studies also focus on the influence of new doping elements in the properties of TiAlN-based coatings.

In this chapter, recent developments made on TiAlN-based coatings are going to be presented, namely:

- The studies being conducted about monolayered TiAlN-based coatings, mentioning the influence of doping elements (such as, Ru, Ta, Y and Mo) on the overall properties of these coatings;
- Studies about multilayered coatings, presenting the new structures and composition combination which are being employed recently and analyzing its influence on the coating's properties;
- Studies being conducted about nanolayered coatings, mentioning the recent developments being made on this topic, presenting the various structures and their benefits when compared to the other types of TiAlN-based coatings.

In total, three coating types are going to be divided into subsections inside this section. Additionally, the information regarding these coating's mechanical properties, especially Hardness and Young's Modulus values, are going to be presented in a subsection for each of the coating types. The wear mechanisms that these coatings suffer for various machining applications are also presented, mentioning the wear behavior of these types of coatings and presenting the obtained values for tool life based on the information provided by the various analyzed articles.

2.1. Monolayered TiAlN-Based Coatings

Regarding monolayered TiAlN-based coatings, recent research has been made on the effects on the coating properties of doping TiAlN coatings with certain elements. It has been found that the addition of certain elements to the coating can improve properties such as corrosion resistance and wear behavior. Additionally, the addition of these elements is also linked to an improvement in mechanical properties such as hardness and Young's Modulus. In the study performed by Yang et al. [59], the influence of Mo content on TiAlMoN films is presented. The authors have produced five types of TiAlMoN coatings, varying the amounts of Mo. Composition of the samples used in this work can be observed in Table 1.

Table 1. Chemical composition of the TiAlMoN coatings developed in the work presented by Yang et al. [59].

Sample ID	Composition (at. %)			
	Ti	Al	Mo	N
S1	27.6	24.1	2.8	45.6
S2	23.9	22.2	6.9	47.0
S3	17.8	18.0	8.3	54.9
S4	18.2	19.4	10.1	54.4
S5	16.4	16.7	12.1	54.8

The authors analyzed these coatings determining the mechanical properties for each one of the produced samples. Furthermore, the microstructure of each of the TiAlMoN films was analyzed. It was noticed that the hardness and Young's Modulus increased with the addition of Mo, reaching peak levels of hardness for 12.1% Mo content (S5), concretely, and 50 GPa and 610 GPa for hardness and Young's Modulus value, respectively.

Authors registered the influence of the Mo content on the microstructure, by obtaining SEM images of the coating's cross-section. This phenomenon can be observed in the following images, starting with Figure 4, depicting the film's structure with a Mo content of 2.8 at. % (a) and 6.9 at. % (b).

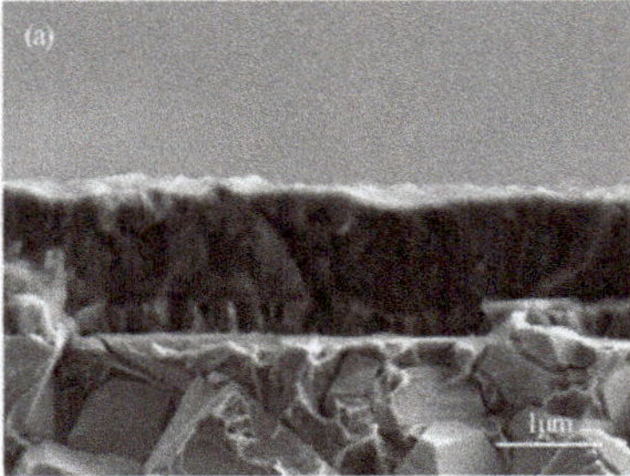

Figure 4. SEM cross-section of the TiAlMoN films with differing Mo contents: 2.8 at. % (**a**); 6.9 at. % (**b**), presented by Yang et al. [59].

It can be observed that at low Mo contents, the structure presents some columnar grains, however, it is not yet uniform. This uniformization was registered at a Mo content of 8.3 at.%. This can be observed in Figure 5.

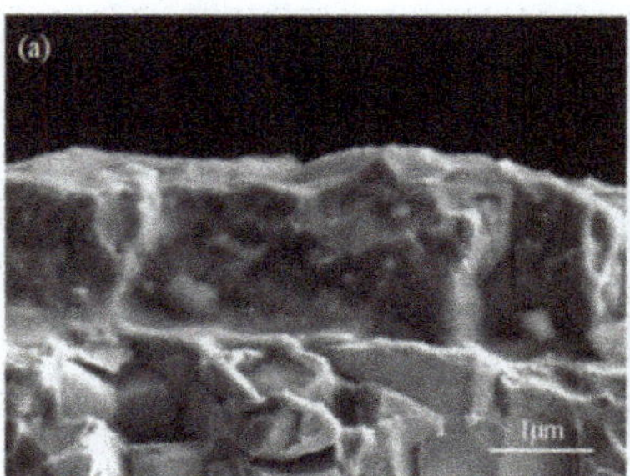

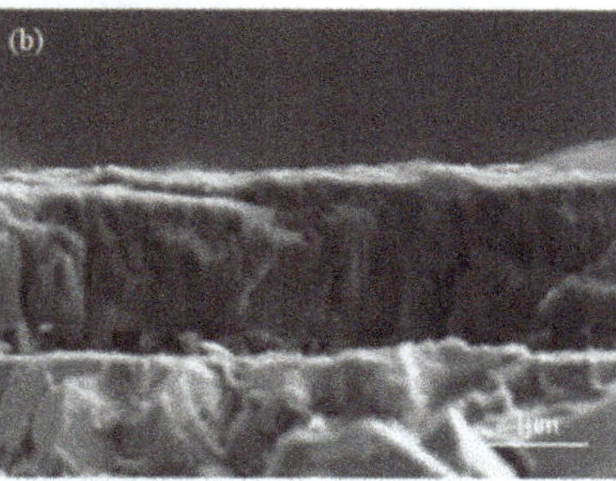

Figure 5. SEM cross-section of the TiAlMoN films with differing Mo contents: 8.3 at. % (**a**); 10.1 at. % (**b**), presented by Yang et al. [59].

There is a change in the film's structure at 10.1 at. % of Mo content, the film's structure starts becoming columnar, finally becoming fully columnar at a Mo content of 12.1 at.%. The latter structure exhibited face-centered cubic TiN-based phases with a preferred orientation. The film's structure with a 12.1 at. % Mo content is displayed in Figure 6.

Figure 6. SEM cross-section of the TiAlMoN films with differing Mo contents: 12.1 at. %, presented by Yang et al. [59].

The addition of Mo can also improve the wear behavior of the coating, however, the wear behavior of the TiAlMoN films did not increase with the Mo content, as seen for the mechanical properties. The coating's that exhibited the best wear behavior was the S3 coating, with a Mo content of 8.3 at. %. This is based on the toughness and mechanical properties of the film. One way to evaluate the wear performance of the film is by the analysis of the H/E (Hardness/Young's Modulus) ratio. This ratio can be related to the film's toughness; however, it provides information regarding the plastic deformation of the film (with high H/E ratio values being tied to high plastic deformation resistance). The S3 coating displaying the highest value for this ratio, among all the other TiAlMoN coatings. Still regarding the influence of Mo addition to TiAlN-based coatings, the work presented by Tomaszewski et al. [60] also evaluated the influence of the addition of this element, concluding that the addition of this element is linked to an improvement of mechanical properties. Indeed, a small addition of up to 7.7 at. % allowed to report a significant improvement on the wear behavior of the coating. The authors also registered an improvement on the corrosion resistance properties of the coating, observing that the coating exhibited an improved resistance to pitting corrosion.

The addition of Nickel to TiAlN-based coatings is also an interesting topic. Similar to Mo, the addition of Ni to coatings also has an influence on the microstructure, as reported by Yi et al. [61] in their study, where three samples of AlTiN-Ni with differing contents of Ni are analyzed and subsequently tested in the turning of Inconel 718. The analyzed AlTiN-Ni coatings had 0%, 1.5% and 3% Ni. The AlTiN-Ni coatings were obtained via PVD cathodic arc evaporation. Figure 7 shows the three cross-sectional images of the coatings.

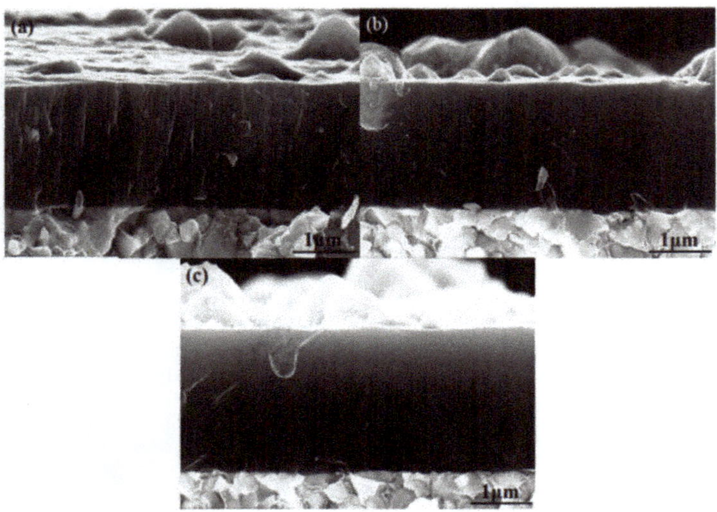

Figure 7. AlTiN-Ni coatings with differing Ni contents: 0% (**a**); 1.5% (**b**); 3% (**c**), presented by Yi et al. [61].

There is an evident change in microstructure depending on the Nickel content. Observing Figure 7a, the microstructure of the AlTiN-Ni coating is columnar. However, it can be noted that the addition of Ni promotes a homogenization of the structure (nanocrystalline structure). The authors registered the hardness and Young's Modulus values for these coatings. AlTiN-Ni with 0% Ni content exhibited the highest values for these properties (26.2 GPa and 315 GPa for hardness and elastic modulus, respectively). These values decreased with the increase in Ni content. For a Ni content of 1.5%, the values registered for hardness and Young's Modulus were 24.3 GPa and 315.8 GPa, respectively, seeing another decrease to 20.9 GPa and 300.5 GPa for a Ni content of 3%. From these presented values, the decrease in mechanical properties was not very accentuated from 0% to 1.5% Ni content. Furthermore, the authors observed that the toughness of the AlTiN—Ni (1.5%) coating was the highest of all the coatings, improving tool life by 160%.

Still regarding the addition of elements to TiAlN-based coatings, in the study carried out by Liu et al. [62], the addition of Ruthenium (Ru) to TiAlN coatings is performed and evaluated. The authors compared the microstructures of base TiAlN coating and two other coatings with differing Ru contents—7% and 15%, respectively. As seen in the work presented above [61], it was noted that the Ru addition promoted a change in the coating's microstructure, and similar to the addition of Ni, it promoted a homogenization of the microstructure. By analyzing Figure 8, the microstructure changes from a columnar structure to a homogenous structure.

Figure 8. SEM cross-sectional image of TiAlN coatings with different Ruthenium contents: 0% (**a**), 7% (**b**), 15% (**c**), presented in the study carried out by Liu et al. [62].

An increase in mechanical properties was also registered for the coating with 7% Ru (33.15 GPa and 498.55 GPa for hardness and Young's Modulus, respectively), however, the coating with 15% Ru content showed a decrease in hardness and Young's Modulus value, even when compared to the base TiAlN coating. Other elements, such as Ytrium (Y) and Tantalum (Ta), are linked to an increase in properties for TiAlN-based coatings, as seen in the study performed by Aninat et al. [63], where the addition of Y is linked to an increase in hardness values. Ta, however, did not have a significant influence on the coating's mechanical properties, influencing the amount of residual compressive stresses of the coating, which is linked to the wear behavior of the coating. Ta coatings exhibited a better wear behavior, whereas Y doped coatings exhibited overall better mechanical properties. Regarding coating residual stress control, this factor has also been linked to coating thickness, with this being analyzed by Chandra et al. [64]. In total, three TiAlN coatings with differing thicknesses were studied. Their mechanical properties suffered low variation, exhibiting a small decrease with a thickness increase. However, the residual stresses are greatly influenced by the coating's thickness, exhibiting higher values for thinner coatings.

Another element related to a significant increase in mechanical properties is Silicon (Si), significantly increasing hardness and Young's Modulus values. TiAlSiN is an example of a TiAlN coating doped with Si. As seen in [65], TiAlSiN coatings have better mechanical properties and wear behavior when compared to TiAlN coatings.

Regarding recent improvements made on the properties and performance of TiAlN-based coatings, in the study carried out by Chaar et al. [66] two TiAlN coatings with high aluminum content and differing structures were studied. One coating had a fine-grain structure, consisting of a mixture of cubic and hexagonal phases (dual-phase coating), the other coating had a coarse-grain structure of cubic phase. It was found that, although the coatings are very similar in terms of composition, their structure greatly impacts the thermal behavior of the coatings, highlighting the advantage of controlling the coating's structure in order to obtain a desired result, especially in terms of thermal stability.

There have also been recent studies conducted on the improvement of TiAlN-based coatings' performance by applying texturing treatments to the substrates. These texturing treatments are linked with a slight increase in wear behavior and mechanical properties. Moreover, coating adhesion improves greatly in textured tools [67,68].

2.1.1. Machining Applications and Coating Wear Behavior

In this subsection the recent studies on machining applications of monolayered TiAlN-based coatings are going to be presented. An analysis of recent studies was conducted on the wear behavior of these coatings. The tool wear mechanisms suffered by monolayered TiAlN-based coatings for milling and turning are going to be presented.

Milling Process

The milling process has a huge presence in the machining industry, with many studies being conducted about the improvement of this process, either by using new tool geometries, coatings or by employing optimization techniques, such as the Taguchi method [43,69] to optimize machining parameters. TiAlN-based coatings are also being researched, in order to try and improve the various machining processes where they are employed. As seen in the beginning of this section, using doping elements to improve the coating's mechanical properties [59–64]. Other approaches to improve the performance of these coatings involve the study of coating behavior under experimental machining conditions, or the study of the wear mechanisms sustained by these coatings during machining [39,43]. Ravi et al. [70] studied the influence of various lubrication methods on TiN and TiAlN-coated tool's performance. The authors conducted milling experiments under, dry, flooded and cryogenic (liquid nitrogen) conditions. Furthermore, the tests were conducted at 75, 100 and 125 m/min of cutting speed. The cutting temperature and cutting force variation were evaluated for all tools. It was observed that the use of the TiAlN-coated tool over the TiN caused an increase in cutting performance, exhibiting a reduction of approximately 13% in cutting force for all cutting conditions. Moreover, the cutting temperature was 18% lower for the TiAlN-coated tools when compared to TiN. Cutting temperature increased with higher cutting speed values, however, cutting force values decrease for higher cutting speeds. This was particularly evident for the cryogenic cutting conditions, where the TiAlN-coated tools benefited greatly from this, exhibiting the lower cutting force value of all tools for all cutting conditions. This force variation for all coated tools under the three lubrication conditions can be observed in Figure 9.

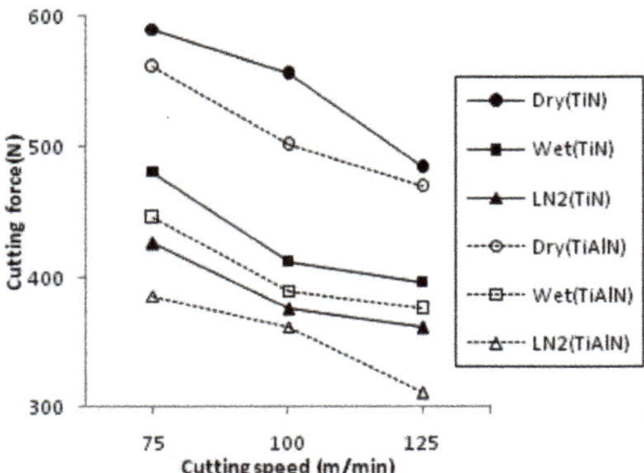

Figure 9. Cutting force variation (N) for different cutting speed values, for TiN and TiAlN-coated tools under the three different machining conditions, presented by Ravi et al. [70].

It was reported by the authors that in this case the main wear mechanism was adhesion and abrasion, for all the cutting conditions, these effects being attenuated by the employment of flooded and cryogenic cutting conditions, especially for the TiN coating. The TiN-coated tool suffered more adhesive and abrasive wear than the TiAlN-coated tool, this can be explained by the Al contained in the coating. This element can confer the coating with better thermal properties, explaining the fact that these are less affected by the temperature dependent adhesive wear. Still regarding the wear analysis of TiAlN-based coatings, Siwawut et al. [71] evaluated the cutting performance and wear characteristics of TiAlSiN and CrTiAlSiN coatings. These coatings were deposited by filtered cathodic arc onto WC (Tungsten Carbide) inserts. The authors tested the coated tools and one uncoated tool in the dry milling of cast iron turbine housings, using a range of 14–300 m/min for cutting speed. Coating's mechanical properties were evaluated, namely hardness and Young's Modulus values. The TiAlSiN coating exhibited the highest hardness value of all tools and the highest Young's Modulus value of both the coated tools, although these values have been very similar to those of the CrTiAlSiN. This produced a higher surface finish quality when using the CrTiAlSiN-coated tool when compared to the CrTiAlSiN coated tool. The H/E ratio of these coatings was also evaluated, as this ratio is strongly correlated with wear performance. The CrTiAlSiN coating exhibited the highest value of all tools (0.112), being followed by the TiAlSiN coating (0.105). The insert wear was analyzed using SEM, whose images can be observed in Figures 10 and 11.

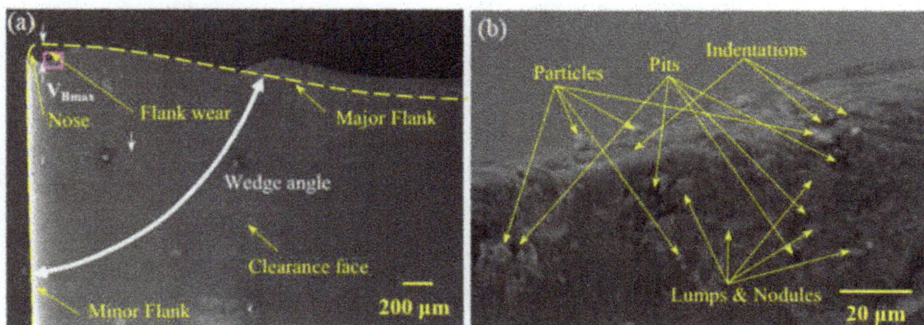

Figure 10. SEM micrographs of the TiAlSiN-coated WC insert, low magnification (**a**), and high magnification (of the zone marked with a pink square) (**b**), presented by Siwawut et al. [71].

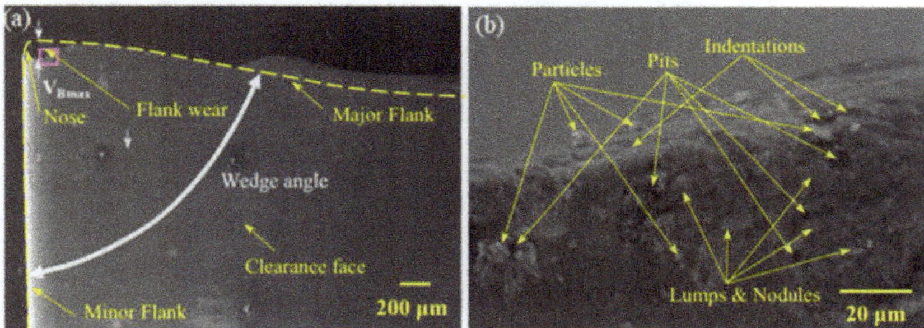

Figure 11. SEM micrographs of the CrTiAlSiN-coated WC insert, low magnification (**a**) and high magnification (of the zone marked with a pink square) (**b**), presented by Siwawut et al. [71].

Both the coatings improved the wear behavior of the WC inserts, however, the CrTiAlSiN coating performed better in this regard, exhibiting lower values for flank wear and suffering overall less damage, when compared to the TiAlSiN-coated insert. The authors observed that TiAlSiN-coated inserts displayed primarily ploughing abrasive wear behavior, while the CrTiAlSiN-coated inserts exhibited wear related to thermal cracking and partial coating delamination.

As mentioned previously, the level of residual stresses inside the coatings impacts their properties and, thus, their machining performance [30,38,39,43]. There are some recent studies made on this topic, as the work presented by Hou et al. [72], where the influence of compressive stresses on the TiAlN-coated tool's performance when milling Ti alloy Ti-6Al-4V was investigated. Both coatings had the same composition and were deposited onto the same substrate material, one of them was unaffected by compressive stresses (Coating 1) and the other was affected (Coating 2). These coatings were characterized, determining hardness and Young's Modulus values, presented in Table 2.

Table 2. Hardness and Young's Modulus values determined for TiAlN coatings (Coating 1 and 2), presented by Hoy et al. [72].

Coating	Hardness (GPa)	Young's Modulus (GPa)
1 (Unaffected)	30.6	482
2 (Affected)	34.9	567

The residual stresses have a clear benefit for the coating's mechanical properties, with Coating 2 having an increase in hardness and Young's Modulus values by 12% and 15%, respectively. These were then subjected to milling tests, having their wear behavior analyzed. In Figures 12 and 13 the wear mechanisms sustained by the two coated tools can be observed.

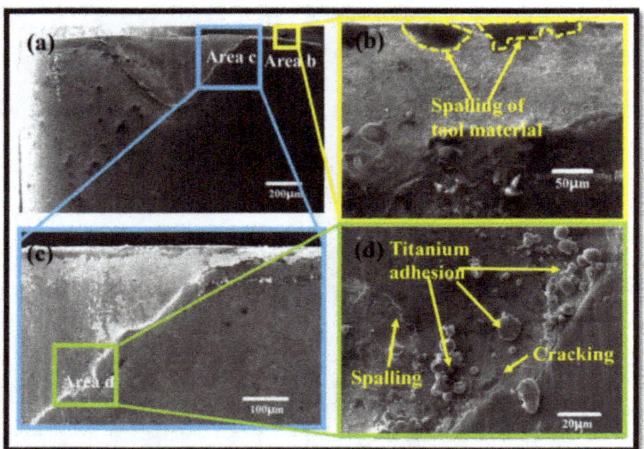

Figure 12. Wear patterns exhibited by Coating 1 after machining: SEM image of flank face (**a**); magnification of area b (**b**); magnification of area c (**c**); magnification of area d (**d**). Presented by Hou et al. [72].

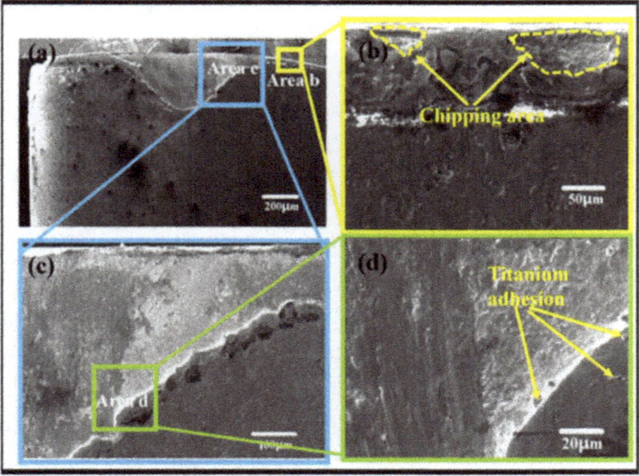

Figure 13. Wear patterns exhibited by Coating 2 after machining: SEM image of flank face (**a**); magnification of area b (**b**); magnification of area c (**c**); magnification of area d (**d**). Presented by Hou et al. [72].

From Figures 12 and 13, it can be observed that the wear sustained by Coating 1 is more intense than that of Coating 2, the former presenting more cracking and spalling than the other coating. There is, however, material (titanium) adhesion on both the tool coatings. The sustained flank wear over the cutting length for both coatings was also analyzed. The authors determined that the wear was less intense on Coating 2. The wear behavior for both coatings can be observed in Figure 14.

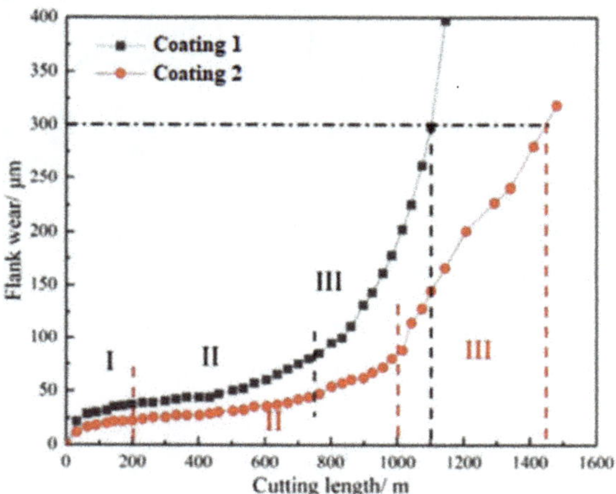

Figure 14. Flank wear measurements (μm) for various cutting lengths, for both coatings, presented by Hou et al. [72].

Compressive stresses can improve the wear behavior greatly and reduce the wear, as seen from Figures 12 and 13, because the coating unaffected by stresses suffered more cracking, and subsequently, coating spalling damage, while the coating affected by compressive stresses exhibited less cracking and less wear damage for equal cutting length values. As the crack propagates deeper into the coating, high stress values can prevent this crack propagation, retarding this damage.

As seen from presented recent works, there seems to be a focus on the optimization/study of milling processes of hard-to-machine materials, as titanium alloys [72] and some other alloys with high strength applied primarily to the aeronautical industry such as aluminum, particularly from the 6000 and 7000 series [69]. Another of these alloys is Inconel, with some recent works being conducted on the study of milling cases of this alloy. TiAlN coatings have seen some application in the machining of this alloy, as seen in the study performed by Sen et al. [73], in which the wear behavior of TiAlN coated carbide tools is analyzed in the milling of Inconel 690. Similar to the study presented by Ravi et al. [70], the authors have studied the influence of different lubricating conditions on the performance of TiAlN coating, using flooding, MQL (Minimum Quantity Lubricant) with palm oil, and MQL with 0.9% alumina enriched palm oil. Due to the high temperatures developed during the machining of this material and its characteristic high hardness, the wear mechanisms that were reported were mainly abrasion and adhesion. The wear related to these mechanisms can be observed in Figure 15. This was registered for all the lubricating conditions, however, the MQL 0.9% alumina enriched condition led to less wear damage on the tool.

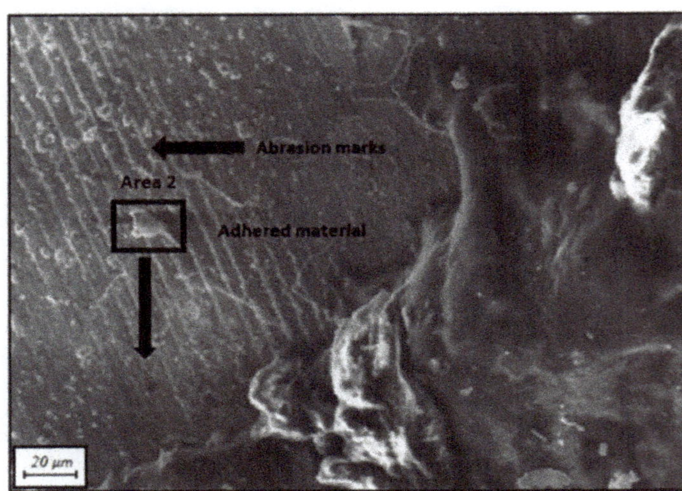

Figure 15. Abrasion marks and adhered material on TiAlN coating after milling of Inconel 690 (under the MQL with palm oil lubricating condition), presented by Sen et al. [73].

The main wear mechanisms that TiAlN-based coatings suffer in milling operations are mainly adhesion and abrasion. These mechanisms are more evident and intense in the milling experiments conducted on hard-to-machine materials, where the high machining temperature and material hardness heavily impacts tools life. High hardness values will induce heavy abrasive wear on the coatings and the high temperatures will promote adhesion of the machined material to the tool. This leads to coating delamination and, ultimately, to tool failure. These wear mechanisms are also registered for micro-milling experiments. There is a recent trend for the application of TiAlN based coatings for micro-milling [74], this because of the properties that make this coating ideal for high-speed machining, whose are also suited for this process, due to high rotational speeds usually used in micro-machining. Recent studies focus on the machining of hard-to-machine alloys [75], employing TiAlN coatings and AlTiN [76] coatings in the machining of titanium alloys [75,76] and nickel-based superalloys, such as Nimonic 75 [77]. The main wear mechanism sustained by coated micro-milling tools, when machining hard-to-machine materials, are adhesion and abrasion, as seen for the milling process, with cutting speed and depth of cut being registered as the main influencer on the development of this wear.

Turning Process

The turning process as also seen some applications of monolayered TiAlN-based coated tools. Studies made on this topic, similarly to the ones conducted on the milling process, evaluate various coatings performances and wear behavior. The studies seem to focus on the turning of hard-to-machine alloys, however, these are focused primarily on the machining of Inconel. Zhao et al. [78] study the influence of coating thickness on the machining performance of TiAlN-coated tools. The authors studied two TiAlN coatings, TiAlN-1 having 1 μm thickness, and TiAlN-2 with 2 μm thickness. Both coatings were deposited onto a WC-Co carbide and were employed in the dry turning of Inconel 718. Turning tests were performed at the cutting speeds of 30, 60, 90 and 120 m/min. The cutting forces developed during turning were evaluated, concluding that using coating with less thickness would result in lower cutting force values. Moreover, the cutting temperature was lower when using the thinner coating. Regarding the wear mechanisms and wear behavior of these coatings, these are presented in Figures 16 and 17. These images depict the wear sustained by both coatings at the tested machining speeds.

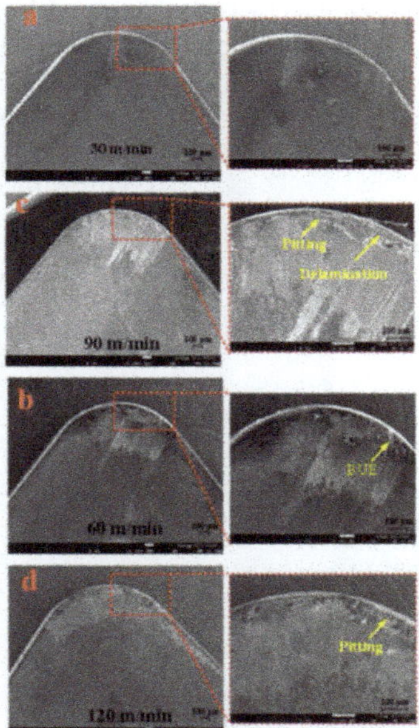

Figure 16. Wear of TiAlN-1 coated tool at different cutting speeds: 30 m/min, with magnification of marked area (**a**); 60 m/min, with magnification of marked area (**b**); 90 m/min, with magnification of marked area (**c**); 120 m/min, with magnification of marked area (**d**), presented by Zhao et al. [78].

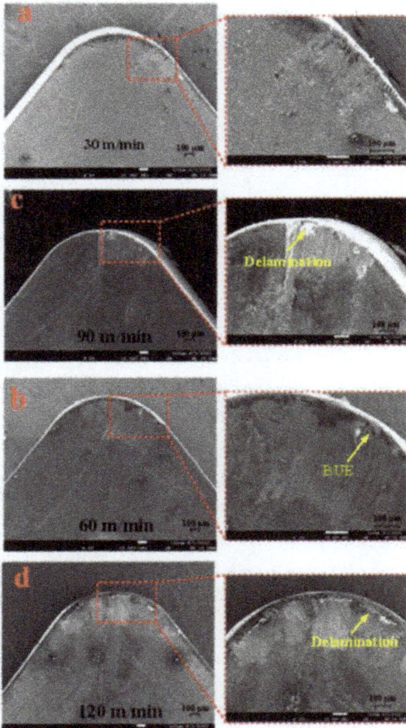

Figure 17. Wear of TiAlN-2-coated tool at different cutting speeds: 30 m/min, with magnification of marked area (**a**); 60 m/min, with magnification of marked area (**b**); 90 m/min, with magnification of marked area (**c**); 120 m/min, with magnification of marked area (**d**), presented by Zhao et al. [78].

The presented coatings both exhibit the same type of wear mechanisms, these being built-up edge (BUE), pitting and coating delamination. From the figures it can be noted that adhesion is also a problem, this last mechanism being responsible for BUE and coating delamination. These wear mechanisms, as in milling, are characteristic of machining hard-to-machine materials such as Inconel, due to material's mechanical properties. Still regarding TiAlN coating's performance when turning Inconel, the study performed by Kurniawan et al. [79] evaluates the machinability of modified Inconel 713C, using a TiAlN-coated WC tool. The cutting characteristics of Inconel 713C are very similar to those of Inconel 718, making it a very hard to machine material. The authors have reported abrasive wear as the main wear mechanism in the tool's flank, being followed either by tool failure or BUE (Build up Edge) formation, as can be observed in Figure 18. High machining temperatures and the ductility of Inconel 713C, caused material adhesion to the tool, which promoted abrasive wear and subsequent coating delamination.

Figure 18. Flank wear of TiAlN-coated WC turning tool at 100 m of cutting distance used in machining Inconel 713C, presented by Kurniawan et al. [79].

Another study performed by Zhao et al. [80], studies the cutting behavior of AlTiN coatings on the turning of Inconel 718 at cutting speeds of 40, 80 and 120 m/min. Machining temperature and cutting forces were evaluated and compared to an uncoated tool. These factors both decreased when using the AlTiN coated tool, proving that this coating is suited for the turning of these alloys. Similar to the studies presented in [78,79], the main wear mechanism suffered by the tools during machining was abrasive wear. Both the TiAlN and AlTiN coatings proved to be very useful when machining hard-to-machine materials, effectively reducing cutting force and machining temperature values. These values are tied to wear rate, with the coated tools exhibiting a significantly lower wear rate than uncoated turning tools [81]. The addition of Si to TiAlN coatings is known to significantly improve their mechanical properties [71], thus making these types of coatings ideal for the machining of materials such as titanium alloys. Lu et al. [82] compares the performance of TiAlN and TiAlSiN-coated tools in the high-speed turning of TC4 titanium alloy. The authors also studied the performance of a gradient TiAlSiN coating. The microstructures of these coatings and the distribution of Si on the gradient coating can be observed in Figures 19 and 20.

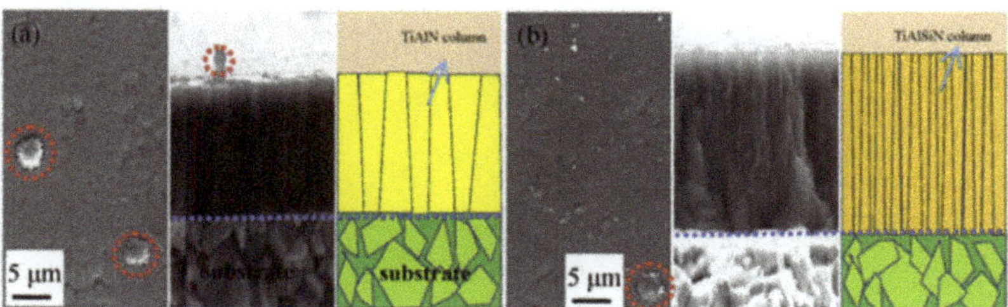

Figure 19. Surface morphology and microstructure of: TiAlN coating (**a**); TiAlSiN coating (**b**), presented by Lu et al. [82].

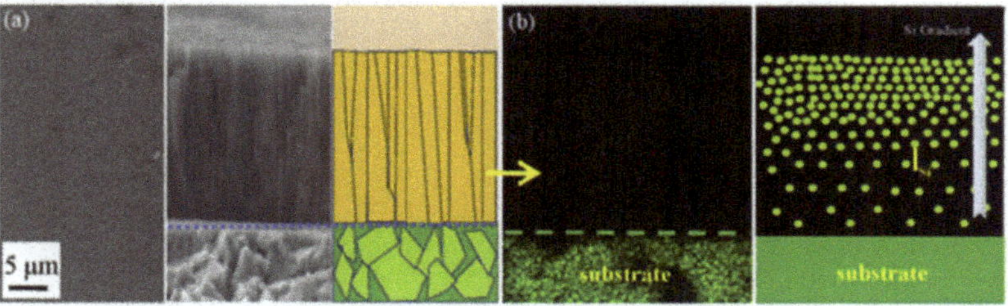

Figure 20. Surface morphology and microstructure (**a**), and Si distribution on TiAlSiN gradient coating (**b**), presented by Lu et al. [82].

The coating's hardness was also evaluated, the TiAlSiN coating exhibiting the highest hardness value (21 GPa) followed by the gradient TiAlSiN coating (15 GPa). The gradient TiAlSiN coating presented an increase in hardness by 47% when compared to the TiAlN coating. An improvement in surface quality and adhesion is also registered for the gradient TiAlSiN coating. All the machining experiments were conducted at 100 m/min cutting speed under flooded lubrication. The coated tools exhibited the same wear mechanism, this being mainly abrasion and BUE, with some adhesion being registered on the TiAlN coated tool. Although the wear mechanism has been the same, their intensity varied, being more intense in the TiAlN coated tool. The coated tool's wear behavior is depicted in Figure 21, where flank wear values (measured in mm) are displayed for the three coatings, for various cutting lengths.

The improved adhesion and surface quality of the gradient TiAlSiN, coupled with the increase in mechanical properties due to the addition of Si provides this gradient coating with great wear performance, suffering overall less wear and at a much later stage in the machining process. Still regarding the addition of elements to improve coating turning performance, in the study performed by Kulkarni et al. [83] the authors study the influence of Cr addition to AlTiN coatings. Here, the authors evaluate the turning performance of AlTiN, AlTiCrN and TiN/TiAlN coated tools, by analyzing cutting forces, wear mechanism and tool-life values. The coating's microstructure was evaluated, with all coatings showing a dense columnar structure, however, in terms of surface morphology, the AlTiCrN coating exhibited a smooth surface. Furthermore, the adhesion strength of this coating was the highest of all three. The coatings were employed in the dry turning of SS 304 steel at a constant feed rate and depth of cut, varying the cutting speed from 140 to 320 m/min (in 60 m/min increments). Regarding the cutting force values registered during the process, these tended to decrease as cutting speed increases, however, the AlTiCrN coating exhibited the lowest cutting force values obtained of all coatings. This can be observed in Figure 22.

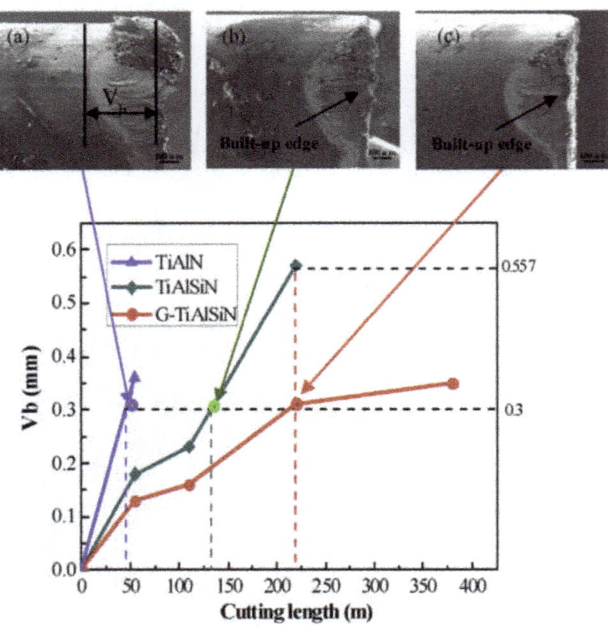

Figure 21. Tool wear of the three coated tools tested in the turning of TC4 titanium alloy: TiAlN (**a**), TiAlSiN (**b**), Gradient TiAlSiN (**c**), presented by Lu et al. [82].

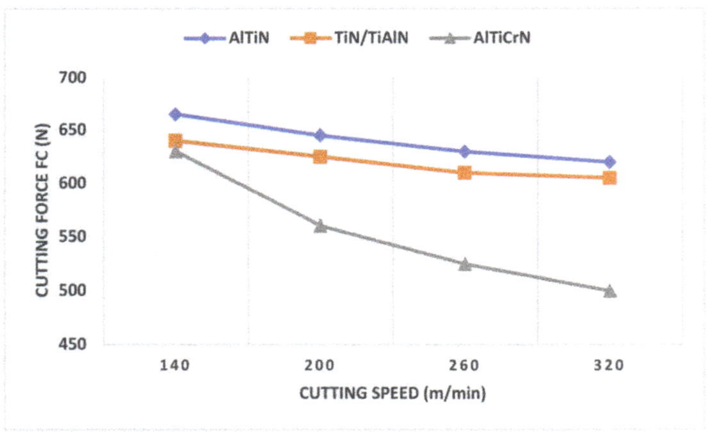

Figure 22. Cutting force variation for the tested cutting speeds, for all coated tools at the following parameters: depth of cut = 1 mm; feed = 0.2 mm/revolution (based on data from [83]).

For the AlTiN tool, the main wear mechanisms were abrasion, chipping and BUE. This was registered for the TiN/TiAlN coating as well, however, there was some adhesive wear. For the AlTiNCrN coating, the main wear mechanism was abrasion, albeit less intense than in the AlTiN coating. Regarding tool-life of these coated tools, it tended to decrease with an increase in cutting speed. Further, as seen for the cutting forces, the AlTiCrN coating outperformed the other coated tools, exhibiting the highest tool life values, reaching 28 min (at 200 m/min). These values can be attributed to the coating's excellent adhesion properties, smooth surface and having high hardness.

From the presented articles conducted about the turning process using TiAlN-based tools, it can be concluded that the main wear mechanisms that coated turning tools are subjected to are, abrasion and adhesion, with the formation of BUE [39]. In the following Figure 23 a clear example of BUE can be observed; in the image the built up material can be seen the coated tool's edge.

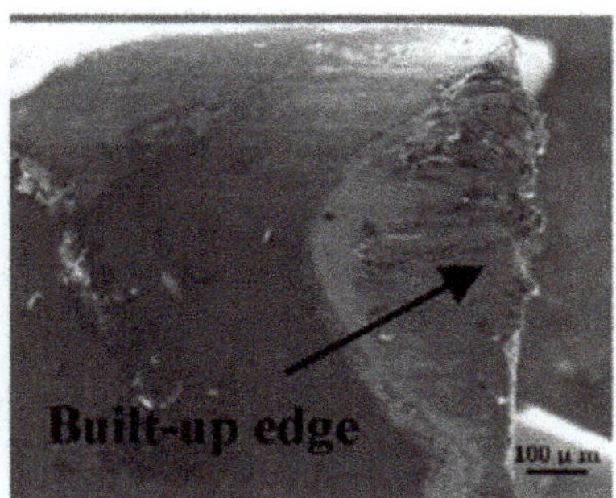

Figure 23. BUE observed at a gradient TiAlSiN-coated tool's tip, after some machining (roughly 125 m of cutting length [82].

2.1.2. Comparison of the Coating's Mechanical Properties

In this chapter an analysis of the mechanical properties of the monolayered TiAlN-based coatings registered in the various works presented in this section is going to be presented. The best hardness and Young's Modulus values for the main coatings, that were obtained by the various authors, were collected, and are presented in Table 3.

Table 3. Monolayered TiAlN-based coatings mechanical properties.

Coating	Hardness (GPa)	Young's Modulus (GPa)
TiAlN [60,64–66,68,84–86]	29.2–34.9	482–511
AlTiN [67,85]	20.0–24.3	337–358
TiAlSiN [65]	25.2–29.8	266–288
CrTiAlSiN [71]	24.0–27.1	200–275
AlTiN–Ni [61]	20.9–24.3	300–316
TiAlMoN [59]	21.58–37.0	510–620
TiAlYN [63]	31.6–35.2	520–575
TiAlTaN [63]	27.0–31.0	542–588
TiAlTaYN [63]	31.7–34.3	526–568
TiAlRuN [62,87]	26.09–33.2	334–492

The addition of elements is a very influential factor on the coating's mechanical properties. However, the coating's deposition method, microstructure and level of residual stresses also impact on the coatings' mechanical properties, with some elements having a greater influence on the hardness, such as Si and Cr. This hardness increase is well documented for these elements, with coatings such as TiAlSiN seeing great use in machining applications, due to its increased wear performance when compared to base TiAlN coatings. New doping elements are being tried, with good results coming from the addition of elements such as Mo, Ru and Y, causing a significant increase in hardness and Young's Modulus values. However, research regarding machining applications for TiAlN-based coatings doped with these elements is quite sparce, as these types of coatings are quite novel. The addition of Ta is also quite novel for monolayered, yet this element was already implemented in the architectures of multilayered TiAlN-based coatings.

Some deposition methods and even variations on the composition of TiAlN-based coatings, such as TiAlN and AlTiN, also influence their mechanical properties (as seen in Chapter 1). One common case study is the evaluation of the coating's residual stresses on their mechanical properties and wear behavior. In most cases, some amount of residual compressive stresses is preferred, as it usually increases hardness and Young's Modulus values. Furthermore, these stresses are also related with better wear performance, lowering coating wear rates and crack propagation.

Regarding these coatings wear behavior, from the works presented above, it can be noted that the H/E ratio heavily influences the coating's wear performance, with a higher value being usually preferred. This ratio is presented in Figure 24, based on the information provided by Table 3. From the analysis of the table, it can be noted that CrTiAlSiN and TiAlSiN coatings are the ones with the highest H/E value. This is especially due to the addition of elements such as Cr and Si, that confer the tool with excellent mechanical properties and wear behavior. The TiAlMoN coatings also show a high ratio value, making them a very promising coating for machining applications. Furthermore, the addition of Mo also promotes a better corrosion resistance of the coated tool. Although these values can be indicative of the coating's wear performance, this is also dependent on the machined material. For example, coatings with low hardness value will experience more abrasion (despite having a higher H/E ratio). Thermal fatigue is also a factor, for example, while the CrTiAlSiN coating is less susceptive to the wear mechanisms such as adhesion and abrasion, which revealed has suffered from thermal cracking due to the machining's high temperatures [71]. This highlights the fact that coating choice is very important for a certain machining operation.

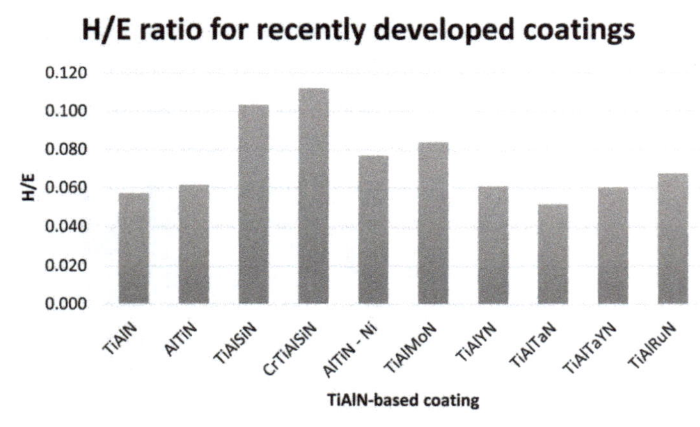

Figure 24. H/E ratios for the analyzed monolayered TiAlN-based coatings.

It is also worth to note that, although some coatings have very similar H/E ratios, their composition influences greatly their wear behavior and it should be factored on the coating choice.

2.2. *Multilayered TiAlN-Based Coatings*

Multilayered coatings are very appealing to the machining industry, as they enable combinations of various coating's properties to best fit a machining application. Further, their multilayered structure prevents crack growth [23,53]. This versatility makes them the most used type of coatings, in terms of coating structure, in the machining industry, for both the turning and milling sectors [39,43]. In terms of multilayer coating developments in general, there seems to be a trend in optimizing layer thickness [39]. Furthermore, there is a focus on studying the behavior of various types of coating combinations. This seems to be the case on the studies conducted around TiAlN-based coatings. Analyzing recent studies made on this regard, new combinations based on the coatings presented in Section 2.1 are being studied. There are also some studies about the improvement of already well-known TiAlN-based multilayered coatings, such as the TiAlN/TiN, Ti/TiAlN and TiN/TiAlN coatings, as seen in this work conducted by Zhang et al. [88], where the authors study the cyclic oxidation of Ti/TiAlN coatings with differing thicknesses of Ti layer, having 0.15 and 0.3 μm for coating 1 and 2, respectively. TiAlN single layer coatings with differing thicknesses were also studied. Regarding the results obtained for the multilayered coatings, it was noted that the residual stresses of the thinner coating would be relieved after the tests. This coating exhibited cracking, as the stresses could not be accommodated. However, for the coating with the thicker Ti layer, the residual compressive stresses increased, with this coating showing no cracking due to the thicker ductile Ti layers. This is a relevant study, as compressive stresses contribute for the tribo-mechanical properties of the coating, leading to a better cutting performance [89]. Regarding the study of TiAlN/TiN coatings, Çomakli [90] compares the mechanical and corrosion properties of TiN, TiAlN and multilayered TiAlN/TiN-coated tools. It was reported that the multilayered coating presented a smoother surface, having a smaller grain than the surfaces of the other coatings. The hardness for the multilayer coating was also higher, due to the number of layers present in the coating [39,43]. Due to the combination of coating surface and mechanical properties, the multilayered coating had a lower friction coefficient, conducing to a lower wear rate. A similar study employs the use of CrN (Chromium Nitride) on a multilayered architecture for TiAlN coatings [91]. The author compares the wear performance and mechanical properties to TiAlN monolayer coating, reaching similar conclusions to those of the previous study, significantly improving the wear behavior by reducing the friction coefficient, while increasing the coating's mechanical properties.

As mentioned in the study presented above [88], the addition of interlayers of Ti can improve the wear performance of multilayered coatings. Similar to this, the work completed by Shugurov et al. [92] studies the influence of TiAl interlayers on a TiAlN-based multilayer coating. The authors study the influence of the number of layers/interlayers and their respective thickness on coating's mechanical properties and their wear behavior. The coating's structure can be observed in Figure 25.

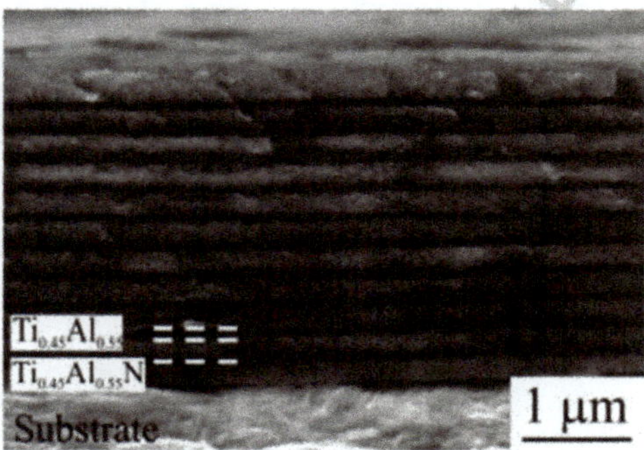

Figure 25. Multilayer structure of one of the TiAlN/TiAl coatings presented by Shugurov et al. [92].

The authors concluded that four layers of TiAlN and three interlayers of TiAl with thicknesses of 0.6 and 0.2 μm, respectively, would produce the best wear performance, exhibiting a wear rate three times inferior to that of the monolithic TiAlN coating. The results of this study provide very useful information regarding coating design, as this method can be used for a wide range of applications, especially machining.

The use of Ta is also being recently researched, as shown in the work presented in Section 1 [27], where various multilayered TiAlTaN-(TiAlN/TaN) with differing layer thicknesses were tested and subsequently characterized. In another recent study carried out by Shang et al. [93], the mechanical properties of a multilayered TiAlN/Ta coating are evaluated. The coating consists of three layers, a TiAl layer followed by a TiAlN layer and finally, a Ta layer (Figure 26). The mechanical properties of the multilayered coating were compared to a TiAlN monolithic coating. TiAlN/Ta coating exhibited a higher value of hardness and elastic modulus (31 GPa and 315 GPa, respectively), showing an increase of 29% and 47%, respectively. Ta is a ductile material and can dissipate energy through deformation, thus minimizing crack propagation. Furthermore, this element confers the coating with a better thermal stability. Studies such as these [27,93] show that Ta is very beneficial to improve TiAlN-based coating's wear behavior.

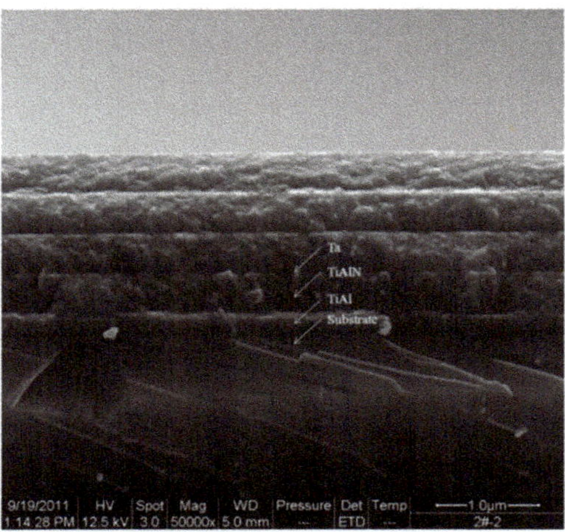

Figure 26. SEM cross-sectional image of the TiAlN/Ta multilayered coating, presented by Shang et al. [93].

It is known that multilayered coatings benefit from all the properties of their layers, as seen in the previous study [93], where a Ta layer promoted a lower wear rate for the coating, based on its ductility (primarily). This can be true for the same coating, yet with different mechanical properties, as seen in this study performed by Zhao et al. [94], where various multilayered TiAlSiN coatings are studied. These coatings consist of alternating layers of TiAlSiN deposited at different chamber pressures, thus causing hardness variation between the two coatings. TiAlSiN-1, obtained at 0.08 Pa chamber pressure presented 32.25 GPa of hardness, while the TiAlSiN-2, obtained at 0.2 Pa, exhibited 37.56 GPa for hardness value. Coatings were produced with 2, 5, 7 and 10 alternating layers of TiAlSiN-1 and TiAlSiN-2. It was observed that the hardness values for coatings with five or more layers were very close (of about 37 GPa). This was also the case for their elasticity modulus, however, the coating with five alternating layers showed the best plastic deformation resistance, and thus, the lowest wear rate. Though, the wear rate for coatings with five or more layers were very similar. The wear behavior of the multilayered coatings was also compared to monolayered TiAlSiN coatings, the former being significantly lower than the latter. There are also some recent coatings being developed, showing promising results, such as the TiAlCN/TiAlN coatings [30], and the TiCrAlCN/TiAlN [95]. These coatings show promising results in terms of hardness and wear behavior, however, the amount of research made around them is still quite sparse.

Although the multilayered coating is the most used in machining, recent study trends show that most of the research made about these types of coatings is about nano-multilayered coatings. This is justified by the ability to obtain very thin layers using recent technologies. These types of coatings will be the focus in Section 2.3.

In the next section, recent machining applications of the TiAlN-based multilayered coatings are going to be presented. This will be done in the same manner as in the previous Section 2.1.1.

2.2.1. Machining Applications and Coating Wear Behavior

Recent machining applications for TiAlN-based coatings were analyzed. As for the development and study of multilayered coatings, recent machining applications tend to be centered on nanolayered and nanocomposite coatings. However, there is still some research being made about the use of regular multilayered coatings in machining operations, namely milling in turning.

Milling Process

Regarding the study of the wear behavior of multilayered coated tools, there has also been some research made on this regard. Based on its versatility, the multilayered coating is very popular in the machining industry. However, in terms of recent research, there are few papers made on the study of the milling performance of new TiAlN-based multilayered coatings.

Due to their versatility and multilayer structure, these types of coatings usually outperform regular monolayered coatings, especially due to their toughness and crack propagation resistance. With recent studies such as the one developed by An et al. [96], where the performance of CVD and PVD coatings on the face milling of Ti-6242S and Ti-555 titanium alloys is evaluated. The PVD coatings is a multilayered TiAlN+TiN coating and the CVD coating is a TiCN+Al_2O_3+TiN coating. The cutting forces were evaluated and during the process, these were lower for the use of the PVD coating, as seen in Figure 27.

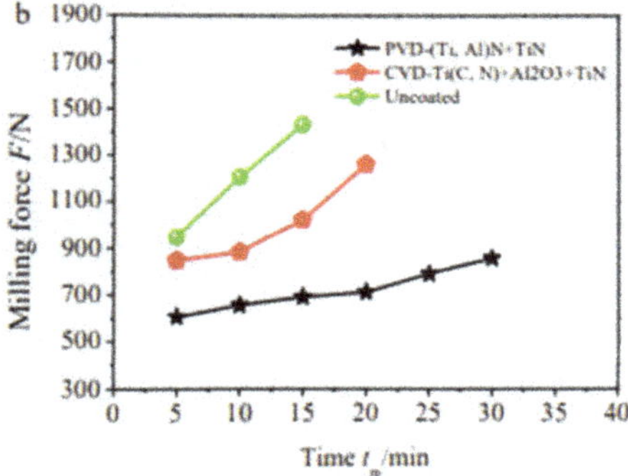

Figure 27. Cutting force value over machining time, for uncoated, PVD and chemical vapor deposition (CVD)-coated tools, presented by An et al. [96].

The wear sustained by the coated tools was also evaluated, reporting for the milling of Ti-624S that the main wear mechanisms for both CVD and PVD coating was microchipping and adhesive wear. Milling tests carried out on the Ti-555 alloy produced severe chipping and adhesive wear on the CVD-coated tool. However, in the PVD-coated tool, the main wear mechanism was adhesion, being less severe than with the other tool. The TiAlN-based coating proved to be better suited for the milling of these titanium alloys, showing better wear resistance and fracture resistance than the CVD-coated tool (the wear sustained by the PVD tools can be observed in Figures 28 and 29). This is not only due to the TiAlN properties for high-speed machining, but also due to the multilayered structure and residual stresses, characteristic of PVD coatings [43]. These characteristics make these types of coating highly resistance to crack propagation. Still regarding TiAlN-based multilayer coating applications in milling processes, the TiAlN/NbN coating is also known for its machining applications, due to their excellent mechanical properties, as shown by Varghese et al. [97], where they determine the coating's properties and employ them in the dry end milling of AISI 304 steel. The authors studied and evaluated the wear suffered by the coated inserts, reporting that abrasion was the main wear mechanism, eventually resulting in coating chipping and breakage.

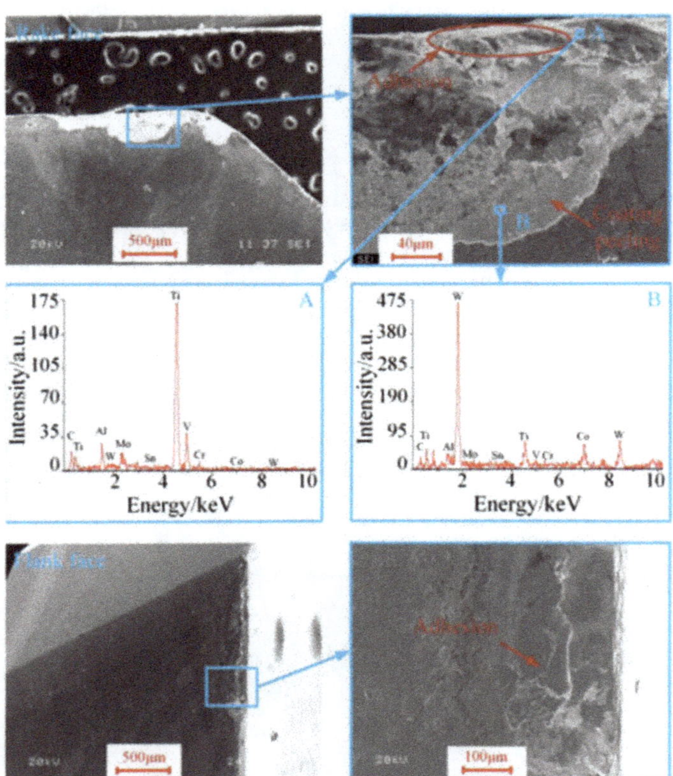

Figure 28. SEM and EDS analysis of the PVD coated tool's wear in milling Ti-555, EDS graphics correspond to the zones A and B, providing information regarding element composition, presented by An et al. [96].

As seen from Figures 28 and 29, and the results reported by the authors, the wear mechanisms sustained by multilayered coated tools in milling are very similar to those sustained by monolayered coated tools. Indeed, the main wear mechanisms present in the milling process are adhesion and abrasion. However, the multilayer architecture contributes to an improvement of coating's properties, such as wear resistance and crack propagation resistance. This causes these mechanisms to manifest at a much later stage of the machining process, albeit, in a similar way to monolayered coatings.

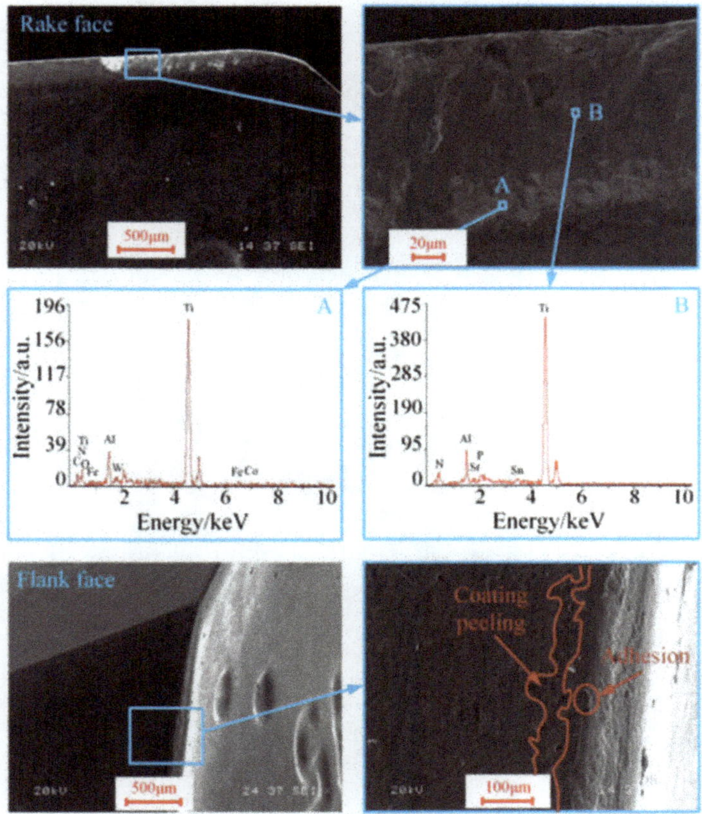

Figure 29. SEM and EDS analysis of the PVD-coated tool's wear in milling Ti-624S, EDS graphics correspond to the zones A and B, providing information regarding element composition, presented by An et al. [96].

Turning Process

Similar to the case of multilayered TiAlN-based coatings for milling applications, there seems to be very few new researches being made on this topic, with the attention being shifted to nanolayered and nanostructured coatings. However, there are some recent studies that focus on the wear performance of well-known TiAlN-based multilayered coatings, such as TiAlN/TiN. Analyzing their wear mechanisms when machining certain materials, such as in the study presented by Zheng et al. [98], in which the wear mechanisms of TiAlN/TiN coated tool are analyzed, for dry turning of 300 M steel. The coated tool suffered was mainly mechanical abrasion and adhesion, leading to chipping and coating delamination. It was also reported the existence of micro-cracks in the tool flank. This is due to the high machining temperatures developed during dry turning, which also promoted adhesion. The TiAlN/TiN-coated tool's wear can be observed in Figure 30.

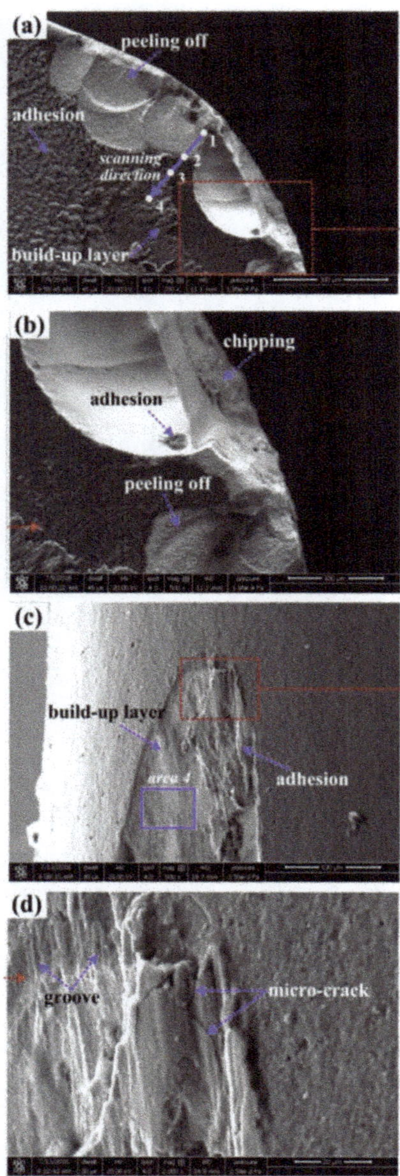

Figure 30. TiAlN/TiN-coated tool wear after dry turning of 300 M steel, rake face with 200× magnification (**a**) and 500× (**b**) and flank face with 200× magnification (**c**) and 500× magnification (**d**) presented by Zheng et al. [98].

The use of these types of multilayer coatings in hard-to-machine materials, and the study of its wear mechanisms is still a relevant subject. As the use coatings reduces machining forces and wear rates, thus making the dry machining method viable for the machining of these alloys. The use of TiN/TiAlN coating in the dry turning of Inconel 825 is studied by Thakur et al. [99], for finishing and roughing conditions. In total, three lubrication techniques were employed; these being dry, flooded and MQL. However, only the coated tool was employed in the dry machining. For the coated tool, the machining forces that were registered were lower, the surface finish of the machined material was better. However, the machining temperature was the highest for coated tool use. Wear was also analyzed for the tested tools, registering main wear mechanisms as abrasion and BUE and some minor coating delamination for the TiN/TiAlN-coated tool.

Multilayered TiAlN coatings are also being studied for turning applications. It is known that compressive residual stresses can improve cutting behavior and wear behavior of the coated tool's, however, to much compressive stresses can be detrimental for the coating's properties. As seen in this paper by Abdoas et al. [100], where the authors deposit three 11 μm thick TiAlN multilayer coating with different levels of residual stress. The three coatings, having 1.3 GPa, 2.1 GPa and 4.5 GPa, showed different cutting performances, with the coatings with fewer residual stresses having the most tool life. However, in terms of surface roughness, these coatings produced very similar results. The increase in tool life for the coating will less stresses, can be explained by the fact that this coating had better adhesion properties than the other coatings, thus promoting a better wear behavior for the coated tool. As for the wear mechanisms, these were the same for all three coatings, albeit in different intensities. The main ones that were registered were: material adhesion and BUE, with some abrasion being registered as well. This follows a similar trend to the wear mechanisms registered in monolayered coatings, similar to the milling cases mentioned in this section.

2.2.2. Comparison of the Coating's Mechanical Properties

As in Section 2.1.2, here the mechanical properties of the studied multilayered TiAlN-based coatings are going to be presented. These values are taken based on the results presented in various articles about the development/study of these types of coatings. Unfortunately, the table is missing some values that were not provided by the authors.

The various multilayered coating's properties are now going to be presented in Table 4.

Table 4. Multilayered TiAlN-based coatings mechanical properties.

Coating	Hardness (GPa)	Young's Modulus (GPa)
TiAlN (multilayer) [89,100]	29.0–37.0	370–462
TiAlSiN (multilayer) [94]	32.5–37.6	400–415
TiAlN/Ta [93]	29.0–33.0	300–325
Ti/TiAlN [88]	25.0–35.0	325–410
TiAlN/CrN [91]	37–41	420–475
TiAlN/TiN [90,96,98]	42–46	N/A
TiAlN/TiAl [92]	22.5–33.1	220–350
TiAlN/NbN [97]	16–30	433–606
TiCrAlCN/TiAlN [95]	23	N/A

The study of multilayered TiAlN-based coatings and their application in machining is not very abundant. However, the studies that are made about this topic focus primarily on the improvement of already existing multilayered TiAlN-based coating. There are some studies on novel coating structures such as the TiCrAlCN/TiAlN, but there is few information regarding its mechanical properties. Furthermore, machining case studies with the application of these are also sparse, these focusing on already existing multilayered coatings. Recent studies also focus on the improvement of coating's mechanical properties by introducing a novel structure, as seen in the case for TiAlSiN recently released [94]. This is also the case for recent research about TiAlN multilayer coatings, where layer thickness and residual stresses are tied to mechanical properties and wear behavior. The use of Ta is also studied for multilayered coatings, with the research made on this topic bearing better results in terms of mechanical properties and wear behavior using a first layer of Ta, it was concluded that coating's hardness and wear performance was significantly improved.

As presented in Section 2.1.2, the H/E ratio of the various analyzed coatings is going to be presented in Figure 31. Only coatings with complete information regarding this ratio will be presented.

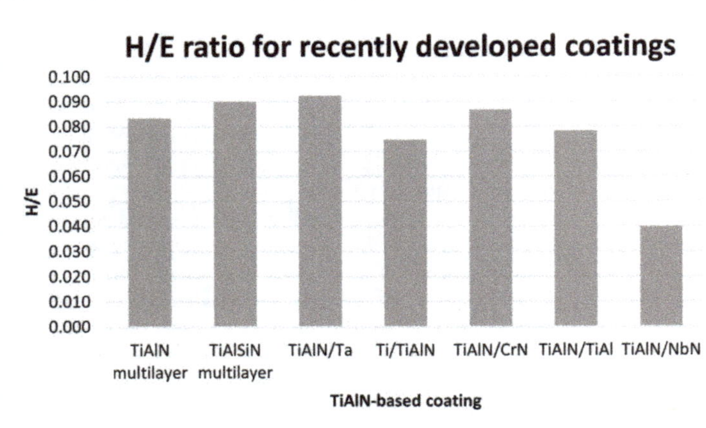

Figure 31. H/E ratio for recently researched multilayered TiAlN-based coatings.

As it was noted for monolayered coatings, the addition of certain elements such as Si, Ta and Cr significantly improve the wear behavior of the TiAlN based coatings. Recent research also shows that residual compressive stresses, microstructure and layer thickness also influence the wear behavior of the coatings, with TiAlN and TiAlSiN multilayered coatings achieving high values of hardness and Young's Modulus values, while also having a good performance in terms of wear.

2.3. *Nanolayered TiAlN-Based Coatings*

Recent studies made on coatings tend, in general, to be about nanostructured and nanocomposite coatings [39,43]. Nanolayered coatings, similarly to the multilayered coatings, have an increased crack propagation resistance. This is more intense on nanolayered coatings, due to the higher number of layers. Hardness values also tend to be higher on these types of coatings.

Regarding TiAlN-based coatings, the research trend is the same for other coating types, being more abundant in the area for nanolayered and nanocomposite TiAlN-based coatings. These recent papers tend to focus on the study of various novel coatings, based on the development of monolayered TiAlN-based coatings, more precisely, using elements that are tied to an improvement in mechanical properties such as Mo, Ta and Cr. In this section the various novel nanolayered/nanocomposite TiAlN-based coatings that are under development/study are going to be presented, mentioning their mechanical properties and wear behavior. Furthermore, as done in previous sections, the machining applications for these types of coatings are going to be presented for turning and milling. The various wear mechanisms that these tools are subject to are also going to be mentioned, presenting a comparison between coatings (when possible).

There is a current focus on the study of these types of coating over regular monolayered and multilayered coatings, with commonly known multilayered structures such as TiAl/TiAN [101], being attempted at a nanometric scale, with thinner layers conferring the coating with improved mechanical properties, such as high hardness, improved corrosion performance and high coating adhesion. As the layers are considerably thinner when compared to regular multilayered coatings, the number of layers is also higher in nanolayered coatings. This not only increases hardness, but also increases the crack propagation resistance of the coating. The influence of the layer thickness in nanolayered coatings is analyzed in the study performed by Wang et al. [102], where a TiN/TiAlN coating (another well-known multilayered architecture) is characterized in terms of mechanical properties and wear behavior. By controlling the rotation speed of coating deposition, the authors were able to control the thickness of deposited TiN and TiAlN layers. They determined the layer thickness value for highest mechanical properties (hardness and Young's Modulus), this value was found to be 13 nm. It was also noted that the wear behavior of the coating was improved for lesser thick layers, as this promoted crack propagation resistance. The coatings microstructure at different deposition rates can be observed in Figure 32.

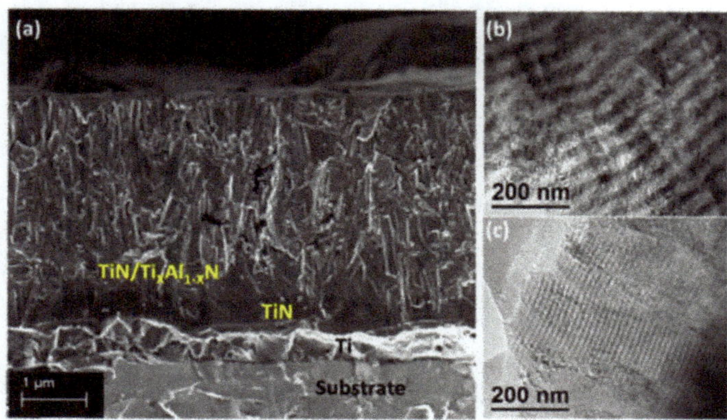

Figure 32. SEM cross-sectional image of TiN/TiAlN coatings for a rotation speed of 1 r.p.m. (**a**); TEM cross-sectional images for rotation of: 1 r.p.m. (**b**), 2 r.p.m. (**c**), presented by Wang et al. [102].

Another multilayered structure that has got some attention is the Al/TiAlN coating. In the study carried out by Liang et al. [103], an Al/TiAlN nanocomposite coating deposited on AZ91D magnesium alloy is analyzed. Similar to the previous study, here the authors studied the influence of different layer thickness on the mechanical properties and microstructure of the coating. In addition to the properties, the authors also analyzed the corrosion resistance of the coating. The authors produced four coatings with a thickness of 5 μm. The phases that form the nanocomposite film are TiN nanocrystal and amorphous AlN. These coatings had different interface periods, 100 nm, 200 nm, 300 nm and 400 nm. Regarding coating microstructure, it is columnar for 400 nm periods, however, it changes to multi-interfaces as the period is thinner. Hardness values reached their highest value for the lowest thickness period (100 nm), reaching 31.3 GPa. The authors also noted that the reduced thickness induced improved corrosion resistance, with the lowest thickness performing better in this regard as well. Layer thickness influence is a highly researched topic in nanolayered coatings, being related with an increase in mechanical properties (primarily hardness) and, as mentioned before, in nanolayered coatings the overall thickness of the coating improves hardness, as the high number of layers promotes a hardness increase. Additionally, this high number of layers in thick nanolayered coatings promotes a compressive stress relief on the coatings. The number of layers and their arrangement in a nanolayered coating also influences coating performance, which can be controlled by changing target arrangement (in deposition) and altering deposition time. This is highlighted in the study presented by Seidl et al. [104], which evaluated the influence of target arrangement in producing various AlCrN/TiAlTaN coatings. The control of the number of layers of AlCrN or TiAlTaN can influence greatly the coating's properties, with AlCrN (in this case) promoting better mechanical properties, while the TiAlTaN coating promoted a better oxidation resistance at higher temperatures. As seen in this study, the authors used a TiAlN coating with Ta addition, an element that is recently being researched for monolayered TiAlN-based coatings, as its addition is tied to improve mechanical properties.

The addition of Si is also very beneficial for the coating's mechanical properties, with some studies evaluating new structures such as the structure presented in [105], consisting of a first layer of TiAlN, followed by a nano multilayered TiAlN/AlCrSiN coating and finally, an outer layer of TiAlN coating. This coating's structure can be observed in Figure 33.

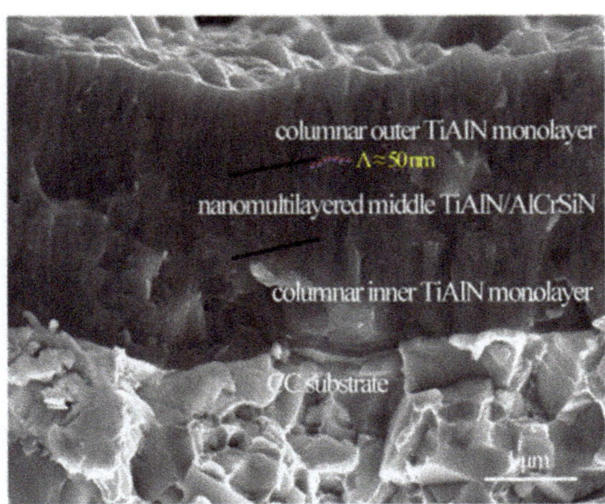

Figure 33. SEM cross-sectional image of TiAlN-(TiAlN/CrAlSiN)-TiAlN coating deposited on a cemented carbide presented by Xian et al. [105].

From the studies presented in Section 2.1, elements such as Mo and Y, when added to TiAlN coatings can significantly improve their hardness and Young's Modulus values, as well as their wear performance and corrosion resistance. In the study performed by Pshyk et al. [106], the novel nanocomposite (TiAlSiY)N and nanoscale (TiAlSiY)N/MoN multilayered coating obtained by arc-PVD are evaluated. The authors analyzed the microstructure, phase composition and mechanical properties of these coatings. Regarding the microstructure, the authors concluded that the multilayered coating had the preferred orientation, when compared to the monolayered nanocomposite coating. Moreover, the mechanical properties of the multilayered coating were better than the nanocomposite coating, showing values of 38.37 GPa and 392.5 GPa for hardness and Young's Modulus values, respectively. The (TiAlSiY)N/MoN coating also exhibited a better fracture toughness and a higher H/E ratio than the monolayered coating, thus having a better wear performance. Another study carried out by Kravchenko et al. [107] compare this novel coating with other similar structured coatings, namely, (TiAlSiY)N/CrN and (TiAlSiY)N/ZrN. These coatings were also obtained by arc-PVD and their mechanical properties were characterized. From that study, the authors concluded that the (TiAlSiY)N/MoN coating had higher values for hardness and Young's Modulus (35.9 GPa and 406.8 GPa, respectively) than the other coatings, with (TiAlSiY)N/CrN coating having 23.4 GPa and 300 GPa, and (TiAlSiY)N/ZrN having 22.1 GPa and 271 GPa, for hardness and Young's Modulus, respectively. However, the H/E and the plastic deformation indexes were very similar for all coatings, which means that these coatings (CrN and ZrN multilayer) have good tribological properties. Studies such as these [106,107] highlight the benefits of using these nano-multilayered coatings, especially for extreme tribological applications. Regarding TiAlN based nanolayered films with ZrN, Wang et al. [108] studied the influence of Zr_3N_4 on a nano multilayered TiAlN/Zr_3N_4. Additionally, the authors also studied the influence of layer Zr_3N_4 thickness in the coating's mechanical properties, and, as seen in the studies previously presented, thinner layers promote higher hardness and H/E ratio values. The authors also report that the Zr_3N_4 causes a significant increase in coating hardness (34.7 GPa) while retaining a very high toughness, conferring this coating with an excellent wear behavior. The hardness,

Young's Modulus and H/E ratio variation of TiAlN/ Zr_3N_4 for different layer thicknesses can be observed in Figure 34.

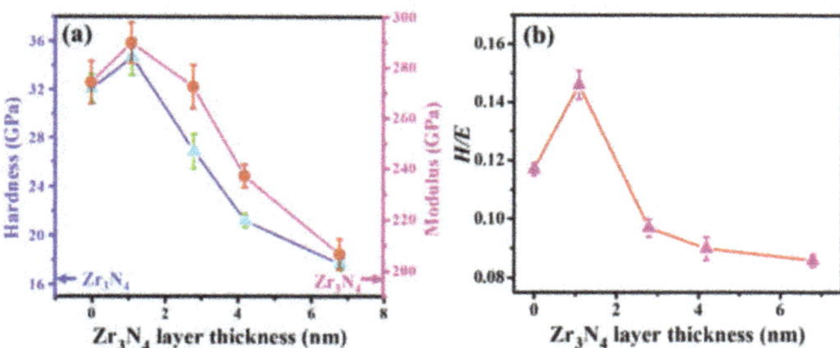

Figure 34. Hardness, Young's Modulus values (**a**) and H/E ratio values (**b**) for TiAlN/ Zr_3N_4 nano multilayered coatings, presented by Wang et al. [108].

The addition of Cu to tool coatings has some advantages, such as the reduction of friction coefficient as it is a soft and ductile material. The addition and use of these elements in nanocomposite coating has got some promising results, as seen in the study carried out by Chen et al. [109], where nanocomposite TiAlN/Cu coatings provided with varying percentages of Cu concentration (0–1.4 at % Cu concentration) are deposited by filtered cathodic arc ion plating. The authors have evaluated the coating's microstructure (Figure 35) and mechanical properties.

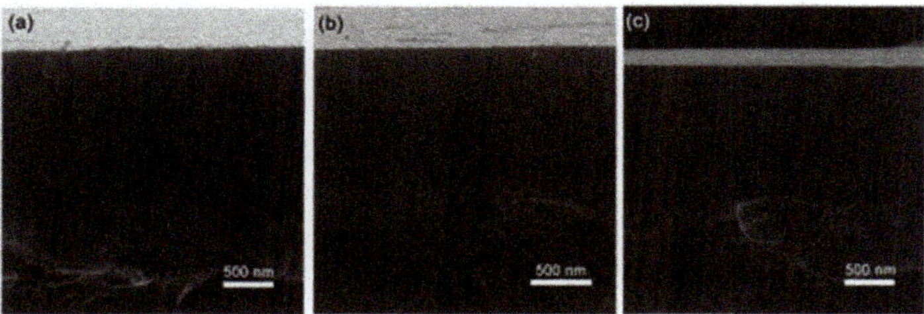

Figure 35. SEM cross-sectional images of TiAlN/Cu coatings with different concentrations of Cu: 0% at. % Cu (**a**); 0.8 at. % Cu (**b**); 1.4 at. % Cu (**c**), presented by Chen et al. [109].

From Figure 35, it can be observed that the addition of Cu results in a reduction in grain size, decreasing from 45 nm (in TiAlN) to 30 nm for the coating containing the highest Cu concentration. TiAlN presents a distinct columnar structure that gradually fades away with the addition of Cu. Regarding the coating's mechanical properties, there is a decrease in hardness value and Young's Modulus value with the increase in Cu concentration. There is also an influence on the friction coefficient, with the lowest value being obtained for a 0.8 at. % Cu concentration. Other Cu concentrations produced a higher friction coefficient, surpassing even the TiAlN coating. The addition of this element is an interesting topic, however, the sacrifice in mechanical properties does not seem to be relevant enough to use it to decrease in friction coefficient. However, the results presented in that paper show potential for the employment of Cu in coatings, as a decrease in friction coefficient is desirable, especially for applications such as machining. Still regarding the additions of softer elements to hard coatings such as TiAlN, in a similar study performed by Mejía et al. [110], the characterization of a TiAlN (Ag,Cu) nanocomposite coating is studied. As in the study previously mentioned here [109], the influence of the addition of different concentrations of (Ag,Cu) nanoparticles on the coating's mechanical properties and microstructure is studied. The addition of this softer element causes a grain refinement in the TiAlN coating's microstructure, changing from a columnar structure to an amorphous one with a smaller grain. Equal to the addition of Cu, the addition of these softer elements causes a decrease in mechanical properties (hardness and Young's Modulus value) and a decrease in friction coefficient. Additionally, the authors noted that with an increasing concentration of (Ag,Cu) the coating's residual compressive stresses would decrease.

2.3.1. Machining Applications and Coating Wear Behavior

In this section, the various studies regarding milling and turning applications of these types of coatings are going to be presented, mentioning the improvements that these types of coatings bring for machining. The wear behavior described in these studies is also going to be analyzed and described in this chapter.

Milling Process

Recent research on TiAlN-based coating's performance seems to be shifting to the use of these nanolayered and nanocomposite tool coatings. There have been many improvements recently on the development of new promising coatings for machining applications. However, the today's studies seem to focus on nanolayered and nanocomposite coatings containing elements such as Si and Cr, known as able to confer excellent mechanical properties and cutting performance to tool coatings. This recent paper presented by Geng et al. [111] studies the milling performance of TiSiN/AlTiN nanolayered composite film. The authors also evaluate de coating's microstructure and mechanical properties, which can be observed in Figure 36.

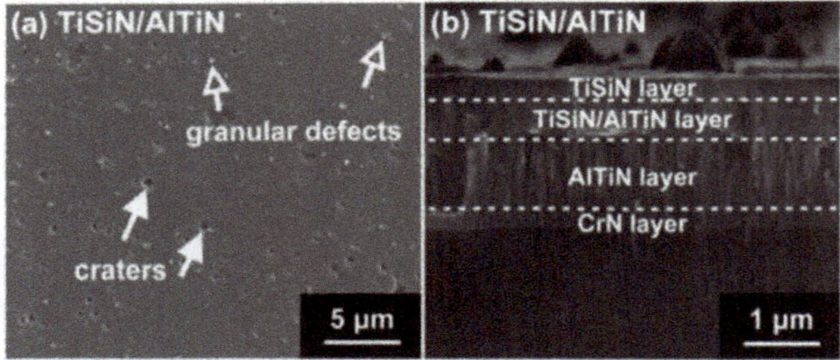

Figure 36. Surface (**a**) and SEM cross-sectional image (**b**) of TiSiN/AlTiN film, presented by Geng et al. [111].

Regarding the coating's mechanical properties, the registered peak values for hardness and Young's Modulus were 41.7 GPa and 340 GPa, respectively. These coatings were deposited onto 4-fluted end-mill with a 6-mm diameter and were subsequently employed in the dry milling of SKD 11 tool steel. The tool's wear was analyzed after the milling operations, and it was reported that the main wear mechanism was abrasion, as seen in Figure 37.

The authors noted that the TiSiN/AlTiN-coated tools showed little to no adhesion of SKD 11 to the coating's surface and the tools' edges. This resulted in a reduction of machining forces and cutting temperature, thus significantly improving the tool's life. Furthermore, there is a slight reduction in wear rate for a machining temperature of 400 °C. This study highlights the wear benefits that come from the employment of these coatings in machining.

As seen in the previous Section 2.3 (Nanolayered TiAlN-based coatings), multi nanolayered coatings with thinner layers exhibit an increase in mechanical properties (primarily hardness) and in wear behavior (low wear rate). This is also related to tool life, as presented by Teppernegg et al. [112]. Here, the authors studied a nano multilayer coating consisting of TiAlN and CrAlN sublayers with different thicknesses (10, 30, 100 and 300 nm), seen in Figure 38. These coatings are deposited onto inserts and their mechanical properties are evaluated; these are then employed in the milling of 42CrMo4 steel.

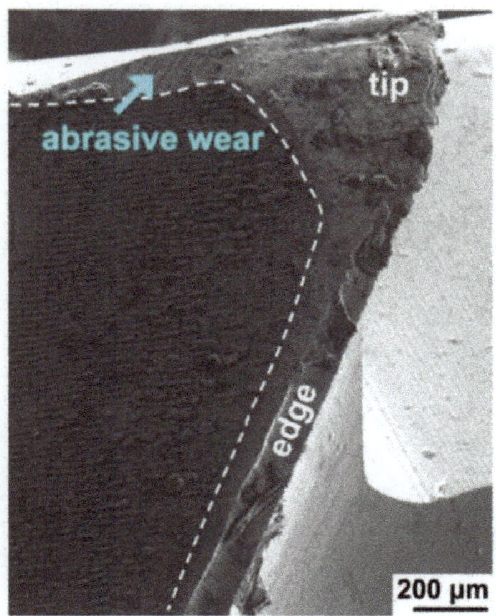

Figure 37. SEM images of the flank wear sustained bu TiSiN/AlTiN-coated tools after dry milling of SKD 11, presented by Geng et al. [111].

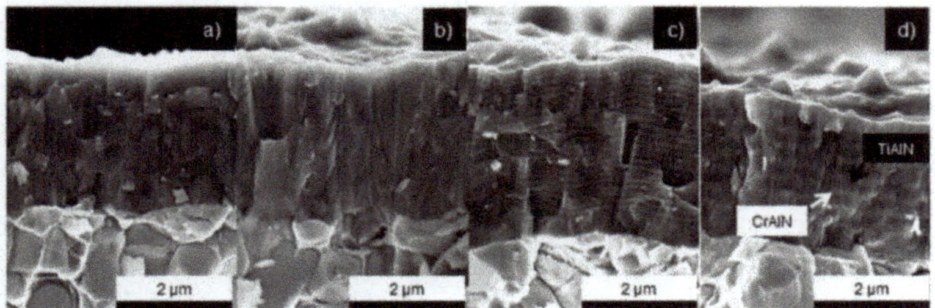

Figure 38. SEM cross-sectional image of the multilayered coatings with differing sublayer. 10 nm (**a**), 30 nm (**b**), 100 nm (**c**), 300 nm (**d**), presented by Teppernegg et al. [112].

It was noticed that an increase in Al content promoted higher hardness values being it related to cutting performance. Thus, the coating's that presented a higher Al content performed better in the milling tests. The sublayer thickness did not influence the hardness greatly, however, in terms of tool life this was not the case. The authors reported that with increased sublayer thickness the tool life would decrease. They were able to determine optimal sublayer thickness as seen in the Figure 39.

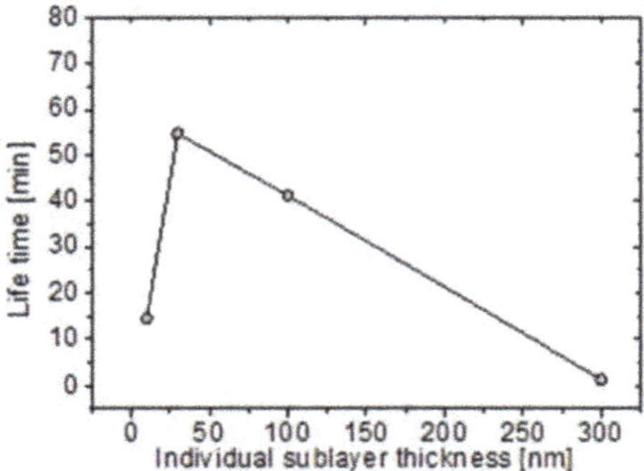

Figure 39. Tool life for the various layer thicknesses tested for the TiAlN/CrAlN nano multilayer coating, presented by Teppernegg et al. [112].

All the tested coatings exhibit the same type of wear. Coating failure occurs due to abrasion on the flank, causing coating erosion and delamination thus exposing the substrate. The other type of wear presented by the tools is thermal fatigue, this generating comb-cracks appearing on the coated inserts. These cracks can be seen in Figure 40.

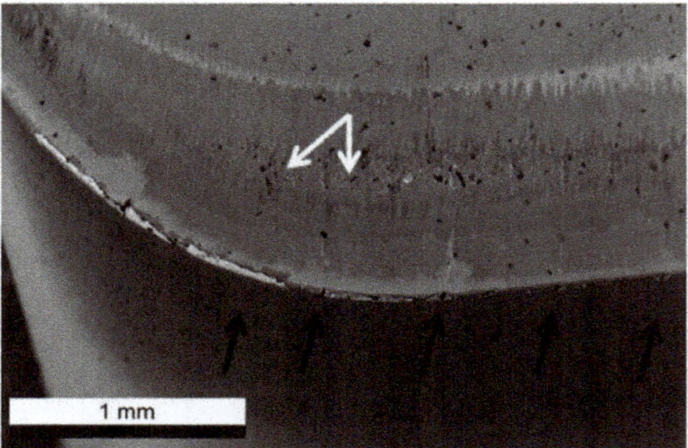

Figure 40. SEM image of the coated insert's cutting edge, the black arrows mark the position of comb-cracks and the white arrows mark the position of cracks that are parallel to the cutting edge, presented by Teppernegg et al. [112].

The employment of TiAlCrSiYN/TiAlCrN has also seen some studies lately, with Chowdhury et al. [113,114] evaluating the mechanical properties of this coating for various coating architectures and differing interlayer thicknesses. The authors also employ these coatings in the dry milling of stainless steel at a machining speed of 600 m/min. As seen in previous studies, the thinner the interlayer thickness is, the higher the hardness is. Chowdhury et al. [113,114] also compare these multilayered coatings with monolithic TiAlCrSiYN and TiAlCrN coatings. It was concluded that the nanolayered coatings, with optimized interlayer thickness produced the best results in terms of cutting performance and wear behavior, indicating that these types of nanolayered coatings are an excellent choice for extreme machining applications.

Turning Process

Similarly, to the research presented for the use of nanolayered and nanocomposite coatings in milling, for the turning process research focuses on coatings with Cr and

Si additions. These elements confer the coatings with excellent mechanical properties, and they improve the cutting performance significantly (especially when combined with TiAlN-based coatings). The use of these elements in nanolayered coatings has even more potential, increasing even more these properties and the cutting performance of coated tools, especially by extending the tool-life [115]. In the study conducted by Sui et al. [116] the performance of TiAlN/CrN coatings with different bilayer periods is evaluated in the high-speed turning of TC4 titanium alloy at 100 m/min. Bilayer periods between 12 nm and 270 nm were tested, and the influence of these periods on coating microstructure can be observed in Figure 41.

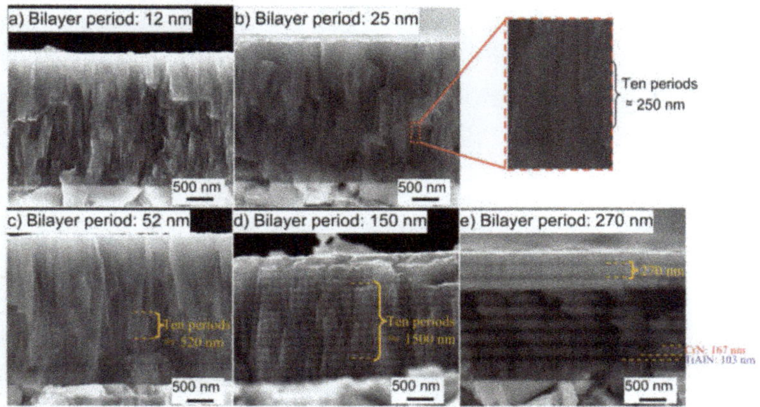

Figure 41. SEM cross-sectional image of the TiAlN/CrN coatings with different bilayer periods: 12 nm (**a**), 25 nm (**b**), 52 nm (**c**), 150 bm (**d**) and 270 nm (**e**), presented by Sui et al. [116].

With an increase in bilayer period, an increase in the coating's hardness was observed. This is due to a grain refinement that occurs with thinner layers. However, for the thinnest bilayer period, the hardness was also very high, such as the values obtained for the bilayer period of 270 nm. The authors also registered a higher wear rate for the thinner bilayer period (12 nm), with the thickest (270 nm) showing the best wear rate values. Regarding the wear mechanisms sustained by the tools, they were primarily abrasion and coating delamination. Although the tools experienced the same type of wear, the tool coated with the coating provided with the thickest layers exhibited less severe wear.

Still regarding the turning performance of nanolayered coatings, in the study developed by Zhang et al. [117], a comparison between AlTiN monolayered coating and AlTiN/AlCrSiN nano multilayered coatings is made. Furthermore, various nanolayered coatings were deposited with differing modulation period, that is, with differing layer thickness. The tested modulation period was between 4.2 nm and 17.8 nm. The various coatings mechanical properties were also determined. It was determined that the period of 8.3 nm (Figure 42) produced the best results in terms of hardness and Young's Modulus values for the coating (37.5 GPa and 486.9 GPa, respectively). Moreover, this coating exhibited the best values of H/E ratio and the best adhesion strength.

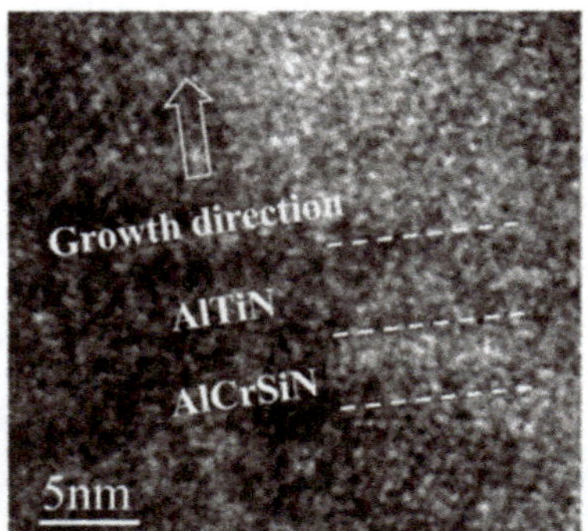

Figure 42. High-resolution TEM micrograph of AlTiN/AlCrSiN with a modulation period of 8.3 nm, presented by Zhang et al. [117].

The turning performance of these coatings was evaluated in the dry turning of SKD 11 tool steel at 250 m/min (cutting speed). The same wear mechanisms for AlTiN/AlCrSiN coated tools were reported, with abrasive and adhesive wear being reported as the main mechanisms. There was also the formation of BUE and some plastic deformation reported on the tool's rake face. The tool wear for the nanolayered coating with a modulation period of 8.3 nm can be observed in Figure 43. This coated tool exhibited the least wear rate of all coated tools, this is due to their high mechanical properties H/E ratio and high adhesion strength.

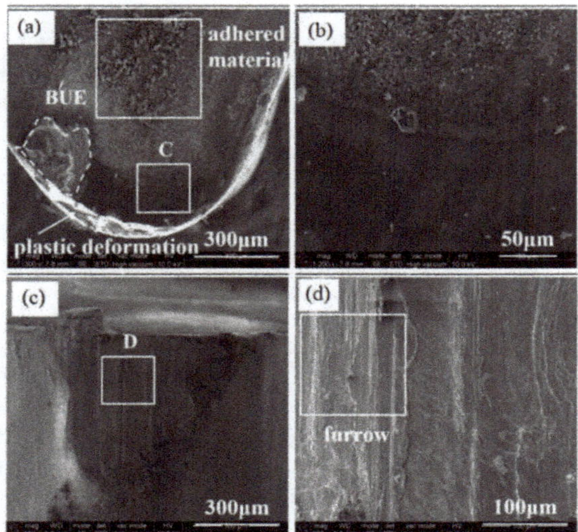

Figure 43. SEM image of the AlTiN/AlCrSiN (modulation period of 8.3 nm) coated tool's wear: rake face (**a**); magnified rake face (**b**), flank face (**c**), amplified flank face (**d**), presented by Zhang et al. [117].

From the studies presented, it can be concluded that abrasion is the common wear mechanisms sustained by nanolayered TiAlN-based coated tools. This is similar to the other types of coating's as well (monolayered and multilayered). However, the nanolayered coatings exhibit the highest tool-life values when compared to other types of coatings. This is due to their high count of very thin layers, conferring the coating with high hardness and high crack propagation resistance.

2.3.2. Comparison of the Coating's Mechanical Properties

The research trend for TiAlN-based coatings is heavily focused on the development and characterization of novel nanolayered and nanocomposite coatings. There have been improvements to already-known coating structures that are heavily employed, such as the TiN/TiAlN and Al/TiAlN. Studies conducted about these known structures focus on the improvement of mechanical properties and wear behavior by taking advantage of the nanolayered structure's benefits, such as an increased hardness value, reduced friction coefficient, reduced wear rate and high crack propagation resistance. These benefits become nanolayered and nanocomposite coatings very appealing, and therefore there are a higher number of papers done about this topic, than about regular multilayered coatings.

In Section 2.1., the addition of Mo to TiAlN-based coatings was covered. This element promotes an increase in the coating's mechanical properties and wear behavior. This was also observed in nanolayered and nanocomposite coatings, showing satisfactory results in creating coatings with high toughness and hardness values. The mechanical properties from the coatings that were mentioned in the previous subsubsection are going to be presented in Table 5.

Table 5. Hardness and Young's Modulus values for the nanolayered and nanocomposite TiAlN-based coatings.

Coating	Hardness (GPa)	Young's Modulus (GPa)
TiN/TiAlN [102]	31.5–42.5	417–520
Al/TIAlN [103]	23.0–31.0	N/A
(TiAlSiY)N/MoN [106,107]	36.0–38.0	395–406
(TiAlSiY)N/ZrN [107]	21.1–22.7	265–275
(TiAlSiY)N/CrN [107]	22.1–23.4	289–302
AlCrN/TiAlTaN [104]	32.0–42.0	N/A
TiAlN/ZrN [107,108]	22.0–36.0	210–290
TiAlN/Cu [109]	24.0–29.0	320–350
TiAlN(Ag,Cu) [110]	6.7–15.2	140–216
AlTiN/AlCrSiN [117]	28.5–31.0	410–450
TiSiN/TiAlN [115]	33.0–39.0	550–570
TiAlN/CrAlN [112]	25.0–30.0	N/A
TiAlCriSiYN/TiAlCiN [113,114]	24.0–33.0	430–475

It can be seen from the number of coatings presented in Table 4 alone, that there is considerably more research being made in novel coating development for these types of coatings when compared to monolayered and multilayered TiAlN-based coatings. Once again, it can be seen that the use of Mo increases significantly the coatings mechanical properties. There is some research, however, made about the improvement of already known structures. In regard to the TiN/TiAlN, a very high value of hardness was achieved, due to the nanolayered coating's properties.

Regarding these coating's wear behavior, (TiAlSiY)N/MoN showed great potential with very high values of hardness and good Young's Modulus values, becoming the types of coating that are very appealing for extreme applications, such as where wear is very intense. Some satisfactory results also come from the use of ZrN on the coatings, especially when coupled with TiAlN, achieving incredible mechanical properties. Figure 44 shows the H/E ratio of the coating analyzed in Section 2.3.

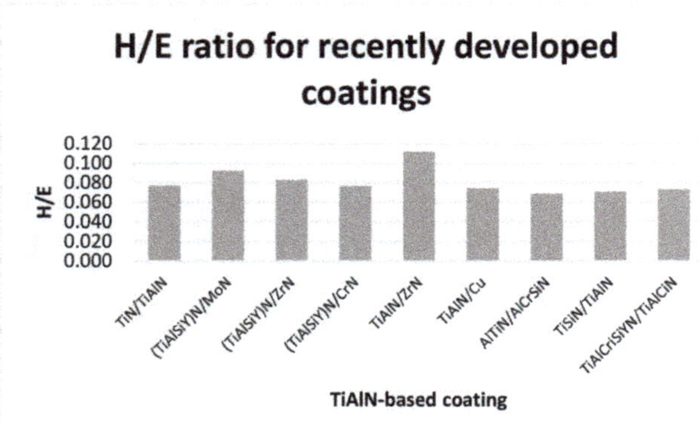

Figure 44. H/E ratio for recently researched nanolayered and nanocomposite coatings.

From analyzing the H/E ratios of the various coatings the TiAlN/ZrN coating has an incredibly high ratio, due to its mechanical properties. Indeed, this coating showed an excellent wear behavior, making it very promising for machining applications. Although there is variation of these ratios, only the TiAlN/ZrN coating ratio does not vary too much, an indicator that the wear performance of the nanolayered and nanocomposite coatings is superior to that of monolayered and multilayered TiAlN based coatings.

3. Machining Conditions and Tool Wear Mechanisms for TiAlN Based Coatings

Most of the research made about the TiAlN-based coatings regarding machining applications, is usually done about hard-to-machine materials. In this section, the various papers presented in Section 2 will be analyzed, presenting the materials that are currently being used in TiAlN based coating's testing. There are two sections, one for milling and another one for turning.

3.1. Milling Process

The Table 6 summarizes the various coatings used to machine a certain material, the range of machining speeds used and the main wear mechanisms of the coating.

Table 6. Materials machined by TiAlN-based coatings, applied in milling.

Material	Coating	Wear Mechanisms	Machining Speed Range [m/min]
Tool steels [70,71,74,97,111–114]	TiAlN	-Adhesion -Abrasion	40–600
	TiAlSIN	-Ploughing -Abrasion	
	CrTiAlSiN	-Thermal cracking -Abrasion	
	TiSiN/AlTiN	-Abrasion	
	TiAlN/CrAlN	-Micro-cracks -Abrasion	
	TiAlN/NbN	-Adhesion -Abrasion	

Table 6. *Cont.*

Material	Coating	Wear Mechanisms	Machining Speed Range [m/min]
Titanium alloys [72,75,76,96]	TiAlN	-BUE -Abrasion -Adhesion	9.6–173
	AlTiN	-Adhesion -Abrasion	
	TiAlN/TiN	-Adhesion -Abrasion -BUE	
Nimonic 75 [77]	TiAlN	-Coating chipping	13
Inconel 690 [73]	TiAlN	-Adhesion -Abrasion	60

As it was already concluded, the main wear mechanisms suffered while milling is abrasion and adhesion. For the milling of tool steel, abrasion is more predominant, however, for titanium alloys adhesion is more frequent. It is also important to note that multilayered and nanolayered tend to not suffer as much from adhesion problems.

3.2. Turning Process

Next, as in the previous section, the various coating used in the turning of various materials are going to be presented (Table 7), as well as the machining speeds that were used during testing.

Table 7. Materials machined by TiAlN-based coatings, applied in turning.

Material	Coating	Wear Mechanisms	Machining Speed Range [m/min]
Tool steels [81,83,98,100,117]	TiAlN	-Adhesion -Abrasion	120–500
	AlTiN	-Abrasion -Chipping	
	AlTiCrN	-Abrasion -Adhesion -BUE	
	TiAlN/TiN	-Abrasion -BUE	
Titanium alloys [82,115,116]	TiAlSiN	-Abrasion -BUE	100
	TiAlN/CrN	-Abrasion -Coating delamination	
	TiSiN/TiAlN	-Abrasion	
Inconel [78–80,99]	TiAlN	-Abrasion -Adhesion -BUE	30–120
	AlTiN	-Abrasion	
	TiN/TiAlN	-Abrasion	

The main wear mechanisms sustained by TiAlN-based coated tools when turning is abrasion and BUE. The studies about the employment of these coatings on turning operations, similarly to the ones for milling, focus on the machining of tool steel, titanium alloys and Inconel alloys.

Regarding the machining of Inconel, there is a clear advantage in having a higher concentration of Al in the coating, as it is related to the reduction of adhesive damage when machining these alloys.

4. Current Research Trends of TiAlN-Based Coatings

An analysis of current research about TiAlN-based coatings has been made and presented in this paper. Developments regarding monolayered, multilayered and nanolayered coating are currently being made, with a clear focus on the study of the addition of certain elements to TiAlN-based coatings in order to improve them. There are also studies made about the development of new coatings, such as self-lubricating coatings. As seen in the first chapter, the use of solid lubricants can significantly improve the performance of the coated tool [5]. This nanocomposite coating proves to have great potential, especially due to its low friction coefficient (due to self-lubricating behavior) which results in reduced wear behavior.

Still regarding recent research trends of TiAlN based coatings, similarly to the other coating type, there seems to be a focus on the study and development of nanolayered and nanocomposite coatings as presented in the previous chapters. These coatings have very high hardness values, due to the high number of layers, conferring super-hardness to these types of coatings (more than 40 GPa). There is also quite a lot of research done about these super-hard coatings (that do not need to be nanolayered as seen in [59]). These TiAlN-based super-hard coatings have better cutting performance when compared to the other regular TiAlN based coatings. This is due to their high hardness, a careful architecture of the coating's microstructure [118] and addition of some elements such as Mo [59] and Ta [119].

Regarding the research made on doping elements, it seems to be a popular topic, there are quite a lot of studies about the influence of certain elements in TiAlN-based coatings, with some yielding very satisfactory results. The use of Mo in coatings is well documented, with these coatings containing Mo, such as MoSeC coatings [120]. Mo addition not only improves the coating's mechanical properties, such as hardness and Young's Modulus [59,106,107], but also improves the friction coefficient of the coating, which promotes a better tool-life (reducing wear rate) [121,122]. There are some studies conducted about the reduction of wear coefficient of coatings, such as the ones using Cu and Ag [109,110] as doping elements. As it was observed, the addition of these elements resulted in a decrease in the coating's wear coefficient, but also hindered its mechanical properties, as it caused a reduction of hardness in the coatings. Mo is quite popular as it offers an increase in the coating's mechanical properties and improves the wear behavior of the coating, as seen in [106,107]. Moreover, it can be used in multilayer architecture, to be applied under the form of nitrides, such as MoN. The development of these low-friction coatings that can perform in rough conditions is collecting some attention, as seen in the paper presented by Bondarev et al. [123], where the MoSeC coating is paired with TiAlN based coating: TiAlSiCN. This super-hard coating (with about 41 GPa of hardness) had a comb-like structure that confers a high-thermal stability [124] and can be tailored to have a high oxidation resistance [125]. When this coating is paired with the MoSeC coating, which presents low friction coefficient [120], a very hard coating with an excellent wear behavior can be achieved. The authors [123] evaluated the wear behavior of TiAlSiCN/MoSeC coatings and have found that the addition of MoSeC caused a reduction in the hardness values registered in [124]. However, this addition highly improved the wear performance of these coatings, reducing the friction coefficient and improving the wear behavior at higher temperatures (up to 300 °C). Studies such as these show that the employment of Mo in coatings is quite beneficial, thus making it a popular research topic. Another element that as seen some research is Si, as the use of this element is quite popular in machining applications, especially for the TiAlSiN coating, which exhibit better mechanical properties and wear behavior than the TiAlN coating [65]. Its employment is quite beneficial, being used in some monolayered coatings, such as the CrTiAlSiN [71], and in some multilayered architectures [94], and nanolayered coatings [106,107,113,114]. As seen from the studies presented above [123–125] the TiAlSiCN coating has a very high thermal stability and excellent mechanical properties. From the study presented in [123], its employment in multilayered architectures also brings some benefits in terms of wear behavior. In the study performed by Golizadeh et al. [126], the authors evaluate the thermal stability and oxidation resistance of SiBCN/TiAlSiCN and AlOx/TiAlSiCN coatings. It was concluded that the base oxidation resistance and thermal stability of the TiAlSiCN coating were improved when using the multilayer coatings, with the AlOx/TiAlSiCN coating exhibiting the highest performance of all the coatings. Sill regarding doping elements of TiAlN-based coatings, some of these that are currently under research are Ta [63,119] and Y [63], yielding satisfactory results in terms of wear behavior and mechanical properties. There are, also, some novel coatings that use Ru [62,84] in their composition, which are also showing some promising results, however the amount of research in this matter is not as abundant as the other elements.

As previously mentioned, the nanolayered and nanocomposite coatings exhibit increased mechanical properties (presenting very high values of hardness generally) and wear behavior making them very appealing for the machining industry. New coatings have been developed, based on the research made about additive elements for monolayered coatings. Once again, the employment of Mo is under study for extreme applications, this time on nanocomposite coatings [106,107]. Still regarding the research made on nanolayered coatings, many studies are being made on the influence of layer thickness in coating's mechanical properties and cutting performance [39]. Furthermore, improvements are being made to already known coating architecture such as the TiAlN [89–100], TiAlSiN [65,94], TiAlN/TiN [90,96,98] and Al/TiAlN [103], with studies presenting increases in the coating's performances by employing methods to control layer thickness or concentration of phases, such as TiN and Al.

Regarding the machining applications that these coatings are applied to, from the analyzed research papers presented in this review, it is noticed that the coatings are mainly used in turning and milling operations. It was noticed that there are many research papers about milling operations using the TiAlN-based coated tools. However, for both machining cases, it was found that there is a large interest in the study of the wear performance of these coatings in the cutting of steels [70,71,74,81,83,97,98,100,111–114,117], titanium alloys [72,75,76,82,96,115,116] and Inconel [73,78–80,99]. This is due to these alloys being employed in the aeronautical industry, making the study of the machinability of these materials quite interesting. The applied coatings are usually novel nanocomposite coatings or already known structures (aforementioned) that have been improved in some way, being noticed that the coatings structure highly improves the wear behavior of these coatings (as previously mentioned). In the presented studies, the wear mechanisms are also evaluated, as there is a great interest in knowing how these coatings perform.

5. Concluding Remarks

In this paper, an analysis of recent research on TiAlN-based coatings applied to machining was performed. The main research topics about monolayered, multilayered and nanolayered and nanocomposite coatings were presented, mentioning the new developments made about these coating types, their benefits for the coating's mechanical properties, wear and cutting behavior. A comparison of these coating's mechanical properties was also made, as a way to link these to the coating's wear behavior. These coating's main applications were also analyzed and presented, mentioning the main wear mechanisms that these coated tools suffered when machining various types of material.

It was found that, regarding monolayered coatings, the research was primarily about the study of doping elements (as previously mentioned), with Mo additions yielding very satisfactory results. These additions promoted an increase in the coating's hardness, elastic modulus and toughness, thus promoting a slow wear rate for TiAlN-based coatings containing Mo. This was also registered for other elements such as Ru, Zr and Ta. Regarding other type of coating architecture, the main focus of recent research appears to be nanolayered and nanocomposite coatings, as their mechanical properties exceed those of the other coating types. Some of these coatings are super-hard, presenting hardness values above 40 GPa. This is very appealing as this promotes the wear performance of the coating, protecting it from abrasive wear. Furthermore, these coatings present very high crack resistance and very high toughness.

These types of coating architecture have an influence on the coating's properties and subsequently on their wear performance and cutting behavior. Properties that are of high importance and improve the coating's wear behavior are:

- Hardness;
- Toughness;
- H/E ratio (plasticity);
- Friction coefficient.

Regarding the coating's wear mechanisms, it was found that the main wear mechanisms present in milling are adhesion and abrasion, however, the employment of nanolayered coatings improves the adhesive damage suffered by the coatings. Regarding the coatings employed in the turning process, these usually suffer abrasive wear and BUE, with some coating's exhibiting adhesive wear. As for milling, the use of nanolayered and nanocomposite coatings improves the cutting behavior and tool life of the coated tools, with these types of coating outperforming regular monolayered TiAlN-based coatings.

Author Contributions: V.F.C.S.: investigation and writing original draft; F.J.G.D.S.: conceptualization, supervision, writing—review and editing; A.B.: supervision, writing—review and editing; G.F.P.: supervision, writing—review and editing; R.A.: supervision and formal analysis. All authors have read and agreed to the published version of the manuscript.

Funding: The present work was done and funded under the scope of project ON-SURF (ANI | P2020 | POCI-01-0247-FEDER-024521, co-funded by Portugal 2020 and FEDER, through COMPETE 2020-Operational Program for Competitiveness and Internationalization. LAETA/INEGI/CETRIB is acknowledge due to the support provided in all research activities.

Institutional Review Board Statement: Not applicable.

Informed Consent Statement: Not applicable.

Data Availability Statement: Not applicable.

Conflicts of Interest: The authors declare no conflict of interest regarding this paper.

References

1. Kulkarni, B.H.; Nadakatti, M.M.; Kulkarni, S.C.; Kulkarni, R.M. Investigations on effect of nanofluid based minimum quantity lubrication technique for surface milling of Al7075-T6 aerospace alloy. *Mater. Today Proc.* **2020**, *27*, 251–256. [CrossRef]
2. Mersni, W.; Boujelbene, M.; Salem, S.B.; Singh, H.P. Machining time and quadratic mean roughness optimization. *Mater. Today Proc.* **2020**, *26*, 2619–2624. [CrossRef]
3. Gabiccini, M.; Bracci, A.; Battaglia, E. On the estimation of continuous mappings from cradle-style to 6-axis machines for face-milled hypoid gear generation. *Mech. Mach. Theory* **2011**, *46*, 1492–1506. [CrossRef]
4. Moriya, T.; Nakamoto, K.; Ishida, T.; Takeuchi, Y. Creation of V-shaped microgrooves with flat-ends by 6-axis control ultraprecision machining. *CIRP Ann.* **2010**, 61–66. [CrossRef]
5. Siddiqui, T.U.; Singh, S.K. Design, fabrication and characterization of a self-lubricated textured tool in dry machining. *Mater. Today Proc.* **2020**. [CrossRef]
6. Ji, W.; Zou, B.; Zhang, S.; Xing, H.; Yun, H.; Wang, Y. Design and fabrication of gradient cermet composite cutting tool, and its cutting performance. *J. Alloys Compd.* **2018**, *732*, 25–31. [CrossRef]
7. Zhou, X.; Wang, K.; Li, C.; Wang, Q.; Wu, S.; Liu, J. Effect of ultrafine gradient cemented carbides substrate on the performance of coating tools for titanium alloy high speed cutting. *Int. J. Refract. Met. Hard Mater.* **2019**, *84*, 105024. [CrossRef]
8. Chowdhury, M.S.I.; Bose, B.; Yamamoto, K.; Shuster, L.S.; Paiva, J.; Fox-Rabinovich, G.S.; Veldhuis, S.C. Wear performance investigation of PVD coated and uncoated carbide tools during high-speed machining of TiAl6V4 aerospace alloy. *Wear* **2020**, *446–447*, 203168. [CrossRef]

9. Hovsepian, P.E.; Luo, Q.; Robinson, G.; Pittman, M.; Howarth, M.; Doerwald, D.; Tietema, R.; Sim, W.M.; Deeming, A.; Zeus, T. Surface and Coatings Technology TiAlN/VN superlattice structured PVD coatings: A new alternative in machining of aluminium alloys for aerospace and automotive components. *Surf. Coat. Technol.* **2016**, *201*, 265–272. [CrossRef]
10. Sánchez, J.M.; Rubio, E.; Álvarez, M.; Sebastián, M.A.; Marcos, M. Microstructural characterisation of material adhered over cutting tool in the dry machining of aerospace aluminium alloys. *J. Mater. Process. Technol.* **2005**, *164–165*, 911–918. [CrossRef]
11. Peng, Z.; Zhang, X.; Zhang, D. Performance evaluation of high-speed ultrasonic vibration cutting for improving machinability of Inconel 718 with coated carbide tools. *Tribol. Int.* **2020**, 106766. [CrossRef]
12. Kumar, C.S.; Majumder, H.; Khan, A.; Patel, S.K. Applicability of DLC and WC/C low friction coatings on Al2O3/TiCN mixed ceramic cutting tools for dry machining of hardened 52100 steel. *Ceram. Int.* **2020**, *46*, 11889–11897. [CrossRef]
13. Lakshmanan, S.; Xavior, M.A. Performance of coated and uncoated inserts during intermittent cut milling of AISI 1030 steel. *Procedia Eng.* **2014**, *97*, 372–380. [CrossRef]
14. Sokovic, M.; Kopac, J.; Dobrzanski, L.A.; Mikula, J.; Golombek, K.; Pakula, D. Cutting characteristics of PVD and CVD-coated ceramic tool inserts. *Tribology Ind.* **2006**, *28*, 3–8.
15. Bobzin, K. High-performance coatings for cutting tools. *CIRP J. Manuf. Sci. Technol.* **2017**, *18*, 1–9. [CrossRef]
16. Bouzakis, K.-D.; Michailidis, N.; Skordaris, G.; Bouzakis, E. Coated Tools. *CIRP Encycl. Prod. Eng.* **2018**, 1–13. [CrossRef]
17. Lim, S.C.; Lim, C.Y.H. Effective use of coated tools—the wear-map approach. *Surf. Coat. Technol.* **2001**, *139*, 127–134. [CrossRef]
18. Arunnath, A.; Masooth, P.H.S. Optimization of process parameters in CNC turning process on machining SCM440 steel by uncoated carbide and TiCN/Al2O3/TiN coated carbide tool under dry conditions. *Mater. Today Proc.* **2020**. [CrossRef]
19. Sousa, V.F.C.; Silva, F.J.G.; Fecheira, J.S.; Lopes, H.M.; Martinho, R.P.; Casais, R.B.; Ferreira, L.P. Cutting Forces Assessment in CNC Machining Processes: A Critical Review. *Sensors* **2020**, *20*, 4536. [CrossRef]
20. Behera, G.C.; Thrinadh, J.; Datta, S. Influence of cutting insert (uncoated and coated carbide) on cutting force, tool-tip temperature, and chip morphology during dry machining of Inconel 825. *Mater. Today Proc.* **2020**. [CrossRef]
21. Zhao, J.; Liu, Z.; Wang, B.; Hu, J.; Wan, Y. Tool coating effects on cutting temperature during metal cutting processes: Comprehensive review and future research directions. *Mech. Syst. Sig. Process.* **2021**, *150*, 107302. [CrossRef]
22. Thakur, A.; Gangopadhyay, S. Influence of tribological properties on the performance of uncoated, CVD and PVD coated tools in machining of Incoloy 825. *Tribol. Int.* **2016**, *102*, 198–212. [CrossRef]
23. Silva, F.J.G.; Baptista, A.P.M.; Pereira, E.; Teixeira, V.; Fan, Q.H.; Fernandes, A.J.S.; Costa, F.M. Microwave plasma chemical vapour deposition diamond nucleation on ferrous substrates with Ti and Cr interlayers. *Diam. Rel. Mater.* **2002**, *11*, 1617–1622. [CrossRef]
24. Caliskan, H.; Panjan, P.; Kurbanoglu, C. 3.16 Hard coatings on cutting tools and surface finish. *Compr. Mater. Finish.* **2017**, *3*, 230–242. [CrossRef]
25. Baptista, A.; Silva, F.J.G.; Porteiro, J.; Míguez, J.L.; Pinto, G.; Fernandes, L. On the physical vapour deposition (PVD): Evolution of magnetron sputtering processes for industrial applications. *Procedia Manuf.* **2018**, *17*, 746–757. [CrossRef]
26. Baptista, A.; Silva, F.; Porteiro, J.; Míguez, J.; Pinto, G. Sputtering physical vapour deposition (PVD) coatings: A critical review on process improvement and market trend demands. *Coatings* **2018**, *8*, 402. [CrossRef]
27. Silva, F.J.G.; Fernandes, A.J.S.; Costa, F.M.; Baptista, A.P.M.; Pereira, E. A new interlayer approach for CVD diamond coating of steel substrates. *Diam. Relat. Mater.* **2004**, *13*, 828–833. [CrossRef]
28. Silva, F.J.G.; Fernandes, A.J.; Costa, F.; Teixeira, V.; Baptista, A.P.; Pereira, E. Tribological behaviour of CVD diamond films on steel substrates. *Wear* **2003**, *255*, 846–853. [CrossRef]
29. Silva, F.J.; Martinho, R.P.; Alexandre, R.J.D.; Baptista, A.P.M. Wear resistance of TiAlSiN thin coatings. *J. Nanosci. Nanotechnol.* **2012**, *12*, 9094–9101. [CrossRef]
30. Sousa, V.F.C.; Silva, F.J.G. Recent Advances in Turning Processes Using Coated Tools—A Comprehensive Review. *Metals* **2020**, *10*, 170. [CrossRef]
31. Seidl, W.M.; Bartosik, M.; Kolozsvári, S.; Bolvardi, H.; Mayrhofer, P.H. Influence of coating thickness and substrate on stresses and mechanical properties of (Ti,Al,Ta)N/(Al,Cr)N. *Surf. Coat. Technol.* **2018**, *347*, 92–98. [CrossRef]
32. Fernández-Abia, A.I.; Barreiro, J.; Fernández-Larrinoa, J.; Lacalle, L.N.L.; de Fernández-Valdivielso, A.; Pereira, O.M. Behaviour of PVD coatings in the turning of austenitic stainless steels. *Procedia Eng.* **2013**, *63*, 133–141. [CrossRef]
33. Klocke, F.; Krieg, T. Coated tools for metal cutting—Features and applications. *CIRP Ann.* **1999**, *48*, 515–525. [CrossRef]
34. Sousa, V.F.C.; Silva, F.J.G. Recent Advances on Coated Milling Tool Technology—A Comprehensive Review. *Coatings* **2020**, *10*, 235. [CrossRef]
35. Vannan, R.R.R.; Moorthy, T.V.; Harihan, P.; Prabhu, P. Effect of physical vapor deposited bilayer (AlCrN + TiAlN) coating on high-speed steel single point cutting tool. *Mater. Manuf. Processes.* **2017**, *32*. [CrossRef]
36. Fox-Rabinovich, G.S.; Kovalev, A.I.; Aguirre, M.H.; Beake, D.B.; Yamamoto, K.; Vekduis, S.C.; Endrino, J.L.; Wainstein, D.L.; Rashkovskiy, A.Y. Design and performance of AlTiN and TiAlCrN PVD coatings for machining of hard to cut materials. *Surf. Coat. Technol.* **2009**, *204*, 489–496. [CrossRef]
37. Czarniak, P.; Szymanowski, K.; Kucharska, B.; Krawczynska, A.; Sobiecki, J.R.; Kubacki, J.; Panjan, P. Modification of tools for wood based materials machining with TiAlN/a-CN coating. *Mater. Sci. Eng. B* **2020**, *257*, 114540. [CrossRef]
38. Mejía, H.D.; Echavarría, A.M.; Calderón, J.A.; Bejarano, G. Microstructural and electrochemical properties of TiAlN(Ag,Cu) nanocomposite coatings for medical applications deposited by dc magnetron sputtering. *J. Alloys Compd.* **2020**, *828*, 154396. [CrossRef]

39. Nunes, V.; Silva, F.J.G.; Andrade, M.F.; Alexandre, R.; Baptista, A.P.M. Increasing the lifespan of high-pressure die cast molds subjected to severe wear. *Surf. Coat. Technol.* **2017**, *332*, 319–331. [CrossRef]
40. Selvam, P.T.; Pugazhenthi, R.; Dhanasekaran, C.; Chandrasekaran, M.; Sivaganesan, S. Experimental Investigation on the Frictional Wear Behaviour of TiAlN Coated Brake Pads. *Mater. Today Proc.* **2020**. [CrossRef]
41. Ma, X.-F.; Wu, Y.-W.; Tan, J.; Meng, C.-Y.; Yang, L.; Dang, W.-A.; He, X.-J. Evaluation of corrosion and oxidation behaviors of TiAlCrN coatings for nuclear fuel cladding. *Surf. Coat. Technol.* **2019**, *358*, 521–530. [CrossRef]
42. Alat, E.; Motta, A.T.; Comstock, R.J.; Partezana, J.M.; Wolfe, D.E. Multilayer (TiN, TiAlN) ceramic coatings for nuclear fuel cladding. *J. Nucl. Mater.* **2016**, *478*, 236–244. [CrossRef]
43. Romero, E.C.; Macías, A.H.; Nonell, J.M.; Canto, O.S.; Botero, M.G. Mechanical and tribological properties of nanostructured TiAlN/TaN coatings deposited by DC magnetron sputtering. *Surf. Coat. Technol.* **2019**, *378*, 124941. [CrossRef]
44. Zauner, L.; Ertelthaler, P.; Wojcik, T.; Bolvardi, H.; Kolozsvári, S.; Mayrhofer, P.H.; Riedl, H. Reactive HiPIMS deposition of Ti-Al-N: Influence of the deposition parameters on the cubic to hexagonal phase transition. *Surf. Coat. Technol.* **2020**, *382*, 125007. [CrossRef]
45. Zhao, B.; Zhao, X.; Lin, L.; Zou, L. Effect of bias voltage on mechanical properties, milling performance and thermal crack propagation of cathodic arc ion-plated TiAlN coatings. *Thin Solid Films* **2020**, *708*, 1381116. [CrossRef]
46. Tillmann, W.; Grisales, D.; Stangier, D.; Thomann, C.-A.; Debus, J.; Nienhaus, A.; Apel, D. Residual stresses and tribomechanical behaviour of TiAlN and TiAlCN monolayer and multilayer coatings by DCMS and HiPIMS. *Surf. Coat. Technol.* **2020**, 126664. [CrossRef]
47. Tillmann, W.; Grisales, D.; Stangier, D.; Jebara, I.B.; Kang, H. Influence of the etching processes on the adhesion of TiAlN coatings deposited by DCMS, HiPIMS and hybrid techniques on heat treated AISI H11. *Surf. Coat. Technol.* **2019**, *378*, 125075. [CrossRef]
48. Liu, L.; Zhou, L.; Tang, W.; Ruan, Q.; Li, X.; Wu, Z.; Qasim, A.M.; Cui, S.; Li, T.; Tian, X.; et al. Study of TiAlN coatings deposited by continuous high power magnetron sputtering (C-HPMS). *Surf. Coat. Technol.* **2020**, *402*, 126315. [CrossRef]
49. Zhirkov, I.; Polcik, P.; Kolozsvári, S.; Rosen, J. Process development for stabilization of vacuum arc plasma generation from a TiB2 cathode. *AIP Adv.* **2019**, *9*, 015103. [CrossRef]
50. Sun, P.L.; Hsu, C.H.; Liu, H.S.; Su, C.Y.; Lin, C.K. Analysis on microstructure and characteristics of TiAlN/CrN nano-multilayer films deposited by cathodic arc deposition. *Thin Solid Films* **2010**, *518*, 7519–7522. [CrossRef]
51. Warcholinski, B.; Gilewicz, A.; Myslinski, P.; Dobruchowska, E.; Murzynski, D.; Kochmanski, P.; Rokosz, K.; Raaen, S. Effect of nitrogen pressure and substrate bias voltage on the properties of Al–Cr–B–N coatings deposited using cathodic arc evaporation. *Tribol. Int.* **2021**, *154*, 106744. [CrossRef]
52. Knotek, O.; Löffler, F.; Krämer, G. Multicomponent and multilayer physically vapour deposited coatings for cutting tools. *Surf. Coat. Technol.* **1992**, *54–55*, 241–248. [CrossRef]
53. Shuai, J.; Zuo, X.; Wang, Z.; Guo, P.; Xu, B.; Zhou, J.; Wang, A.; Ke, P. Comparative study on crack resistance of TiAlN monolithic and Ti/TiAlN multilayer coatings. *Ceram. Int.* **2020**, *46*, 6672–6681. [CrossRef]
54. Man, B.Y.; Guzman, L.; Miotello, A.; Adami, M. Microstructure, oxidation and H2- permeation resistance of TiAlN films deposited by DC magnetron sputtering technique. *Surf. Coat. Technol.* **2004**, *180–181*, 9–14. [CrossRef]
55. Hsieh, J.H.; Liang, C.; Yu, C.H.; Wu, W. Deposition and characterization of TiAlN and multi-layered TiN/TiAlN coatings using unbalanced magnetron sputtering. *Surf. Coat. Technol.* **1998**, *108–109*, 132–137. [CrossRef]
56. Kale, A.N.; Ravindranath, K.; Kothari, D.C.; Raole, P.M. Tribological properties of (Ti, Al)N coatings deposited at different bias voltages using the cathodic arc technique. *Surf. Coat. Technol.* **2001**, *145*, 60–70. [CrossRef]
57. Mitsuo, A.; Uchida, S.; Nihira, N.; Iwaki, M. Improvement of high-temperature oxidation resistance of titanium nitride and titanium carbide films by aluminum ion implantation. *Surf. Coat. Technol.* **1998**, *103–104*, 98–103. [CrossRef]
58. Du, H.; Zhao, H.; Xiong, J.; Xian, G. Effect of interlayers on the structure and properties of TiAlN based coatings on WC-Co cemented carbide substrate. *Int. J. Refract. Met. Hard Mater.* **2013**, *37*, 60–66. [CrossRef]
59. Yang, K.; Xian, G.; Zhao, H.; Fan, H.; Wang, J.; Wang, H.; Du, H. Effect of Mo content on the structure and mechanical properties of TiAlMoN films deposited on WC–Co cemented carbide substrate by magnetron sputtering. *Int. J. Refract. Met. Hard Mater.* **2015**, *52*, 29–35. [CrossRef]
60. Tomaszewski, L.; Gukbinski, W.; Urbanowicz, A.; Suszko, T.; Lewandowski, A.; Gulbinski, W. TiAlN based wear resistant coatings modified by molybdenum addition. *Vacuum* **2015**, *121*, 223–229. [CrossRef]
61. Yi, J.; Chen, S.; Chen, K.; Xu, Y.; Chen, Q.; Zhu, C.; Liu, L. Effects of Ni content on microstructure, mechanical properties and Inconel 718 cutting performance of AlTiN-Ni nanocomposite coatings. *Ceram. Int.* **2018**, *45*, 474–480. [CrossRef]
62. Liu, Z.R.; Chen, L.; Du, Y.; Zhang, S. Influence of Ru-addition on thermal decomposition and oxidation resistance of TiAlN coatings. *Surf. Coat. Technol.* **2020**, *401*, 126234. [CrossRef]
63. Aninat, R.; Valle, N.; Chemin, J.-B.; Duday, D.; Michotte, C.; Penoy, M.; Bourgeois, L.; Choquet, P. Addition of Ta and Y in a hard Ti-Al-N PVD coating: Individual and conjugated effect on the oxidation and wear properties. *Corros. Sci.* **2019**, *156*, 171–180. [CrossRef]
64. Chandra, N.G.P.S.; Otsuka, Y.; Mutoh, Y.; Yamamoto, K. Effect of coating thickness on fatigue behavior of TiAlN coated Ti-alloys. *Int. J. Fatigue* **2020**, *140*, 105767. [CrossRef]
65. Das, S.; Guha, S.; Ghadai, R.; Swain, B.P. A comparative analysis over different properties of TiN, TiAlN and TiAlSiN thin film coatings grown in nitrogen gas atmosphere. *Mater. Chem. Phys.* **2021**, *258*, 123866. [CrossRef]

66. Chaar, A.B.B.; Rogstrom, L.; Johansson-Joesaar, M.P.; Barrirero, J.; Aboulfadl, H.; Schell, N.; Ostach, D.; Mucklich, F.; Oden, M. Microstructural influence of the thermal behavior of arc deposited TiAlN coatings with high aluminum content. *J. Alloys Compd.* **2021**, *854*, 157205. [CrossRef]
67. Jacob, A.; Gangopadhyay, S.; Satapathy, A.; Mantry, S.; Jha, B.B. Influences of micro-blasting as surface treatment technique on properties and performance of AlTiN coated tool. *J. Manuf. Processes* **2017**, *29*, 407–418. [CrossRef]
68. Zhang, K.; Deng, J.; Meng, R.; Lei, S.; Yu, S. Influence of laser substrate pretreatment on anti-adhesive wear properties of WC/Co-based TiAlN coatings against AISI 316 stainless steel. *Int. J. Refract. Met. Hard Mater.* **2016**, *57*, 101–114. [CrossRef]
69. Masooth, P.H.S.; Jayakumar, V.; Bharathiraj, G. Experimental investigation on surface roughness in CNC end milling process by uncoated and TiAlN coated carbide end mill under dry conditions. *Mater. Today Procc.* **2020**, *22*, 726–736. [CrossRef]
70. Ravi, S.; Gurusamy, P. Experimental investigations on performance of TiN and TiAlN coated tools in cryogenic milling of AISI D2 hardened steel. *Mater. Today Procc.* **2020**, *33*, 3612–3615. [CrossRef]
71. Siwawut, S.; Saikaew, C.; Wisitsoraat, A.; Surinphong, S. Cutting performances and wear characteristics of WC inserts coated with TiAlSiN and CrTiAlSiN by filtered cathodic arc in dry face milling of cast iron. *Int. J. Adv. Manuf. Technol.* **2018**. [CrossRef]
72. Hou, M.; Mou, W.; Yan, G.; Song, G.; Wu, Y.; Ji, W.; Jiang, Z.; Wang, W.; Qian, C.; Cai, Z. Effects of different distribution of residual stresses in the depth direction on cutting performance of TiAlN coated WC-10wt%Co tools in milling Ti-6Al-4V. *Surf. Coat. Technol.* **2020**, *397*, 125972. [CrossRef]
73. Sen, B.; Gupta, M.K.; Mia, M.; Mandal, U.T.; Mondal, S.P. Wear behaviour of TiAlN coated solid carbide end-mill under alumina enriched minimum quantity palm oil-based lubricating condition. *Tribology Inter.* **2020**, *148*, 106310. [CrossRef]
74. Tansukatanon, S.; Tangwarodomnukun, V.; Dumkum, C.; Kruytong, P.; Plauchum, N.; Charee, W. Micromachining of Stainless steel using TiAlN-coated tungsten carbide end mill. *Procedia Manuf.* **2019**, *30*, 419–426. [CrossRef]
75. Ziberov, M.; Oliveira, D.; Silva, M.B.; Hung, W.N.P. Wear of TiAlN and DLC coated microtools in micromilling of Ti-6Al-4V alloy. *J. Manuf. Process.* **2020**, *56*, 337–349. [CrossRef]
76. Bandapalli, C.; Sutaria, B.M.; Bhatt, D.V.P.; Singh, K.K. Tool Wear Analysis of Micro End Mills—Uncoated and PVD TiAlN and AlTiN in High speed Micro Milling of Titanium alloy-Ti-0.3Mo-0.8Ni. *Procedia CIRP* **2018**, *77*, 626–629. [CrossRef]
77. Swain, N.; Venkatesh, V.; Kumar, P.; Srinivas, G.; Ravishankar, S.; Barshilia, H.C. An experimental investigation on the machining characteristics of Nimonic 75 using uncoated and TiAlN coated tungsten carbide micro-end mills. *CIRP J. Manuf. Sci. Technol.* **2017**, *16*, 34–42. [CrossRef]
78. Zhao, J.; Lie, Z. Influences of coating thickness on cutting temperature for dry hard turning Inconel 718 with PVD TiAlN coated carbide tools in initial tool wear stage. *J. Manuf. Process.* **2020**, *56*, 1155–1165. [CrossRef]
79. Kurniawan, R.; Park, G.C.; Park, K.M.; Zhen, Y.; Kwak, Y.I.; Kim, M.C.; Lee, J.M.; Ko, T.J.; Park, C.S. Machinability of modified Inconel 713C using a WC TiAlN-coated tool. *J. Manuf. Process.* **2020**, *57*, 409–430. [CrossRef]
80. Zhao, J.; Liu, Z.; Wang, B.; Hu, J. PVD AlTiN coating effects on tool-chip heat partition coefficient and cutting temperature rise in orthogonal cutting Inconel 718. *Int. J. Heat Mass Transf.* **2020**, *163*, 120449. [CrossRef]
81. Hao, G.; Liu, Z. Thermal contact resistance enhancement with aluminum oxide layer generated on TiAlN-coated tool and its effect on cutting performance for H13 hardened steel. *Surf. Coat. Technol.* **2020**, *385*, 125436. [CrossRef]
82. Lu, W.; Li, G.; Zhou, Y.; Liu, S.; Wang, K.; Wang, Q. Effect of high hardness and adhesion of gradient TiAlSiN coating on cutting performance of titanium alloy. *J. Alloys Compounds* **2020**, *820*, 153137. [CrossRef]
83. Kulkarni, A.P.; Sargade, V.G. Characterization and Performance of AlTiN, AlTiCrN, TiN/TiAlN PVD Coated Carbide tools While Turning SS 304. *Mater. Manuf. Process.* **2015**, *30*, 748–755. [CrossRef]
84. Beake, B.D.; Endrino, J.L.; Kimpton, C.; Fox-Rabinovich, G.S.; Veldhuis, S.C. Elevated temperature repetitive micro-scratch testing of AlCrN, TiAlN and AlTiN PVD coatings. *Int. J. Refract. Metal. Hard Mater.* **2017**, *69*, 215–226. [CrossRef]
85. Gong, M.; Chen, J.; Deng, X.; Wu, S. Sliding wear behavior of TiAlN and AlCrN coatings on a unique cemented carbide substrate. *Int. J. Refract. Metal. Hard Mater.* **2017**, *69*, 209–214. [CrossRef]
86. Vettivel, S.C.; Jegan, R.; Vignesh, J.; Suresh, S. Surface characteristics and wear depth profile of the TiN, TiAlN and AlCrN coated stainless steel in dry sliding wear condition. *Surf. Interf.* **2017**, *6*, 1–10. [CrossRef]
87. Zou, H.K.; Chen, L.; Chang, K.K.; Pei, F.; Du, Y. Enhanced hardness and age-hardening of TiAlN coatings through Ru-addition. *Scriptia Materialia* **2019**, *162*, 382–386. [CrossRef]
88. Zhang, M.; Cheng, Y.; Xin, L.; Su, J.; Li, Y.; Zhu, S.; Wang, F. Cyclic oxidation behaviour of Ti/TiAlN composite multilayer coatings deposited on titanium alloy. *Corrosion Sci.* **2020**, *166*, 108476. [CrossRef]
89. Sprute, T.; Tillmann, W.; Grisales, D.; Selvadurai, U.; Fischer, G. Influence of substrate pre-treatments on residual stresses and tribo-mechanical properties of TiAlN-based PVD coatings. *Surf. Coat. Technol.* **2014**, *260*, 369–379. [CrossRef]
90. Çomakli, O. Improved structural, mechanical, corrosion and tribocorrosion properties of Ti45Nb alloys by TiN, TiAlN monolayers, and TiAlN/TiN multilayer ceramic films. *Ceram. Inter.* **2021**, *47*, 4149–4156. [CrossRef]
91. Çomakli, O. Influence of CrN, TiAlN monolayers and TiAlN/CrN multilayer ceramic films on structural, mechanical and tribological behavior of β-type Ti45Nb alloys. *Ceram. Inter.* **2020**, *46*, 8185–8191. [CrossRef]
92. Shugurov, A.R.; Kazachenok, M.S. Mechanical properties and tribological behavior of magnetron sputtered TiAlN/TiAl multilayer coatings. *Surf. Coat. Technol.* **2018**, *353*, 254–262. [CrossRef]
93. Shang, H.; Li, J.; Shao, T. Mechanical properties and thermal stability of TiAlN/Ta multilayer film deposited by ion beam assisted deposition. *Applied Surf. Sci.* **2014**, *310*, 317–320. [CrossRef]

94. Zhao, F.; Ge, Y.; Wang, L.; Wang, X. Tribological and mechanical properties of hardness-modulated TiAlSiN multilayer coatings fabricated by plasma immersion ion implantation and deposition. *Surf. Coat. Technol.* **2020**, 126475. [CrossRef]
95. Sukuroglu, E.E.; Baran, O.; Totik, Y.; Efeoglu, I. Investigation of high temperature wear resistance of TiCrAlCN/TiAlN multilayer coatings over M2 Steel. *J. Adhesion Sci. Technol.* **2018**. [CrossRef]
96. An, Q.; Chen, J.; Tao, Z.; Ming, W.; Chen, M. Experimental investigation on tool wear characteristics of PVD and CVD coatings during face milling of Ti-6242S ad Ti- 555 titanium alloys. *Refrac. Metals Hard Mater.* **2019**. [CrossRef]
97. Varghese, V.; Chakradhar, D.; Ramesh, M.R. Micro-mechanical characterization and wear performance of TiAlN/NbN PVD coated inserts during Eng milling of AISI 304 Austenitic Stainless Steel. *Mater. Today Proc.* **2018**, *5*, 12855–12862.
98. Zheng, G.; Zhao, G.; Cheng, X.; Xu, R.; Zhao, J.; Zhang, H. Frictional and wear performance of TiAlN/TiN coated tool against high-strength steel. *Ceram. Inter.* **2018**, *44*, 6878–6885. [CrossRef]
99. Thakur, A.; Gnagopadhyay, S. Dry machining of nickel-based super alloy as a sustainable alternative using TiN/TiAlN coated tool. *J. Clean. Product.* **2016**, *129*, 256–268. [CrossRef]
100. Abdoos, M.; Bose, B.; Rawal, S.; Fazal, A.; Arif, M.; Veldhuis, S.C. The influence of residual stress on the properties and performance of thick TiAlN multilayer coating during dry turning of compacted graphite iron. *Wear* **2020**, *15*, 203342. [CrossRef]
101. Bonu, V.; Jeevitha, M.; Kumar, V.P.; Srinivas, G.; Siju; Barshilia, H.C. Solid particle erosion and corrosion resistance performance of nanolayered and multilayered Ti/TiN and TiAl/TiAlN coatings deposited on Ti6Al4V. *Surf. Coat. Technol.* **2020**, *387*, 125531. [CrossRef]
102. Wang, J.; Yazdi, M.A.P.; Lomello, F.; Billard, A.; Kovács, A.; Schuster, F.; Guet, C.; White, T.J.; Sanchette, F.; Dong, Z. Influence of microstructures on mechanical properties and tribology behaviors of TiN/TiAlN multilayer coatings. *Surf. Coat. Technol.* **2016**. [CrossRef]
103. Liang, F.; Shen, Y.; Pei, C.; Qiu, B.; Lei, J.; Sun, D. Microstructure evolution and corrosion resistance of multi interfaces Al-TiAlN nanocomposite films on AZ91D magnesium alloy. *Surf. Coat. Technol.* **2019**, *15*, 83–92. [CrossRef]
104. Seidl, W.M.; Bartosik, M.; Kolozsvári, S.; Bolvardi, H.; Mayrhofer, P.H. Mechanical properties and oxidation resistance of AlCrN/TiAlTaN multilayer coatings. *Surf. Coat. Technol.* **2018**, *347*, 427–433. [CrossRef]
105. Xian, G.; Xiong, J.; Zhao, H.; Fan, H.; Li, Z.; Du, H. Evaluation of the structure and properties of the hard TiAlN-(TiAlN/CrAlSiN)-TiAlN multiple coatings deposited on different substrate materials. *Int. J. Refract. Met. Hard Mater.* **2019**, *85*, 105056. [CrossRef]
106. Pshyk, A.V.; Kravchenko, Y.; Coy, E.; Kempinski, M.; Iatsunskyi, I.; Zaleski, K.; Pogrebnjak, A.D.; Jurga, S. Microstructure, phase composition and mechanical properties of novel nanocomposite (TiAlSiY)N and nano-scale (TiAlSiY)N/MoN multifunctional heterostructures. *Surf. Coat. Technol.* **2018**, *350*, 376–390. [CrossRef]
107. Kravchenko, Y.; Coy, L.E.; Peplinska, B.; Iatsunskyi, I.; Zaleski, K.; Kempinski, M.; Beresnev, V.M.; Kanarski, P.; Jurga, S.; Pogrebnjak, A.D. Nano-multilayered coatings of (TiAlSiY)N/MeN (Me=Mo, Cr and Zr): Influence of composition of the alternating layer on their structural and mechanical properties. *J. Alloys Compd.* **2018**, *767*, 483–495. [CrossRef]
108. Wang, L.P.; Qi, J.L.; Cao, Y.Q.; Zhang, K.; Zhang, Y.; Hao, J.; Ren, P.; Wen, M. N-rich Zr3N4 nanolayers-dependent superhard effect and fracture behavior in TiAlN/Zr3N4 nanomultilayer films. *Ceram. Inter.* **2020**, *46*, 19111–19120. [CrossRef]
109. Chen, L.; Pei, Z.; Xiao, J.; Sun, C. TiAlN/Cu Nanocomposite Coatings Deposited by Filtered Cathodic Arc Ion Plating. *J. Mater. Sci. Technol.* **2017**, *33*, 111–116. [CrossRef]
110. Mejía, H.D.; Perea, D.; Bejarano, G.G. Development and characterization of TiAlN (Ag, Cu) nanocomposite coatings deposited by DC magnetron sputtering for tribological applications. *Surf. Coat. Technol.* **2020**, *381*, 125095. [CrossRef]
111. Geng, D.; Zeng, R.; Wu, Z.; Wang, Q. An investigation on microstructure and milling performance of arc-evaporated TiSin/AlTiN film. *Thin Solid Film* **2020**, *709*, 138243. [CrossRef]
112. Teppernegg, T.; Czettl, C.; Michotte, C.; Mitterer, C. Arc evaporated Ti-Al-N/Cr-Al-N multilayer coating systems for cutting applications. *Int. J. Refract. Metal. Hard Mater.* **2018**, *72*, 83–88. [CrossRef]
113. Chowdhury, S.; Bose, B.; Yamamoto, K.; Veldhuis, S.C. Effect of Interlayer Thickness on Nano-Multilayer Coating Performance during High speed dry milling of H13 Tool Steel. *Coatings* **2019**, *9*, 737. [CrossRef]
114. Chowdhury, S.; Beake, B.D.; Yamamoto, K.; Bose, B.; Aguirre, M.; Fox-Rabinovich, G.S.; Veldhuis, S.C. Improvement of Wear Performance of Nano-Multilayer PVD Coatings under Dry Hard End Milling Conditions Based on Their Architectural Development. *Coatings* **2018**, *8*, 59. [CrossRef]
115. Zha, C.; Chen, F.; Jiang, F.; Xu, X. Correlation of the fatigue impact resistance of bilayer and nanolayered PVD coatings with their cutting performance in machining T-6Al-4V. *Ceram. Inter.* **2019**, *45*, 14704–14717. [CrossRef]
116. Sui, X.; Li, G.; Jiang, C.; Wang, K.; Zhang, Y.; Hao, J.Q. Improved toughness of layered architecture TiAlN/CrN coatings for titanium high speed cutting. *Ceram. Inter.* **2018**, *44*, 5629–5635. [CrossRef]
117. Zhang, Q.; Xu, Y.; Zhang, T.; Wu, Z.; Wang, Q. Tribological properties, oxidation resistance and turning performance of AlTiN/AlCrSiN multilayer coatings by arc ion plating. *Surf. Coat. Technol.* **2018**, *356*, 1–10. [CrossRef]
118. Fernandes, C.; Carvalho, S.; Rebouta, L.; Vaz, F.; Denannot, M.F.; Pacaud, J.; Riviére, J.P.; Cavaleiro, A. Effect of the microstructure on the cutting performance of superhard (Ti,Si,Al)N nanocomposite films. *Vacuum* **2008**, *82*, 1470–1474. [CrossRef]
119. Glechner, T.; Hahn, R.; Wojcik, T.; Holec, D.; Kolozsvári, S.; Zaid, H.; Kodambaka, S.; Mayrhofer, P.H.; Riedl, H. Assessment of ductile character in superhard Ta-C-N thin films. *Acta Materialia* **2019**, *179*, 17–25. [CrossRef]
120. Vuchkov, T.; Yaqub, T.B.; Evaristo, M.; Cavaleiro, A. Synthesis, microstructural and mechanical properties of self.lubricating Mo-Se-C coatings deposited by closed-field unbalanced magnetron sputtering. *Surf. Coat. Technol.* **2020**, *394*, 123889. [CrossRef]

121. Gilmore, R.; Baker, M.A.; Gibson, P.N.; Gissler, W.; Stoiber, M.; Losbichler, P.; Mitterer, C. Low-friction TiN-MoS_2 coatings produced by dc magnetron co-deposition. *Surf. Coat. Technol.* **1998**, *108–109*, 345–351. [CrossRef]
122. Ding, X.; Zeng, X.T.; Goto, T. Unbalanced magnetron sputtered Ti–Si–N:MoSx composite coatings for improvement of tribological properties. *Surf. Coat. Technol.* **2005**, *198*, 432–436. [CrossRef]
123. Bondarev, A.V.; Kiryukhantsev-Korneev, V.; Sheveyko, A.N.; Shtansky, D.V. Structure, tribological and electrochemical properties of low friction TiAlSiCN/MoSeC coatings. *Applied Surf. Sci.* **2015**, *327*, 235–261. [CrossRef]
124. Shtansky, D.V.; Kuptsov, K.A.; Kiryukhantsev-Korneev, V.; Sheveyko, A.N. High thermal stability of TiAlSiCN coatings with "comb" like nanocomposite structure. *Surf. Coat. Technol.* **2012**, *206*, 4840–4849. [CrossRef]
125. Kuptsov, K.A.; Kiryukhantsev-Korneev, V.; Sheveyko, A.N.; Shtansky, D.V. Surface modification of TiAlSiCN coatings to improve oxidation protection. *Applied Surf. Sci.* **2015**, *347*, 713–718. [CrossRef]
126. Golizadeh, M.; Kuptsov, K.A.; Svyndina, N.V.; Shtansky, D.V. Multilayer SiBCN/TiAlSiCN and AlOx/TiAlSiCN coatings with high thermal stability and oxidation resistance. *Surf. Coat. Technol.* **2017**, *319*, 277–285. [CrossRef]

Review

Recent Advances in Turning Processes Using Coated Tools—A Comprehensive Review

Vitor F. C. Sousa and Francisco J. G. Silva *

ISEP—School of Engineering, Polytechnic of Porto, 4200-072 Porto, Portugal; vcris@isep.ipp.pt
* Correspondence: fgs@isep.ipp.pt; Tel.: +35-122-834-0500

Received: 13 December 2019; Accepted: 20 January 2020; Published: 23 January 2020

Abstract: Turning continues to be the largest segment of the machining industry, which highlights the continued demand for turned parts and the overall improvement of the process. The turning process has seen quite an evolution, from basic lathes using solid tools, to complex CNC (Computer Numerical Control) multi-process machines, using, for the most part, coated inserts and coated tools. These coatings have proven to be a significant step in the production of high-quality parts and a higher tool life that have captivated the industry. Continuous improvement to turning coated tools has been made, with many researches focusing on the optimization of turning processes that use coated tools. In the present paper, a presentation of various recently published papers on this subject is going to be made, mentioning the various types of coatings that have recently been used in the turning process, the turning of hard to machine materials, such as titanium alloys and Inconel, as well as the interaction of these coatings with the turned surfaces, the wear patterns that these coatings suffer during the turning of materials and relating these wear mechanisms to the coated tool's life expectancy. Some lubrication conditions present a more sustainable alternative to current methods used in the turning process; the employment of coated tool inserts under these conditions is a current popular research topic, as there is a focus on opting for more eco-friendly machining options.

Keywords: machining; turning process; turning tools; solid tools; cemented carbide; coated tools; coated cemented carbide; Physical Vapor Deposition (PVD); Chemical Vapor Deposition (CVD); multilayered coatings; nanolayered coatings; wear mechanism; tool life; minimum quantity Lubricant (MQL); cutting forces

---

1. Introduction

The machining industry has seen a significant growth in the past 5–6 years, and it is projected to be a 100 billion USD industry by 2025 [1]; this is primarily due to the high demand for higher quality products and computer numerical control machines (CNC), that enable manufacturers to develop these high quality and complex products at higher speeds [2]. Nowadays, there are 6-axis CNC machines, capable of turning feedstock bars into a complete product, which are quite appellative to the machining industry [3]. Moreover, the lathe and milling segment are still leading the market. In 2018, the lathe segment was leading the market, valued at 17.65 billion USD, followed by the milling segment [4]. The lathe market has grown significantly and is expected to keep growing in the following years [5].

The turning process can be described as the machining of a piece of material that is rotating, using a single pointed tool that is stationary, to produce a smooth and straight outside or inside radius on the piece. When turning, some considerations need to be taken into attention, such as the type of turning that is being made, for example longitudinal turning, and external or internal turning; quality demands, such as surface finish or tolerance must also be taken into consideration, because these are factors that depend on the material that is being machined, as well as the tools and the machine that is being used [6]. As previously stated, the tools used for turning are single-pointed, and can be categorized into solid tools and insert tools; these tools can be coated to improve the process. Turning was initially carried out on a lathe, but the processes have seen significant evolution, particularly due to the development of computer numerical control (CNC) machines from the 1980s onwards, as well as turning centers, and later machining centers that could employ different processes such as turning and milling. This led to an increase in their use in the manufacturing industry [7], evolving from basic lathes, to multiple-process machining centers.

As the machines for turning evolved, so did the tools that were being used in the process, from solid tools with simple geometries, usually made out of steel, to coated cemented carbide tools, that make up to 80% of all tool inserts used in today's machining. These coated cemented carbide tools have improved the machining process significantly by enabling the machining of materials at faster rates than using conventional, uncoated tools [8,9]. These coatings are deposited on the surface of the tool to provide more wear resistance or a lower friction coefficient—in summary, these coatings provide a way to machine materials at higher speeds, while maintaining overall good surface quality and, especially, improving tool life, by reducing cutting forces, temperature and tool wear. There have been recent developments in cemented carbide tools, fabricating these tools with gradient layers, where the outer layers are, for example, harder than the substrate [10]. The fabrication of these gradient composite tools will provide tools with more versatility, as the desired properties can be applied on the base tool and improved on the surface, increasing their performance. Studies have been made to analyze the way the thickness of these gradient layers affect the properties of these tools [11]. A study carried out by Zhou et al. [12], tested different gradient cemented carbides, with layers of differing thicknesses, and has tested these with different coatings in the high speed cutting of a titanium alloy. Thus, Zhou et al. found that the thickness of the gradient layer's influences cutting performance, and that this thickness can be controlled by altering the contents of the cemented carbide constituents. Additionally, the authors concluded that the carbide with the thickest layer had the best cutting performance, making the control of these thicknesses very appealing when dealing with this type of substrate.

Coatings can be obtained using two different processes, either by Chemical Vapor Deposition (CVD), or by Physical Vapor Deposition (PVD). CVD films are achieved by having a precursor pumped inside a reactor; this precursor is regulated by control valves. The precursor molecules pass by the substrate and are deposited on its surface, achieving a thin, hard coating. This process runs at high temperatures, reaching temperatures of up to 900 °C and, additionally, the film thickness is usually uniform throughout the substrate surface. PVD consists of different methods, such as evaporation, sputtering and molecular beam epitaxy (MBE). In the sputtering technique, the applied coating is achieved by placing a magnetron near the target, in a vacuum. Then, an inert gas is introduced, then a high voltage is applied between the target and the substrate, releasing atomic size particles from the target. These particles are projected onto the substrate and they start to form a solid film. In the evaporation technique, the target acts as an evaporation source, having the material to be deposited, which works as a cathode. The material is heated at a high vapor pressure, which causes the release of the particles. The pumped gas, inserted in the reactor, clashes with the nano particles, which causes the acceleration of these particles, which in turn creates a plasma. This plasma proceeds trough the deposition chamber, thus depositing the coating's layers onto the substrate. Usually, PVD processes, when compared to CVD, run at a lower temperature (under 500 °C), and are more environmentally safe due to the type of precursors used in the CVD process, which are toxic. Additionally, the energy consumption of the PVD process is considerably lower than that of the CVD process [13–16].

Regarding the PVD process, there are various methods that can be used to obtain different coatings, either in composition or with different properties. Some of these methods are quite novel, or are seeing a new use in order to obtain certain coatings. As previously mentioned, the various methods that are applied in the coating industry are either sputtering or evaporation methods. Magnetron sputtering and arc evaporation methods have seen recent use in the deposition of coatings. Figure 1 shows the various PVD methods.

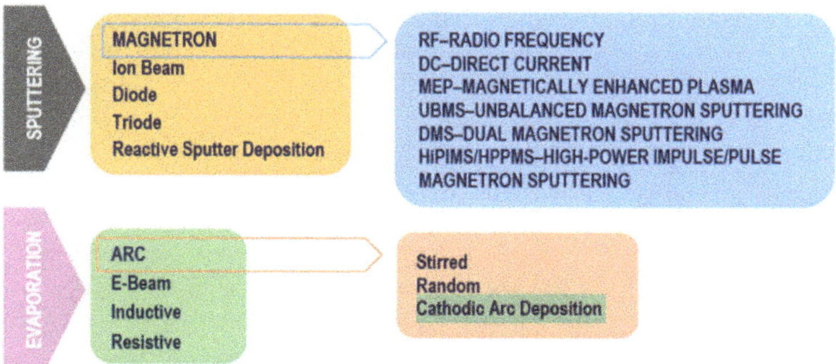

Figure 1. Different physical vapor deposition (PVD) techniques that are used currently in the production of advanced coatings [16].

The most used method, for magnetron sputtering PVD coating, is the direct current method (DC), however there are many more (as seen in the previous image), such as unbalanced magnetron sputtering (UBMS) and the novel high-power impulse/pulse magnetron sputtering (HiPIMS/HPPMS). Regarding these sputtering methods, there is a recent paper, that presents a method on how to control the boron-to-titanium ratios of TiBx thin films. The authors show, in this paper, that the addition of an external magnetic field during the strongly magnetically unbalanced magnetron sputtering of a TiB_2 target in Ar, enables the ability to control the ratio of B/Ti. This research paves the way for synthesizing stoichiometric single-crystal transition-metal diborides [17]. Still regarding magnetron sputtering, a paper by Romero et al. [18] studies the properties of TiAlN/TaN nanostructured coatings deposited by DC magnetron sputtering. These coatings were deposited at different substrate rotation speeds. These different rotation speeds enable the control of the coating's architecture and mechanical properties. Recent studies on the HiPIMS reveal interesting results, such as having better coating adhesion or even having better coating mechanical properties. This paper by Zauner et al. [19], studies the influence of HiPIMS parameters on the properties of Ti–Al–N thin films, by using a Ti–Al composite target in mixed Ar/N_2- atmospheres. The parameters that were studied were both the pulse frequency and duration, and the N_2 flow ratio, substrate bias voltage and target composition. The optimal parameters for obtaining good values of hardness and moderate compressive stresses were determined, and it was found that the regulation of these parameters enables the control of the coating's structure. These variations can promote the formation of a highly preferred cubic phase, though altered gas-to-metal ratios arriving at the film surface.

Regarding evaporation methods, the arc evaporation has seen some use in recent research, especially the cathodic arc deposition method. In this paper, coatings obtained by this method will be mentioned. In a recent paper by Zhirkov et al. [20], a stable and reproductible arc plasma generation from a TiB_2 cathode is presented. The authors show that the use of a Mo (molybdenum) cylinder around the boride cathode limits the movement of the arc spots to within the rim. This, coupled with a TiB_2 cathode containing 1wt% of carbon, generated a stable arc with high reproducibility. These borides are not usually synthesized using DC arc evaporation, although the authors show with these results that the cathodic arc is an efficient method for the synthesis of these metal borides

When selecting the type of coating desired for a tool, the machining process being implemented must be taken into consideration, as there are some advantages and disadvantages to coatings that are applied to cutting tools. For example, in the paper presented by Hovsepian and Ehiasarian [21], the production of coatings using different PVD techniques is explored, namely the conventional DC magnetron sputtering and the HiPMS technique. The produced coatings were tailored for the applications that were chosen, while investigating the properties of the produced coatings. Studies such as these highlight the necessity of good planning when choosing the right coating for the right application. Furthermore, some parameters, such as coating properties (i.e., chemical composition, hardness or coating design), influence tool performance when applied to different machining processes (for example, roughing or finishing), but the deposition method also influences the cutting behavior of these coatings. CVD coatings are usually applied to cemented carbide cutting tools due to the good behavior of these materials under elevated temperatures, whereas, due to the overall low temperature of the PVD process, this also means that this method can be used to coat tool steel, with some studies having been done on the preparation and evaluation of these coatings [22]. However, there are some studies that propose an approach to solve the problem of diamond deposition on steel substrates using the CVD process. The authors propose a Ni/Cu/Ti interlayer for the diamond coating to adhere. The coating showed good adhesion to the multi-metal interlayer, and good wear resistance [23,24]. The study by Silva et al. [25] evaluates the wear resistance of Ti–Al–Si–N coatings deposited by PVD on a steel substrate; the coating was evaluated, including the adhesion of the coating, and the authors reported that a good adhesion of the coating to the tool steel was achieved. PVD, being a line-of-sight process, has some disadvantages, for example, it is harder to apply PVD coatings on complex geometries, although this can be achieved by using, for example, pulsed high-power sputtering [26]. Another disadvantage of PVD is that the coating thickness is harder to control throughout the substrate surface, however, using the mentioned pulsed high-power sputtering, the thickness distribution of the PVD films is more homogenous [26]. PVD coatings are usually thinner than CVD coatings. A thin PVD coating is more suited for finishing operations, as the thinner PVD coatings, confer the tool with a sharp cutting edge (when compared to thicker CVD coatings), also, the compressive stresses exhibited by PVD coatings (CVD coatings exhibit tensile stresses, contrary to the PVD coatings) are favorable; these stresses, coupled with the small layer thickness, make for a stronger and sharper cutting edge, making these coatings ideal for finishing operations [27–29]. Still regarding coating thickness, although generally PVD coatings are thinner than CVD coatings, there is a novel method that enables the deposition of thick PVD coatings, using a state of the art arc-evaporated PVD technique, that can grow coatings up to 24.5 μm [30].

As previously stated, coatings provide tools with properties that best fit machining applications, and, as in the case of gradient cemented carbide tools, that have different layers with different properties; coatings can use the same principle as these tools. This means that a coated tool may have a multi-layered coating, in which, for example, the outer layer has elevated wear resistance and the subsequent layer has a main thermal dissipation function. This versatility makes coated tools very appealing.

Tool coatings are classified by their architecture type (number of layers/layer arrangement) and by their chemical composition (layer composition). They are also characterized by their mechanical properties, such as hardness, indentation modulus and their stress state (stresses induced during the deposition of these coatings). Additionally, the microstructures of these coatings are often analyzed, as these are linked to the mechanical properties of the coatings. There is recent research that studies the altering of the microstructures of some coatings, in order to improve their performance.

Coatings can have various designs, from single layer to multi-layered coating. The types of coating architecture are as follows:

(a) Single layer coating;
(b) Double layer coating;
(c) Gradient coating;
(d) Multilayer coating;
(e) Nanolayered coating;
(f) Nanocomposite coating.

Different coatings with different designs have different types of architecture; the coating can have, for example, a multilayered architecture. The architecture of the coating is linked to coating design, meaning different architectures are chosen to deal with different problems.

Multilayer coating is a type of coating that shows more application in the industry by combining more than one appealing characteristic of each layer of coating. The higher number of layers also contributes to a hardness increase and to a higher crack propagation resistance. In this way, the tool performance can be enhanced, making this type of coating very attractive [14,18,30,31]. Additionally, in a study carried out by Kainz [32], multilayered CVD coating (TiN/TiBN) is compared to single-layered TiN and TiBN, concluding that performance is better with the multilayered coating.

Figure 2 shows, in a scheme, how the different types of coating look when applied on the substrate.

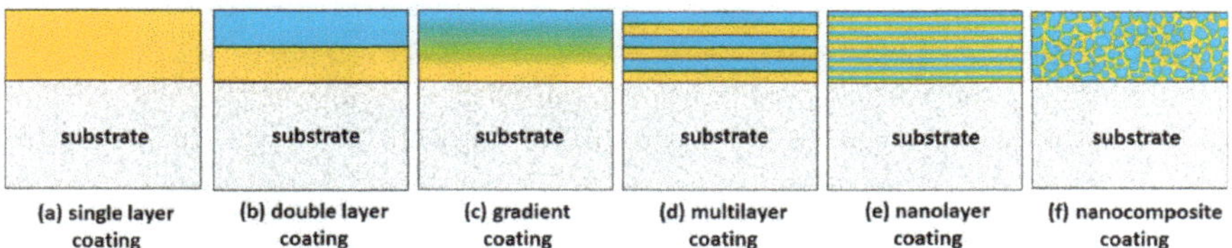

Figure 2. Different designs of hard coatings. Reproduced from [14], with permission from Elsevier, 2017.

Coatings are also characterized by their chemical composition, for example, different types of coatings with different chemical compositions are obtained through different deposition processes. The most common coatings obtained by PVD processes are TiN, Ti(C, N), (Ti, Al)N, while the most common coatings obtained by CVD are Ti(C, N), Al_2O_3, and TiN. Notice that these coatings are employed in different types of architecture, in order to obtain a better-suited cutting tool for a certain machining operation [27]. It is important to note that the chemical composition is important, as different chemical compositions have different hardnesses, friction coefficients and different thermal conductivities, which directly influences coating wear patterns. Regarding coating microstructure, it varies with the coating's chemical composition. Images will be presented regarding the microstructure of various coatings, especially those that have recently been researched. There are various studies regarding coating microstructure, either analyzing coatings obtained by a novel process or analyzing microstructural changes that might occur after machining. In a study by Longt et al. [33], the cutting performance of TiAlN- and CrAlN-coated silicon nitride inserts is analyzed in the dry turning of cast iron (Figure 3). The microstructure of these coatings is also analyzed and is displayed in the following images.

The mentioned coatings were obtained by PVD cathodic arc evaporation, and their thickness were about 4 μm for the TiAlN coating and about 2 μm for the CrAlN coating.

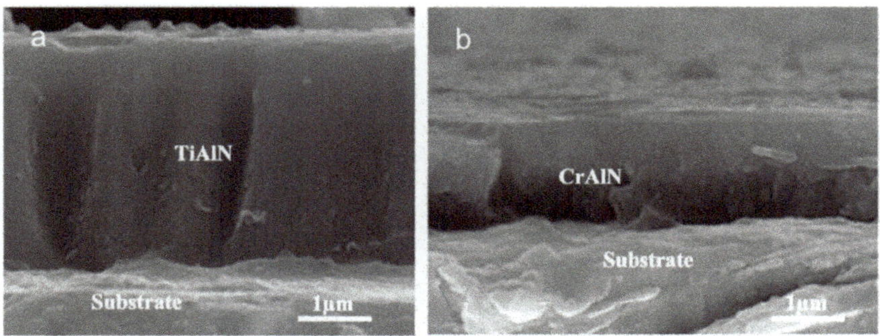

Figure 3. (**a**) TiAlN coating microstructure (**b**) CrAlN coating microstructure. Reproduced from [33], with permission from Elsevier, 2014.

The authors reported that the TiAlN coating had a dense and smooth structure with small grains, while the CrAlN coating's structure had formed columnar crystallites.

Still regarding coating microstructure, another study [34] analyzes the influence of Ni on the microstructure of a PVD cathodic arc evaporation obtained using AlTiN coating (Figure 4). The cutting performance is also analyzed in this study. Three samples were analyzed, one coating composed of AlTiN, another with 1.5% Ni and the third one with 3% Ni content. The following images are of the microstructures of these coatings.

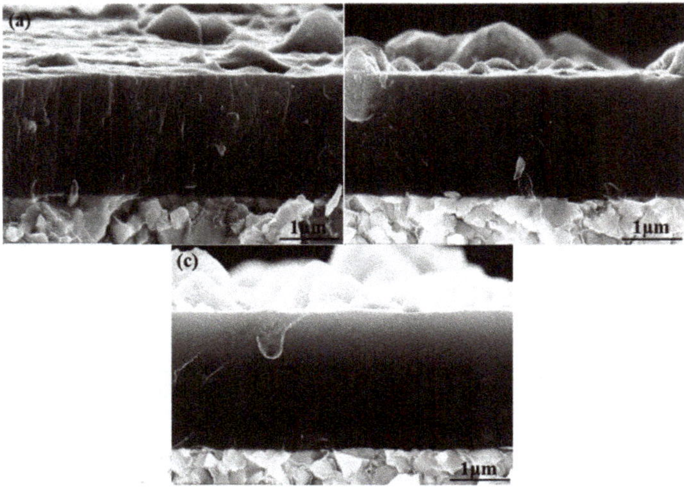

Figure 4. (**a**) AlTiN coating microstructure with 0% Ni; (**b**) AlTiN coating microstructure with 1,5% Ni (**c**) AlTiN coating microstructure with 3% Ni. Reproduced from [34], with permission from Elsevier, 2019.

Notice the microstructural change that is occurring: while the coating with 0% Ni (a) exhibits a columnar microstructure, the addition of Ni promotes a more compact nanocrystalline structure. Although the nanohardness and elastic modulus for the AlTiN coatings had the highest values (26.2 GPa and 315.8 GPa respectively), the authors report that the coating with 1.5% Ni (b) has the best cutting performance. The lowest hardness values and elastic modulus came from the coating with 3% Ni (c) (20.9 GPa and 300.5 GPa respectively). It was also reported that the coating adhesion was worst for the 3% Ni (c) coating, while the 1.5% Ni (b) coating was practically tied with the AlTiN coating in terms of adhesive strength. However, the 1.5% Ni (b) coating was the coating with the greatest toughness, improving tool life by 160%.

Regarding multilayered coatings, these are very appellative to the machining industry, as they confer tools with various properties that can be combined in order to achieve a satisfactory machining process. The trend seems to be reducing layers' thickness, in order to achieve a greater combination of various properties, such as high hardness, a low friction coefficient, thermal conductivity, diffusion barrier and corrosion resistance [35]. Figure 5 shows the overall structure of multilayered coatings.

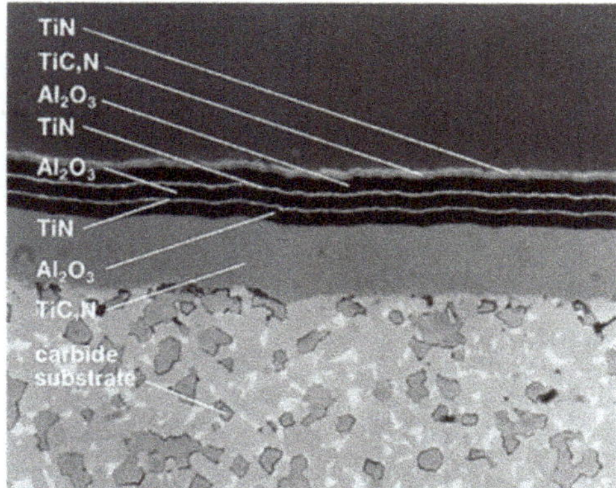

Figure 5. Example of a complex CVD-obtained multilayer coating microstructure. Reproduced from [35], with permission from John Wiley and Sons, 1969.

The nanolayered coatings are a type of multilayered coating, the difference being the layer thickness, which is in the nanometer range. Figure 6 shows the structure of a nanolayer TiB_2 coating.

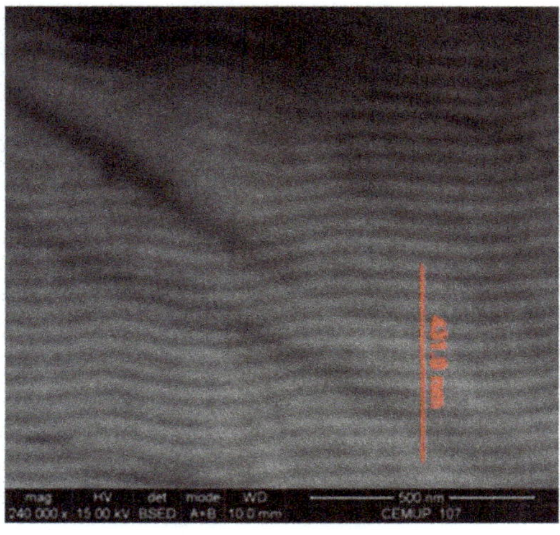

Figure 6. TiB2 nanolayer coating structure. Reproduced from [35], with permission from John Wiley and Sons, 1969.

The coating's structure influences the overall coating strength and adhesion to the substrate. It is an important aspect of fabricating new coatings or finding the correct application to a certain coating.

As previously mentioned, the coatings are also characterized by their mechanical properties, such as hardness, indentation modulus and the stress state of the coated tool (residual stresses that were induced during the deposition process). These mechanical properties can be altered by changing the coating's microstructure, as seen in the previous study [34], or by changing the coating's architecture or structure. As previously mentioned, a multilayered coating has more compressive stresses (e.g., a thin PVD coating); this will make for a stronger and more tough cutting edge. Additionally, their high number of interfaces confers coating strength. However, these residual stresses may provoke coating adhesion problems. In a study by Dai et al. [36], the properties of TiB_2/Cr multilayered coatings with double periodical structures are studied. Various multilayered architectural structures were studied, changing the thickness of the Cr layer in order to see the changes to the microstructure and mechanical properties. It was reported that this double periodical multilayer structure can refine the growth structure of the TiB_2 grains, resulting in a betterment of the coating's mechanical properties. The authors also reported that the residual stresses can be decreased through the deformation of the metallic Cr layers, which, in turn, causes a dramatical improvement to the coating adhesion. Additionally, keeping these residual stresses low can also boost the peel resistance of the coatings [37].

Coatings are applied depending on the machining process and based on the process's pre-requisites, since their performance is heavily tied to the coating's properties (chemical composition, architecture, microstructure and mechanical properties). Because of this, coating design is very important [21,38]. To know which coating to apply and where to apply it, studies have been conducted to evaluate cutting tool behavior. There are many parameters able to affect the cutting performance of a tool, such as cutting speed, feed, depth of cut and lubrication regimen, and tool performance issues which affect the tool's life and the overall finish quality of the workpiece. Cutting tool behavior knowledge has proven to be key in optimizing the turning process. By changing these parameters accordingly, a process can be optimized to meet the expected results [39], such as in the study performed by Krishnan [40], which used the Taguchi method to predict the best parameters to attain the lowest surface roughness when turning IS2062 E250 Steel. The parameters that were varied (input parameters) were the cutting speed, depth of cut and feed rate. The Taguchi loss function was used to compare the experimental values to the desired ones and to predict the output results (surface roughness and material removal rate (MRR)). An analysis of variance (ANOVA) method was used to determine the parameters that most influenced the desired result or the output parameters, in this case the surface roughness and MRR. The process could then be optimized to achieve the good surface roughness that was desired. Similarly, a study conducted by Durga [41], used the Taguchi method to predict the best machining parameters for turning AISI 304 stainless steel with a TiAlN nano-coated tool. The variable parameters of this study were the cutting speed, depth of cut and feed rate, then the authors developed regression equations based on the results for the surface roughness and material removal rate obtained from the empirical tests to develop equations to determine the surface roughness and material removal rate when using coated or uncoated tools for this type of experiment. Recently, studies on the parameters that influence tool performance are focusing on the minimum quantity lubricant (MQL), studying the effect of using this method in the machining of various materials using different tools. The study performed by Khan and Maity [42], which employs MQL using vegetable oil and compares it to dry cutting and flood cooling when turning commercially pure titanium, obtained satisfactory results, reducing cutting forces and cutting temperature when using this approach. Tool geometry also affects the cutting performance, because, by analyzing the cutting tool behavior more adequately, geometries can be created to achieve the desired results. Regarding the study carried out by Harisha et al. [43], cutting tool geometry is analyzed in order to minimize the cutting force when turning hardened steels, this is because tool geometry, when incorrect, leads to energy loss which results in loss of productivity.

In this paper, the recent advances in coated turning tool behavior are analyzed, looking at the different coatings that have been used in recent years. The different wear mechanisms which the coated turning tools are subjected are also going to be analyzed and presented, relating the different wear mechanisms to the coated tool life. There is also a chapter in this paper that will be reserved to the behavior of coated tools under advanced lubrication or cutting conditions, such as, turning using MQL conditions or using cryogenic conditions. Finally, the paper will conclude with a summary of each chapter and there will be mention of what are the recent trends in the turning process using coated tools.

2. Coatings for Turning Tools

In coatings applied to tools, or hard coatings, generally, nitrides, carbides, borides and oxides of transition metals are used. The nitrides used as coatings for cutting tools are TiN, TiAlN, CrN, ZrN, TiSiN, TiAlSiN, CrAlN, TiAlCrN, and cBN [8,9,14,41]; carbides for coatings are TiC, CrC, and WC. For boride coatings, TiB_2 is used due to its chemical inertness, high hardness and good wear resistance.

Additionally, they can be deposited in tool steel with good adhesive capabilities [44,45]. One of the most widely used oxide coatings is the Al_2O_3. Other somewhat common coatings for cutting tools are DLC, MoS_2 and WC-C [13,15].

Due to the drastic tool life reduction, and overall unsatisfactory surface roughness values of the workpiece when using high speed steel (HSS) as a tool material, the industry has been using tool coatings to overcome the issues arising from the use of HSS. In a study by Gupta et al. [46], the cutting characteristics of PVD-coated turning tools were analyzed. The test involved the turning of C45 steel using solid tools coated with TiN, AlCrN and TiAlN. Cutting forces were measured, while cutting speed and feed rate were varied throughout the testing process. The TiAlN coating proved to be the most efficient, relative to tool life, due to its hardness and self-lubricating ability, almost avoiding the adhesion of the workpiece material to the tool surface, and therefore increasing tool life. Research like this continues to be important, as the choice of the right coating can prove to be a profitable choice for manufacturers, a fact that explains the vast amount of studies performed in this area, from finding optimal conditions to machine a certain material, to comparing various types of coatings with different structures or coating methods. In the following paragraphs, recent studies made on the comparison of various coatings, and studies that use coated tools with novel/complex geometries in the turning process, are going to be presented.

The work developed by Koyilada [47], testing the machinability characteristics of Nimonic C-263, used coated cemented carbide for turning that material under dry machining conditions. Carbides were coated with commercially available CVD bilayer coating (TiCN (bottom layer)/Al_2O_3 (top layer)), and PVD multilayer coating (TiAlN/TiN), consisting of alternating layers of TiN and TiAlN on the substrate. Both these coating types were compared. The results show that the PVD coating presented a remarkable improvement in the surface finish of the workpiece when compared to the CVD-coated tool. Cutting forces measured in the tests were also lower when the PVD coating was in use. Additionally, the PVD coating outperformed the CVD bilayer coating in terms of tool life. This is due to the PVD-coated tool presenting a superior compressive strength when compared to the CVD-coated tool (due to deposition technique and multilayer configuration), making it more suitable to work under a fluctuating load. Due to the microstructure of Nimonic C-263, the tool is subjected to dynamic fatigue, even in continuous machining, which explains the underperformance of the CVD coating in this case. Although the PVD multilayer coating is preferred for higher cutting speeds, ranging from 50–90 m/min (further cutting speed augmentation would require the use of a cutting fluid paired with the coated cutting tools), the CVD bilayer coating should be used to turn this material at speeds below 60 m/min, because the top layer of Al_2O_3 has a low thermal conductivity, which results in a higher machining temperature at higher cutting speeds. This eventually leads to wear problems, especially material adhesion and coating disaggregation.

Another study that compares the characteristics of CVD and PVD-coated carbide tools, is a work presented by Ginting [48] which studies the productivity of AISI 4340 hard turning using multilayered CVD (TiN (top layer)/Al_2O_3/TiCN (bottom layer)) and monolayered PVD coatings (TiCN). Productivity was characterized by the material removal rate (MRR) and volume of material removal (VMR). Otherwise, the variables analyzed in the tests were cutting speed, feed rate and depth of cut. The upper limits set to the coated tools reveal that the CVD multilayer-coated carbide can achieve a slightly higher feed rate and depth of cut value. In terms of tool life, the PVD monolayer coated tool endured for longer than the CVD-coated tool. In terms of surface roughness, the PVD coating was more effective as well. As mentioned above, PVD coatings have more sharp edges and are thinner, providing a better quality surface finish, however, in terms of productivity the CVD-coated carbide achieved values higher by about 78–125% than the monolayer PVD-coated tool. This highlights the fact that choosing the right coating method and type of coating can prove an advantage. As seen in that study, the PVD-coated tool performs better regarding tool life and overall surface finish quality, but in terms of MRR the CVD-coated tool is more effective. This is due to the coating properties: while the PVD coating (TiCN) is harder than the outer layer of the CVD coating (TiN), its friction coefficient is also lower when compared to the TiN. This, coupled with the fact that thin PVD coatings confer the tool with sharper cutting edges and have more compressive stresses (caused during the PVD process [29]), give the PVD coating superior finishing capabilities when compared to this CVD coating.

In the work presented by Kumar et al. [49], the performances of PVD coated carbides using TiAlN, AlCrN and TiAlN (top layer)/AlCrN (bottom layer) were tested in the turning of Inconel 825, as well as uncoated carbide tools. The performance of each tool was evaluated, taking into consideration the flank wear of the tool, work-piece surface roughness, cutting force generated during the cutting process and chip formation. The optimal machining parameters were analyzed using grey relational analysis under multiple response optimization, and the results showed that the bilayer coating (TiAlN/AlCrN) outperformed the single layered coatings, TiAlN and AlCrN, in the machining of this alloy.

Still regarding the comparison between CVD- and PVD-coated tools, the study performed by Koseki et al. [50] compares TiN-coated tools obtained by different deposition methods in the continuous turning of Ni-based super-alloys. These high-strength, low-conducting alloys require higher cutting forces and temperatures than other materials during the machining process. The damage suffered during the tests was investigated, and the CVD coating proved to be more efficient in the machining of these alloys, suffering almost no change in coating hardness and overall less plastic deformation in the process.

A method that has seen some use is the coating of a textured tool in order to promote better chip removal, better adhesion of the coating to the tool and even an improved tool life. In the work developed by Mishra [51], the machining performance of laser-textured chevron shaped tools, and untextured tools was evaluated. These tools were coated with AlTiN and AlCrN using PVD. Cutting forces and tool wear were analyzed for textured and untextured cutting tools. Coating growth on textured tools was better, presenting a reduced number of microcavities and macroparticles for both coatings. The value of the cutting forces was lower for the textured tools, resulting in less tool wear, and the texture on the tools improved the tool-coating adhesion. The chips formed by the textured tools were thinner when compared to untextured tools, however, chip fragments were embedded in the textures. Thus, machining parameters need to be adjusted in order to find a balance between these two phenomena in order to produce favorable machining conditions.

As previously stated in Chapter 1, discussing developments in gradient cemented carbide tools, studies on the influence of this gradient and on how to control the grain size of the cemented carbide, have been conducted [11,12]. These gradient cemented carbides already represent a significant improvement when using uncoated cemented carbide tools. Additionally, these gradient cemented carbides may improve the quality of the coating, making it more adherent to the substrate, and even improve tool performance. A paper presented a study regarding CVD coating application on gradient cemented carbide tools of different grain sizes, where it was observed that the coating thickness is related to the carbide grain size, coating thickness increases with smaller grain sizes [52]. It was also concluded that the adhesion strength of the coating is overall better on gradient cemented carbides when compared to regular cemented carbides, however, the adhesion strength drops when the grains of the cemented carbide are finer. A thicker coating may not be ideal for finishing operations, however, for roughing operations, having a thicker coating may be beneficial.

Cutting tool coatings may also provide a sustainable eco-friendlier alternative when machining certain materials, as some of these materials, for example, nickel-based super-alloys, already mentioned in this chapter, require higher cutting speeds, which results in higher cutting forces and a higher temperature in the cutting area. To counteract these problems, lubrication is employed, sometimes by flooding the cutting area. Practices like these have proven to be unsustainable and damaging to the environment.

A recent tendency is to employ methods able to reduce the high usage of these lubricants and make the overall machining process more eco-oriented. These methods (Figure 7) sometimes use biodegradable oils, such as vegetable oils, as lubricants. Dry machining is also a sustainable option, as it removes the use of cooling fluid altogether, having as its alternative the minimum quantity lubrication (MQL) method, in which small amounts of lubricant are employed. MQL presents itself as being a viable alternative to dry machining when having problems with cutting temperature or surface finish quality. Other sustainable methods, such as cryogenic cooling, or the use of a high-pressure coolant, in which the coolant is applied to a select area of the tool, are employed as an alternative to the conventional flood cooling method [53].

Figure 7. Wheel of sustainable machining. Reproduced from [53], with permission from Elsevier, 2019.

The work produced by Thakur and Gangopadhyay [54] proposes a sustainable alternative to the machining of nickel-based super alloy, by employing TiN/TiAlN PVD-coated tools and dry turning of the Incoloy 825. Tests were conducted employing different lubrication methods, such as dry machining, flood machining and MQL. Cutting forces, temperature of the cutting area, tool wear and surface integrity were evaluated. While the temperature in the cutting area was higher for dry machining when using a PVD-coated tool, the overall surface finish was of better quality when compared to flood cooling and MQL. Moreover, cutting forces also presented lower values when using a PVD coating, even in dry machining. This means that the cutting force necessary would be lower, therefore promoting a more environment-friendly alternative (dry machining); an overall better surface finish was obtained using the PVD-coated tool under an MQL environment.

Regarding nanolayered coatings, they are quite appealing, mainly due to their high hardness value, for example, research was done on nanolayer TiN/VN coatings, and results found that there was a very high hardness increment [55]. This hardness increase is related to the nanolayer thickness, as see in Figure 8.

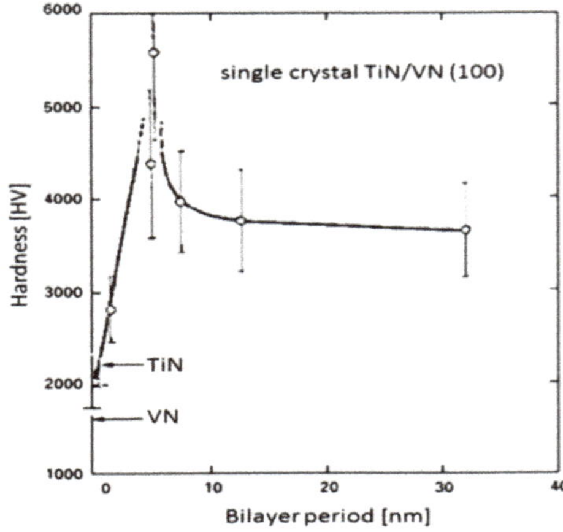

Figure 8. Hardness increase of a TiN/VN coating, over the thickness of the bilayer. Reproduced from [14], with permission from Elsevier, 2017.

The reason for this hardness increase was attributed to the large number of interfaces between layers, characteristic of nanolayered coatings. In Figure 9, the structure of a CrN/TiAlN nanolayered coating deposited on steel is shown [14].

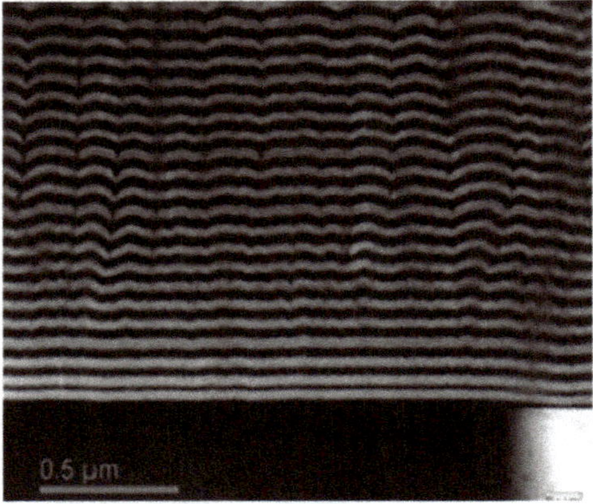

Figure 9. CrN/TiAlN nanolayered coating structure. Reproduced from [14], with permission from Elsevier, 2017.

The structure shown in Figure 9, characteristic of nanolayered coatings, confers the coatings a very high crack propagation resistance, as the cracks tend to not propagate as deep as in a monolayered coating, thus risking damage to the substrate. These coatings are very similar to the multilayered coatings, the main difference being the thickness of each layer and that the hardness value for a nanolayered coating is not equal to the average hardness of its constituents—however, this is the case for regular multilayered coatings [14]. A recent study about the influence of the thickness of these nanolayers in coatings found that, for hardness values at room temperature, there were no significant changes between the coatings with differing thickness layers. However, the cutting properties of the coatings were different, with the coating with the thinnest nanolayers exhibiting a higher tool life than the coating with thicker nanolayers [56].

As mentioned previously, understanding cutting tool behavior is the key to correctly optimizing a machining process. There are many studies on coatings' behavior while cutting, for example, in the study presented by Gassner et al. [57], the thermal crack network on CVD TiCN/Al_2O_3-coated cemented carbide cutting tools is analyzed. The theme of coating tools and coating performance is heavily researched, comparing coating methods to achieve overall better machining results, or even finding an eco-friendlier alternative to machining certain materials.

3. Coating Influence on Turned Surface Quality

There are many factors that influence surface quality in the turning process, as seen in Chapter 1. A turning process can be optimized in order to have the best possible surface quality, by changing certain parameters, such as rotation speed, feed rate and depth of cut. These parameters are mainly dependent on the machine, as the machine also influences cutting performance and the overall surface finish of the machined piece [58]. There have been studies that relate the chip formation thickness to the overall surface roughness of the machined piece [59]. Using a prediction method to determine chip formation thickness can serve as a monitor of the surface roughness of certain materials. However, there are more factors that influence the surface roughness of turned pieces, such as lubrication method and tool geometry, and the coating that is being employed also affects the overall surface quality of the workpiece. As previously mentioned, thin PVD coatings are very well suited to surface finish operations, and provide an overall better surface quality than thicker coatings. This is mainly due to the sharp edges that are conferred to the substrate, and compressive stresses that confer the tool edge strength [27–29].

In this chapter, the influence of coated tools on the surface quality of various materials are presented. The materials selected are mainly titanium alloys and nickel-based super-alloys, as these are very appealing for structural and engineering applications, especially due to their strength-to-weight ratio. Although these alloys have some processing problems associated with them, for example low machinability rating, their poor machinability may be attributed to material properties, such as high hardness at high temperatures, low thermal conductivity and high chemical reactivity [60]. There have been some studies conducted on the turning of hardened steels, which will also be presented in this chapter, highlighting the coating's influence on turned surface quality.

In the aforementioned study [40], a comparison of the machining performance of coated tools, using a monolayer PVD coating TiCN and a multilayer CVD coating TiN/Al_2O_3 in the hard turning of AISI 4340 steel. The authors found that, in this case (and as elaborated above), the PVD coating would be best suited for finishing operations, especially due to the hardness values of TiCN and the lower friction coefficient (when compared to the TiN top layer of the CVD coating). The surface roughness values obtained for the PVD-coated carbide and for the CVD-coated tool, are (0.8–1.6) micron and (1.6–3.2) micron, respectively. However, for material removal rate the CVD coating is preferred, although the turned surface quality is poorer than those using PVD coatings.

The study by Fernández-Abia et al. [28], presents a comparison of four coatings (and an uncoated tool) in the turning of austenitic stainless steel, AISI 304L. The cutting behavior of these coated tools were analyzed. The authors mention that these types of PVD coatings are best suited for achieving low values of surface roughness, especially due to the sharp edges conferred by the PVD process. The coatings used were, AlTiN; AlTiSiN; AlCrSiN; and, finally, TiAlCrN. The first three coatings are nano-structured coatings. The graph of Figure 10 shows the results obtained from this study, regarding the surface roughness of the machined material.

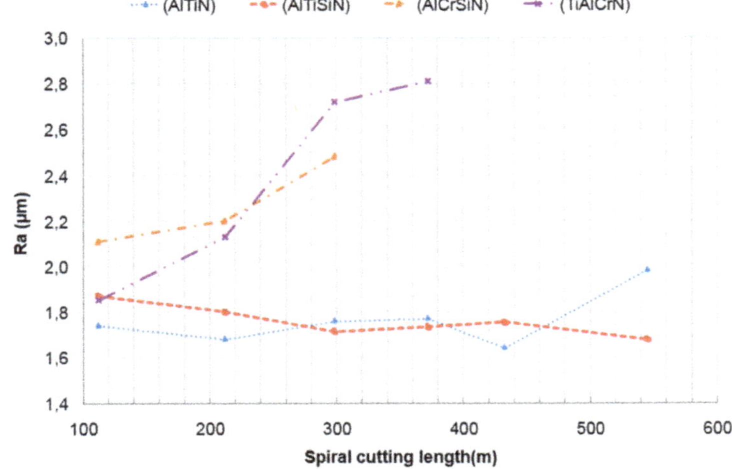

Figure 10. Graphic of surface roughness value (Ra) for different PVD coatings [28].

The best coatings for the machining of this material are the AlTiN and AlTiSiN coatings; this is due to their nano-crystalline structure. Although AlCrSiN also has this beneficial nanostructure, the presence of chromium in its chemical composition favors the creation of an oxide protective layer that is inferior to the AlTiN and AlTiSiN coatings.

Regarding the turning of nickel-based super-alloys, in this study [54], the influence of coating and lubrication/cooling method was observed in the turning performance of Incoloy 825, using a TiN/TiAlN multilayer coating. The results regarding surface roughness obtained from these tests can be interpreted from the graphs in Figure 11.

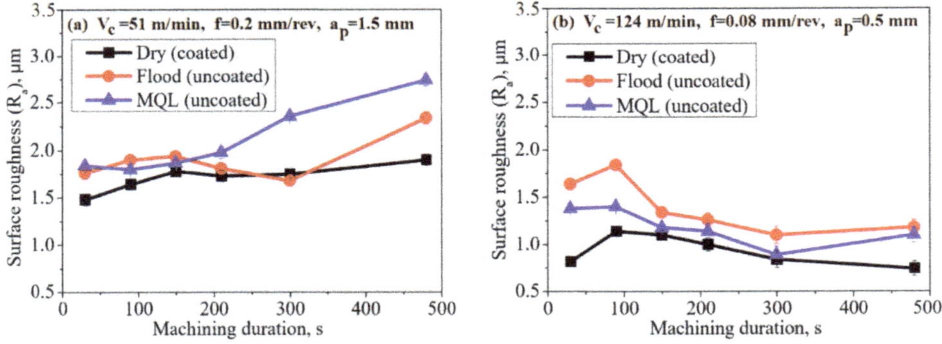

Figure 11. Different values of surface roughness for the machining duration. (**a**) and (**b**) have different cutting parameters. Reproduced from [54], with permission from Elsevier, 2016.

The coated tool provides the better surface roughness quality from the tests that were carried out. The fact that dry machining with coated tools provides a better surface quality is quite an interesting finding, as this type of machining is sustainable and eco-friendly, and these coatings enable the machining of high-quality parts at a lower price/environmental impact. The authors also added that the cutting temperature was lower for the dry machining using PVD TiN/TiAlN-coated tools. A study was also mentioned before [49], in which a PVD and CVD coating were used in the machining of Nimonic C-263; the findings of these authors report that the PVD TiN/TiAlN multilayered coated tool is better for the surface finish. Due to the sharper edges, this coating provided a 14.3% reduction in surface roughness value when compared to the CVD TiCN/Al_2O_3 counterpart, in which a higher edge radius contributed to the lower turned surface quality. As stated previously, the PVD coating is preferred for higher machining speeds than the CVD coating.

Without a doubt, coatings improve the overall surface finish quality of turned parts, however, tools with the correct geometry can rival the low roughness values obtained, especially those with thinner PVD coatings. By analyzing the literature, a trend can be seen, with thin PVD coatings usually being employed in finishing operations, explained by the residual stresses that thin PVD coatings have (compressive stresses), and by the sharp cutting edge that these coatings confer to tools. Their chemical composition is also a contributing factor, as, depending on the coating chemical composition, these will react differently with the material that is being turned, sometimes even forming hard protective layers that lower the friction coefficient, thus improving overall coated tool performance [25].

4. Tool Wear Mechanisms

Coated cutting tools significantly improved the tool life of conventional tools, as coated tools suffer overall less wear in the same lifetime as an uncoated one, particularly at high machining speeds [61], however, coated tools eventually give out, due to the fact that a lot of these coated tools are used in dry machining conditions, which means that the machining temperature is overall higher. A coated tool has different wear mechanisms, such as abrasive wear, thermal cracks, adhesive wear, build up edge (BUE), or coating structure failure, resulting in spalling or cracks appearing on the coating. Understanding the different wear mechanisms for each coating type helps one make a better coating or machining parameters choice to achieve the desired results. These wear mechanisms are related to some parameters; for example, when there is an increase in the cutting force, it can be assumed that the coating is suffering abrasive wear, or that there is a problem with the coated tool's edge. Wear is also related to the coating properties, different coatings (or coating layers) have different hardnesses, friction coefficients and thermal conductivities, all factors that also contribute to the wear patterns of these coated tools. Coating microstructures can influence factors, such as coating adhesion, that might cause fracture wear later. Mechanical properties, such as hardness, indentation modulus and the stress state of the coating itself, affect the wear patterns that these will suffer. In some cases, high residual stresses can cause problems with coating adhesion, that is, the coating's adhesive strength is lower, and the coating is more likely to suffer spalling or delamination, although, as mentioned before, different coatings obtained by different deposition methods have different stress states. Hardness is primarily tied with abrasive wear, as well as the coefficient of friction (COF); the latter being related to coating design (layer thickness or layer chemical composition), as seen in the study by Dai et al. [36], where it is reported that the increase in the thickness of a Cr layer, in a multilayered TiB_2/Cr coating with double periodical structures, would result in a decrease in coating hardness and an increase in coefficient of friction, which will affect the coated tool's wear rate.

In this chapter, some studies regarding different wear mechanisms and coating degradation will be addressed, presenting images for some types of wear mechanisms, as well as a summary of the findings of these studies.

In a work mentioned in the previous chapter [50], there is a study on the wear of TiN coatings obtained by different deposition methods. In addition to abrasive wear and fracture wear, a common wear mechanism is adhesive wear, where the material adheres to the coated tool. In Figure 12, the wear mechanisms for the TiN coating, obtained by different deposition methods, can be seen on the tool's cutting edge.

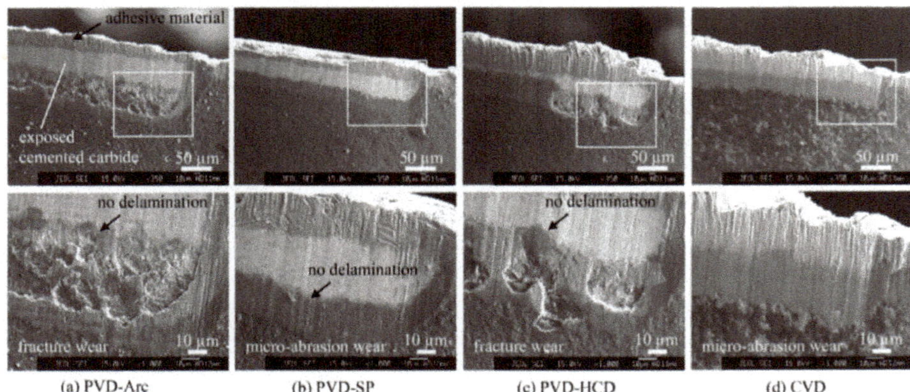

Figure 12. Wear mechanisms on TiN coating, in the continuous turning of a nickel-based super-alloy using different coating techniques: (**a**) PVD-arc, (**b**) PVD-SP, (**c**) PVD-HCD and (**d**) CVD. Reproduced from [50], with permission from Elsevier, 2015.

In Figure 12, in addition to noticing the adhesive material on the cutting edge, other wear mechanisms can be seen, such as fracture wear, where the cemented carbide substrate is exposed. There is also micro-abrasion wear that can be noted on all the samples.

Still regarding the same study [60], some plastic deformation can be observed in the coating (Figure 13). In this image, a broken coating area and adhesive material can also be seen.

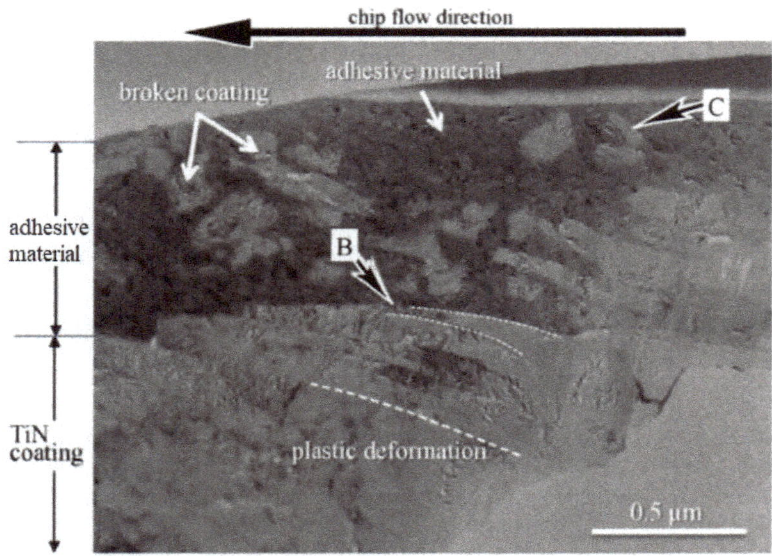

Figure 13. High resolution TEM image regarding a TiN coating obtained by PVD-Arc, noticing wear on the surface of the tool. Reproduced from [50], with permission from Elsevier, 2015.

Also due to temperature, thermal cracking may occur on the coating. Another study regarding thermal cracking of CVD TiCN/Al_2O_3-coated cemented carbides [57] analyzes various methods to close the cracks that occur due to excessive cutting temperature, such as wet blasting or filling the cracks with TiO_2. This wear mechanism can be observed in Figure 14, taken from the same work.

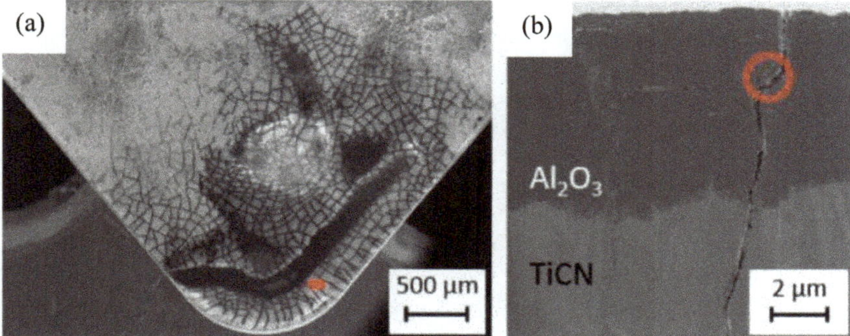

Figure 14. (**a**) SEM image of rake face of a coated CVD insert after face turning (**b**) highlighted section of the rake face. Reproduced from [57], with permission from Elsevier, 2019.

Regarding crack defects on cutting tools, the study performed by Vereschaka et al. [56] highlights the influence of the PVD Ti-TiN-(Ti,Al,Cr,Si,)N nanolayer coating thickness, which was tested in the turning of AISI 321 steel. The coatings differed in the number of nanolayers and nanolayer thickness; the coating with the ticker nanolayers had a total of 33 nanolayers (Coating A), and the other coating, with the thinner nanolayers, had a total of 57 of these layers (Coating B). These coatings were also tested against monolayered Ti-(Ti,Al)N-coated and uncoated tools, and both the nanolayered coatings presented a higher tool life than these last two. However, Coating A has less wear resistance and, with the increase in cutting speed, the cutting temperature also increased, inducing thermal stresses in the superficial layers of the tool. This can be observed in Figure 15, where (a) is the image regarding the thickest nanolayered coating and (b) is the thinnest one. The thinner layers provide the coating with a higher wear resistance, which means that thinner layers better resist thermal stress crack formations.

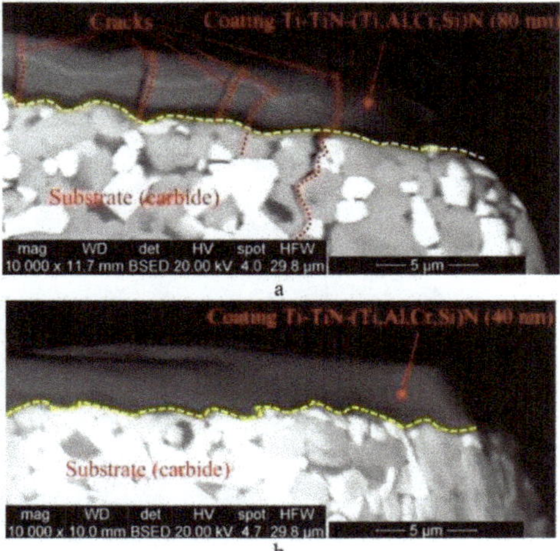

Figure 15. Crack formation on PVD nanolayered coating surface, area directly adjacent to the cutting edge; (**a**) coating with nanolayers of 80 nm and (**b**) coating with 40 nm nanolayer thickness [56].

Coatings, when deposited, mirror the substrate surface, resulting sometimes in an uneven coating surface, with imperfections such as cracks, and even residual stresses that may have resulted from the coating process. These superficial imperfections may cause material transfer and have a negative impact on the coated tool performance, by promoting tool wear or by not conferring the desired surface finish.

Some methods have been proposed to minimize these defects, such as a post-deposition polishing of the coated tool, in order to lower its overall surface roughness and minimize the potential for material transfer [62]. Another method proposed to deal with cracks is shown in the paper presented by Faksa et al. [63], where the authors study the effect of shot peening on residual stresses and crack closure in CVD-coated hard metal cutting inserts. Due to the stresses seen on the deposition process of the CVD coating on hard metal, the surface layer of the coating exhibits cracks. The authors conclude that well placed and calculated shots can close these cracks and prevent crack nucleation and growth in CVD coatings. These residual stresses also have an influence on PVD films, as shown by Skordaris et al. [64], who studied the effects of PVD films' residual stresses on their mechanical properties, brittleness, adhesion and cutting performance. The coatings used are PVD TiAlN, and different coatings were used, these having different levels of residual stress, obtained by heat treatment. The authors concluded that there is a significant contribution of the film's compressive stresses to increasing the mechanical properties of the coating and adhesion, consequently improving tool life.

The wear mechanisms of a MTCVD–TiCN–Al_2O_3-coated cemented tool was also analyzed in another study [65], where wear patterns were observed after 142 min of turning 300 M steel, at a cutting speed of 300 m/min. Crater spalling could be observed, as well as evidence of molten metal particles, which means that the cutting temperature was very high. Signs of adhesion, matrix exposure and build up layer (BUL) were also observed. These wear patterns can be observed in Figure 16.

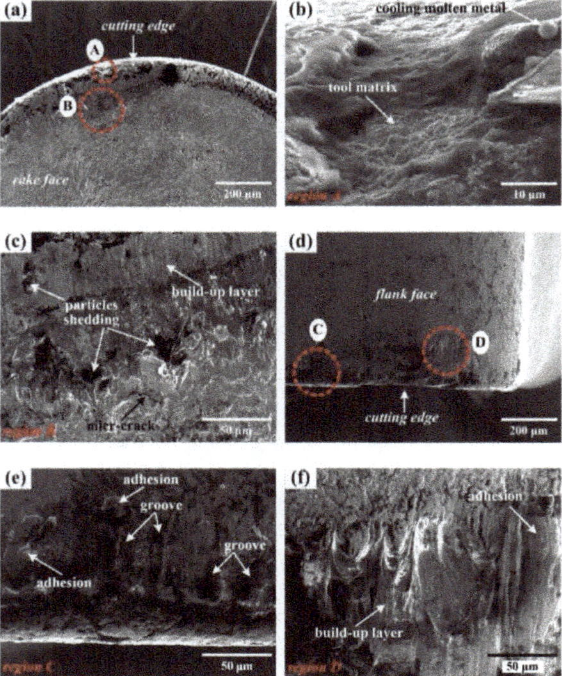

Figure 16. Different wear patterns on the MTCVD coating after 142 min of turning 300 M steel. The rake face (**a**) was analyzed in two regions A (depicted in (**b**) presenting signs of cooling molten metal), and region B (depicted in subfigure (**c**), where signs of BUL and micro-cracks can be noticed). In subfigure (**d**) the flank face is displayed, and zones **C** and **D** are analyzed. Zone **C** is displayed in subfigure (**e**), where adhesion damage can be noted, and some grooves. Lastly, zone **D** is displayed in (**f**), where adhesion damage is the predominant type of wear mechanism. Reproduced from [65], with permission from Elsevier, 2017.

An analogous study [66] using similar conditions (maintaining the same coating and machined material) was conducted in order to analyze the main wear mechanisms of the coated tool. The wear patterns are like those observed in the previous study [65]: adhesive wear, build-up layer (BUL), molten material and crack wear patterns were detected in the coated tool. The authors conclude that the main wear mechanisms are adhesive wear, abrasive wear, oxidation wear and diffusion wear. During the tests, cutting parameters were varied, such as cutting speed and feed rate. Cutting forces were also analyzed, because this is an important step when optimizing a process.

In another work [67], the wear mechanism of PVD-, CrAlN- and TiAlN-coated Si_3N_4 ceramic cutting tools was studied. The conducted tests consisted of the turning of GT250 gray cast iron using these coated tools and characterizing the wear mechanisms. It was found that the adhesive strength of the TiAlN was stronger than the CrAlN coating; during the turning of the material, the CrAlN coating suffered spalling at a cutting speed of 400 m/min, due to low adhesive strength. During the dry turning of the material, abrasive wear and minor adhesive wear were found to be the main wear mechanisms. It was possible to observe that coated tools suffered more adhesive wear than the uncoated inserts. Still regarding the coated Si_3N_4 ceramic cutting tools, studies made on the wear mechanism of these tools coated with diamond were conducted [68,69], the authors observed the machining performance of coated and uncoated tools; it was found that the cutting force was higher during machining with diamond-coated tools, due to the surface roughness of the rake face. Additionally, the parameter that influenced these cutting forces the most was the feed rate, contributing more to the increase in cutting force than cutting speed. Regarding wear mechanisms, it was observed that the high machining temperature promoted the graphitization of the diamond coating, which resulted in its removal from the tool; however, there was no delamination observed in the coating after machining.

As previously stated, a recent shift to dry turning as an alternative to some machining methods has been observed, due to economic and environmental reasons. Naskar et al. [70] in their investigation, compared the flank wear mechanism of CVD and PVD hard coatings in the high-speed dry turning of low- and high-carbon steel. The steels in question are C20 and C80, and they were turned at a cutting speed of 300 m/min and 600 m/min with CVD Al_2O_3 (top layer)/TiC (bottom layer), TiCN (top layer)/TiC (bottom layer) bilayer-coated, and PVD TiAlN single-layer-coated, inserts. The authors found that the main wear mechanisms were abrasive wear, however, plastic deformation-induced necking and dissolution-diffusion were also contributing to the acceleration of tool wear. They concluded that when designing a coating material for high speed machining, the solubility of coating materials has to be taken into account.

Figure 17, taken from the paper presented by Naskar et al. [70], exhibits the wear of coated tools (coatings presented in the above paragraph), in the machining of C80 steel.

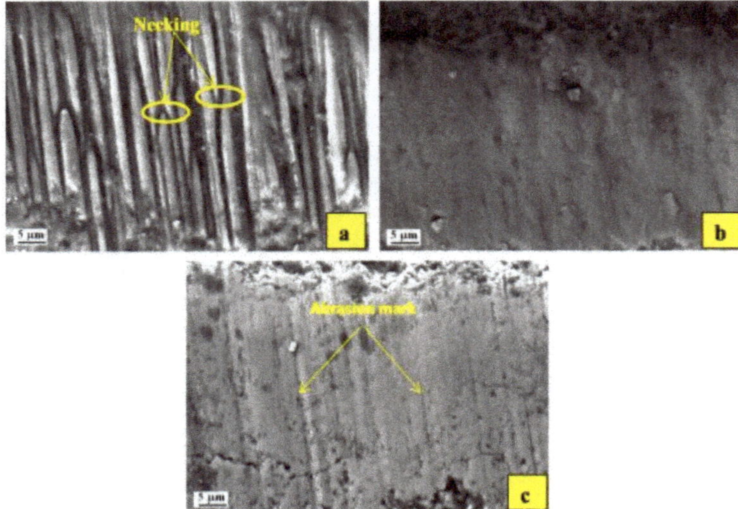

Figure 17. Wear patterns for necking (caused by plastic deformation) (**a**) and abrasion marks (**c**). In image (**b**) the coating's surface is smooth after wear, with no evident abrasion marks or plastic deformation induced defects. Reproduced from [70], with permission from Elsevier, 2018.

Studies like these are very important to understanding and finding ways to improve tool life, by understanding how the wear patterns are displayed. Moreover, machining parameters can be found/calculated in order to optimize the process and improve tool life. Additionally, by observing these wear patterns, new ways of fabricating novel coatings can be achieved. The development of nanostructured composite coatings is still quite novel, and studies such as these help to gain a better understanding of how they behave in certain conditions, such as, for example, in the study carried out by Vereschaka et al. [71], in which the behavior of a nanostructure multilayered composite coating is tested under the high speed turning of steel. It was found that there was adhesive wear from the steel that was being turned. As a result of the turning process, a top layer was "destroyed", no longer exhibiting a nanostructure.

5. Tool Life

Improving tool life has been a strong focus of the machining industry, as having cutting tools last longer and not underperform is quite appealing. Because of this, many studies have been conducted to increase understanding how to improve cutting tool life. As seen in the previous chapter, the knowledge of the cutting tool wear mechanisms is crucial when wanting to improve tool life [70]. There are many factors that directly influence tool life, such as the coating's properties, i.e., mechanical properties, coating architecture and microstructure, and chemical composition. These factors influence the coated tool's wear patterns, thus influencing tool life. Machining parameters also influence the tool's life, such as cutting speed, feed rate and even tool geometry, as there are some papers that study the influence of the micro-textures of cutting tools, relating their surface geometry with tool life [72]. Regarding the influence of machining parameters on cutting tool life, the study carried out by Asha et al. [73], analyzes the effect of these parameters on cutting temperature and tool life while turning EN24 and HCHCr Grade alloy steel. The authors carried out tests where the cutting speed and feed rate were varied, and the depth of the cut was kept constant. The coatings used in the carbide inserts were an M15 grade multilayer coating and an M20 grade. It was determined from the results of the turning tests, that the cutting temperature was higher when machining the HCHCr grade alloy steel, when compared to the EN24; the cutting speed increase applied during tests caused tool life to decrease, however, feed rate did not have a high impact on the tool life. It was also found that the tool life was lower when machining the HCHCr alloy steel; this was possibly due to the presence of high alloying elements in the steel and the hardness of the HCHCr steel.

In this chapter, some studies focused on improving tool life are presented, and the various methods for determining tool life for different tools are also mentioned.

In the study carried out by Boing et al. [74], the tool life of PVD- and CVD-coated tools is evaluated when turning AISI 4340-, 52-, 100- and D2-hardened steels. The authors found that the TiAlN PVD-coated tool promoted better results when turning AISI 4340 steel. Otherwise, the MTCVD TiCN/Al_2O_3//TiN proved to be better at turning the other steels. The authors also found that the hardness and microstructure of these steels were the limiting factor, meaning that the carbide fraction that is present in the steel microstructure limits tool life, due to the impact on the cutting edge, similar to the study presented in the beginning of the chapter by Asha et al. [73].

Another paper presented by Vereschaka et al. [75] relates the coating thickness of a composite nanostructured coating Ti-TiN-(Ti,Al,Cr)N to its tool life, similar to the results from another work [56]. The coating with thinner layers presents overall better mechanical properties and wear resistance properties, having a longer tool life than its thicker competitor.

Another method to improve tool life involves ANOVA analysis, determining the best machining parameters in order to obtain the desired effects, from a better surface finish to a longer tool life. This method is presented in the work by Ranjan Das et al. [76], in which a process that involves the hard turning of AISI 4340 using a CVD TiN/TiCN/Al_2O_3//TiN coating is optimized.

There are some methods proposed to predict tool life based on certain parameters, such as the study that proposes a new model for the prediction of a time-varying heat partition coefficient at the coated tool–chip interface in continuous turning [77]. The heat partition coefficient at the tool-chip interface is important, as it helps to accurately estimate the distribution of heat flux and temperature while machining, therefore this is important, as it gives insight on what influences the heat distribution on the tool-chip; as, for example, coating thickness and substrate material influence the heat partition coefficient, the type of coating and cutting parameters also influences this coefficient, for example, increased cutting speed resulted in higher temperatures, and all of this has an influence on tool life. With this proposed method, there is a new way to design better and more optimized coated tools, and even help the selection of these tools for machining applications.

The study by Zhang et al. [78], proposes the prediction of tool wear, using a 2D Fractal analysis of the cutting force and surface profile, in the turning of an Iron-based super-alloy. Coated carbide tools and cermet tools were used to turn the material, and a dynamometer was used to measure the cutting force. MATLAB (MathWorks, Natick, MA, USA.) was used to calculate the fractal dimension. The results from these tests showed that the cutting force curve and the machined surface profile had fractal characteristics; the authors determined that by using this method, tool wear and machined surface finish quality could be predicted. This study yielded additional results, such as the coated carbide tools having a higher tool life than the cermet tools, and demonstrated that by increasing cutting speed the surface finish quality would also increase, however, with the increase in tool wear the surface quality would deteriorate. Having a prediction method such as this for tool life is very appealing for the industry.

There are also some studies on tool life that focus on the substrate. For example, in the study carried out by Uhlmann et al. [79], the substitution of commercially coated tungsten carbide tools in the dry cylindrical turning process with HiPIMS-coated niobium carbide cutting inserts is proposed. These niobium coatings have shown potential in the machining of iron-based materials. The authors found that, although the cutting performance of the HiPIMS-coated niobium inserts was higher than that of the uncoated inserts, when compared to the commercially coated tungsten carbide tool performance was not improved noticeably. However, it was found that the adhesion of this coating to the substrate was good, providing an alternative to regular coated tungsten carbide tools. Research like this is important in finding new ways to optimize the machining of certain materials, as these new coatings may be a reliable option when wanting to improve cutting tool life.

6. Tool Coatings Under Advanced Cutting and Lubrication Conditions

Since using coated cutting tools for turning has its limitations, especially regarding the lubrication method, there have been some studies on employing some alternative lubrication methods in order to achieve better results in the finished product, and even optimizing the process, making it cheaper by improving tool life or even reducing power usage and, additionally, making the overall process safer for the environment.

In this chapter, recent studies regarding alternative lubrication methods are presented, drawing attention to the minimum quantity lubricant (MQL) method and cryogenic lubrication methods, paying attention to the overall process efficiency and tool behavior while turning. For example, in some cases MQL regimens can have a good impact on surface finish quality when compared, for example, to dry turning. Additionally, the dry turning of some alloys may cause a high temperature in the cutting area, provoking more work hardening when compared to MQL regimen [80]. As mentioned in Chapter One, there are papers that study the influence of extreme pressure anti-wear additives (EP/AW), in the MQL regimen, obtaining good results when compared to other MQL (without additives) regimen and dry turning. These types of lubrication method not only affect the overall finished quality of the product and tool life/performance, but they affect the microstructure as well. Studies show that MQL is quite advantageous when the best surface finish is one of the goals [81,82].

In the work presented by Marques et al. [83], the turning of Inconel 718, applying a vegetable-base cutting fluid mixed with solid lubricants by MQL, is proposed. The authors studied the turning process of this super-alloy under dry machining conditions and under MQL while using graphite solid lubricant. The authors found that under MQL conditions the tool life was improved, because the addition of solid lubricants reduced the cutting forces during the process, making it a good option when intending to extend tool life when turning Inconel 718.

Regarding the use of vegetable-based coolants in the turning process, in the study carried out by Elmunafi et al. [84] the tool life of a tool coated with TiAlN is analyzed under MQL using castor oil. The authors achieve satisfactory results by reducing the overall cutting temperature and cutting forces, suggesting that MQL method would prove to be useful in the turning of hard stainless steels. It was also found that the tool life is inversely proportional to both cutting speed and feed, with the effect of the first being more significant. Additionally, the use of castor oil is more environmentally friendly when compared to other coolants.

In the work presented by Chetan et al. [85], the wear behavior of PVD TiN-coated carbide inserts during the machining of Nimonic 90 and Ti6Al4V super-alloys under dry and MQL conditions is studied. The authors determined that the main mechanism for the wear of the coating during the machining of Nimonic 90 alloy was the abrasive wear and nose fracture, which caused the catastrophic failure of the tool. However, due to the wettability of Ti6Al4V under MQL mode, this provided less intense flank wear at high cutting speeds. Studies such as these help to understand when to apply certain lubrication regimens, depending not only on tool but the workpiece material as well.

There have been some recent studies on cryogenic pre-treated coated tool performance, such as the study performed by Kumar and Senthil [86]. In this work, PVD-coated TiN/AlTiN tungsten carbide inserts were used in the dry turning of a Ti6Al4V titanium alloy. The authors concluded that this treatment increased the hardness of the inserts. Additionally, the cutting forces obtained were lower when using cryogenic treated tools; overall better surface finish was reported, as well as less significant tool wear on the cryogenically treated tools.

Cryogenic cooling methods are a recent focus of attention as well, concerning coated tool behavior. In the study carried out by Dhananchezian et al. [87], the effects of cryogenic cooling on the turning of 2205 duplex stainless steel, using a PVD-coated nano multilayered TiAlN cutting tool, were analyzed and compared to dry turning. From this study, it was found that cryogenic cooling reduced cutting temperature by more than 50% when compared to dry turning and decreased cutting forces by up to 40%. An improvement on roughness was also registered, of about 20%. These results contribute to finding better alternatives, especially when machining hard materials such as duplex steel. Regarding machining under cryogenic conditions, the Taguchi Method can also be used to optimize the machining parameters for these conditions, as shown in the study by Khare et al. [88]. The optimal parameters were chosen when machining AISI 4340. These parameters were: cutting speed, depth of cut and feed rate, while under cryogenic conditions. By optimizing a process such as cryogenic turning, it makes the process more viable, from a financial standpoint, making it more likely to be used in the industry. A similar method, the Taguchi incorporated Gray relational analysis (TGRA), is shown to be implemented with success in another study, this one regarding the cryogenic machining of 17-4 PH stainless steel. As in the previous study, the optimal parameters were predicted, these parameters being the cutting speed, feed rate and depth of cut [89].

There are some additives that can be used in lubricants to improve machining performance, reduce tool wear and even improve tool life. In this study, performed by Gutnichenko et al. [90], a study of the influence of adding graphite nanoplatelets (GnP) to vegetable oil on the MQL-assisted turning of Alloy 718 was performed. From the turning tests, the additives impact the machining performance in a positive manner; by adding the GnP to the vegetable oil, an increase in terms of tool life, surface finish and overall process stability was noted. In addition, adding GnP particles to the oil results in a significant reduction in friction in the cutting area. It was also noted that these particles influence the chip formation process, whereas a pure oil lubrification would act as a coolant. The authors also reported that using the vegetable oil without the nanoparticles would improve the machining process, however, these nanoparticles would contribute heavily to overall process stability.

7. Concluding Remarks

As the machining industry grows, the interest in making it more profitable grows as well. The turning segment still has an important presence in the machining industry and, because of this, is an important focus of research in order to optimize the process to achieve more and more satisfactory results. Tool behavior knowledge is necessary to understand the process, with a vast amount of literature existing in this field. In this paper, the recent studies on and advances in coated tool technology were presented.

Regarding the development of new coatings, the focus seems to be the development of nanolayered and nanocomposite coatings, however, there are many studies on the behavior of the most common CVD and PVD coatings. Many of these studies focus on the comparison of PVD- and CVD-coated tool performance, evaluating the influence of these coatings on the machining of certain materials. The coating's chemical composition, architecture and deposition method are all factors that contribute to the cutting performance of coated tools. Researches show that CVD- and PVD-obtained coatings are used for different types of operations, for example, thin PVD coatings provide a suitable option when the finishing quality is the most desired parameter, and, in turn, CVD coatings prove to be useful in most materials for roughing operations. Of course, this depends heavily on the material that is being machined and the coating properties.

Wear mechanisms were also analyzed, showing how different wear patterns present themselves on the coated cutting tool. There are many comparative studies as well in this field, as understanding how different coatings develop their wear mechanisms and how they develop is key when wanting to improve tool life. These wear patterns usually occur due to abrasion failure, adhesive wear or coating destruction. Properties, such as coating hardness, residual stress and chemical composition, directly influence these wear patterns.

One focus that remains the same is wanting to improve tool life. If a tool can function normally for longer, the machining process would be cheaper, as cutting tools wear out considerably quickly and are quite expensive. Attempts to lower these costs comprises a field with a large amount of research. In addition to tool life improvement, the recent trend is to increase eco-friendliness, with studies that employ alternative lubrication methods in order to reduce the amount used in some current turning processes. This, coupled with the fact that some of these methods reduce the cutting forces of the process, means that the overall power usage will also be lower, lowering the price of the process, albeit slightly, thus making it more environmentally friendly.

Author Contributions: Conceptualization, F.J.G.S. and V.F.C.S.; methodology, V.F.C.S.; investigation, V.F.C.S.; resources, F.J.G.S.; writing—original draft preparation, V.F.C.S.; writing—review and editing, F.J.G.S.; project administration, F.J.G.S. All authors have read and agreed to the published version of the manuscript.

Funding: This research received no external funding.

Conflicts of Interest: The authors declare no conflict of interest.

References

1. CNC Machining Projected to be $100B Industry by 2025. Available online: https://www.thomasnet.com/insights/cnc-machining-projected-to-be-100b-industry-by-2025/ (accessed on 25 October 2019).
2. Global Milling Machines Market 2017–2021. Available online: https://www.maschinenmarkt.international/global-milling-machines-market-2017-2021-a-586045/ (accessed on 28 October 2019).
3. Milling Machines Market—Trends and Forecasts by Technavio Business Wire. Available online: https://www.businesswire.com/news/home/20170201005667/en/Milling-Machines-Market-%E2%80%93-Trends-Forecasts-Technavio (accessed on 28 October 2019).
4. CNC Machining Industry Trends 2019-3ERP. Available online: https://www.3erp.com/blog/cnc-machining-industry-trends-2019/ (accessed on 25 October 2019).
5. Lathe Machines Market Size, Key Drivers and Winning Strategies. 2017. Available online: https://www.alliedmarketresearch.com/lathe-machines-market. (accessed on 28 October 2019).
6. General Turning. Available online: https://www.sandvik.coromant.com/en-gb/knowledge/general-turning/pages/default.aspx (accessed on 31 October 2019).
7. Childs, T.; Maekawa, K. *Metal. Machining*; Arnold: London, UK, 2000; pp. 12–18, ISBN 0-340-69159-X.
8. Lakshmanan, S.; Xavior, M.A. Performance of coated and uncoated inserts during intermittent cut milling of AISI 1030 steel. *Procedia Eng.* **2014**, *97*, 372–380. [CrossRef]
9. Sokovic, M.; Kopac, J.; Dobrzanski, L.A.; Mikula, J.; Golombek, K.; Pakula, D. Cutting characteristics of PVD and CVD-coated ceramic tool inserts. *Tribol. Ind.* **2006**, *28*, 3–8.
10. Ji, W.; Zou, B.; Zhang, S.; Xing, H.; Yun, H.; Wang, Y. Design and fabrication of gradient cermet composite cutting tool, and its cutting performance. *J. Alloys Compd.* **2018**, *732*, 25–31. [CrossRef]
11. Ji, W.; Zou, B.; Huang, C.; Huang, C.; Liu, Y.; Guo, P. Microstructure and mechanical properties of self-diffusion gradient cermet composite tool materials with different characteristics of surface layer. *Ceram. Int.* **2016**, *42*, 19156–19166. [CrossRef]
12. Zhou, X.; Wang, K.; Li, C.; Wang, Q.; Wu, S.; Liu, J. Effect of ultrafine gradient cemented carbides substrate on the performance of coating tools for titanium alloy high speed cutting. *Int. J. Refract. Metals Hard Mater.* **2019**, *84*, 105024. [CrossRef]
13. Chemical Vapor Deposition vs. Physical Vapor Deposition. Available online: https://documents.indium.com/qdynamo/download.php?docid=1958 (accessed on 10 November 2019).
14. Caliskan, H.; Panjan, P.; Kurbanoglu, C. 3.16 Hard coatings on cutting tools and surface finish. *Compr. Mater. Finish.* **2017**, 230–242. [CrossRef]
15. Baptista, A.; Silva, F.J.G.; Porteiro, J.; Míguez, J.L.; Pinto, G.; Fernandes, L. On the physical vapour deposition (PVD): evolution of magnetron sputtering processes for industrial applications. *Procedia Manuf.* **2018**, *17*, 746–757. [CrossRef]
16. Baptista, A.; Silva, F.; Porteiro, J.; Míguez, J.; Pinto, G. Sputtering physical vapour deposition (PVD) coatings: A critical review on process improvement and market trend demands. *Coatings* **2018**, *8*, 402. [CrossRef]
17. Petrov, I.; Hall, A.; Mei, A.B.; Nedfors, N.; Zhirkov, I.; Rosen, J.; Reed, A.; Howe, B.; Greczynski, G.; Birch, J.; et al. Controlling the boron-to-titanium ratio in magnetron-sputter-deposited TiBx thin films. *J. Vac. Sci. Technol. A Vac. Surf. Films* **2017**, *35*, 050601. [CrossRef]

18. Contreras Romero, E.; Hurtado Macías, A.; Méndez Nonell, J.; Solís Canto, O.; Gómez Botero, M. Mechanical and tribological properties of nanostructured TiAlN/TaN coatings deposited by DC magnetron sputtering. *Surf. Coat. Technol.* **2019**, *378*, 124941. [CrossRef]
19. Zauner, L.; Ertelthaler, P.; Wojcik, T.; Bolvardi, H.; Kolozsvári, S.; Mayrhofer, P.H.; Riedl, H. Reactive HiPIMS deposition of Ti-Al-N: Influence of the deposition parameters on the cubic to hexagonal phase transition. *Surf. Coat. Technol.* **2019**, *382*, 125007. [CrossRef]
20. Zhirkov, I.; Polcik, P.; Kolozsvári, S.; Rosen, J. Process development for stabilization of vacuum arc plasma generation from a TiB2 cathode. *AIP Adv.* **2019**, *9*, 015103. [CrossRef]
21. Hovsepian, P.E.; Ehiasarian, A.P. Six strategies to produce application tailored nanoscale multilayer structured PVD coatings by conventional and high power impulse magnetron Sputtering (HIPIMS). *Thin Solid Films* **2019**, *688*, 137409. [CrossRef]
22. Díaz-Parralejo, A.; Macías-García, A.; Díaz-Díez, M.A.; Encinas Sánchez, V.; Carrasco-Amador, J.P. Preparation and characterization of multilayer coatings on tool steel. *Ceram. Int.* **2019**, *45*, 16934–16939. [CrossRef]
23. Silva, F.J.G.; Fernandes, A.J.S.; Costa, F.M.; Baptista, A.P.M.; Pereira, E. A new interlayer approach for CVD diamond coating of steel substrates. *Diam. Relat. Mater.* **2004**, *13*, 828–833. [CrossRef]
24. Silva, F.J.G.; Fernandes, A.J.; Costa, F.; Teixeira, V.; Baptista, A.P.; Pereira, E. Tribological behaviour of CVD diamond films on steel substrates. *Wear* **2003**, *255*, 846–853. [CrossRef]
25. Silva, F.J.; Martinho, R.P.; Alexandre, R.J.D.; Baptista, A.P.M. Wear resistance of TiAlSiN thin coatings. *J. Nanosci. Nanotechnol.* **2012**, *12*, 9094–9101. [CrossRef]
26. Kouznetsov, V.; Macák, K.; Schneider, J.M.; Helmersson, U.; Petrov, I. A novel pulsed magnetron sputter technique utilizing very high target power densities. *Surf. Coat. Technol.* **1999**, *122*, 290–293. [CrossRef]
27. Cutting Tool Materials. Available online: https://www.sandvik.coromant.com/en-gb/knowledge/materials/pages/cutting-tool-materials.aspx (accessed on 30 October 2019).
28. Fernández-Abia, A.I.; Barreiro, J.; Fernández-Larrinoa, J.; Lacalle, L.N.L.; de Fernández-Valdivielso, A.; Pereira, O.M. Behaviour of PVD coatings in the turning of austenitic stainless steels. *Procedia Eng.* **2013**, *63*, 133–141. [CrossRef]
29. Klocke, F.; Krieg, T. Coated tools for metal cutting—features and applications. *CIRP Ann.* **1999**, *48*, 515–525. [CrossRef]
30. Seidl, W.M.; Bartosik, M.; Kolozsvári, S.; Bolvardi, H.; Mayrhofer, P.H. Mechanical properties and oxidation resistance of Al-Cr-N/Ti-Al-Ta-N multilayer coatings. *Surf. Coat. Technol.* **2018**, *347*, 427–433. [CrossRef]
31. Knotek, O.; Löffler, F.; Krämer, G. Multicomponent and multilayer physically vapour deposited coatings for cutting tools. *Surf. Coat. Technol.* **1992**, *54–55*, 241–248. [CrossRef]
32. Kainz, C.; Schalk, N.; Tkadletz, M.; Mitterer, C.; Czettl, C. Microstructure and mechanical properties of CVD TiN/TiBN multilayer coatings. *Surf. Coat. Technol.* **2019**, *370*, 311–319. [CrossRef]
33. Long, Y.; Zeng, J.; Yu, D.; Wu, S. Microstructure of TiAlN and CrAlN coatings and cutting performance of coated silicon nitride inserts in cast iron turning. *Ceram. Int.* **2014**, *40*, 9889–9894. [CrossRef]
34. Yi, J.; Chen, S.; Chen, K.; Xu, Y.; Chen, Q.; Zhu, C.; Liu, L. Effects of Ni content on microstructure, mechanical properties and Inconel 718 cutting performance of AlTiN-Ni nanocomposite coatings. *Ceram. Int.* **2018**, *45*, 474–480. [CrossRef]
35. Silva, F.J.G. Nanoindentation on tribological coatings. In *Applied Nanoindentation in Advanced Materials;* Tiwari, A., Natarajan, S., Eds.; John Wiley and Sons, Ltd.: Hoboken, NJ, USA, 2017; pp. 111–133, ISBN 9781119084495.
36. Dai, W.; Li, X.; Wang, Q. Microstructure and properties of TiB2/Cr multilayered coatings with double periodical structures. *Surf. Coat. Technol.* **2019**, *382*, 125150. [CrossRef]
37. Wang, D.; Lin, S.; Liu, L.; Yang, H.; Shi, J.; Jiang, B.; Zhou, K.; Zhang, X. Micro-nano multilayer structure design and solid particle erosion resistance performance of CrAlNx/CrAlN coating. *Vacuum* **2019**, *172*, 109064. [CrossRef]
38. He, Q.; Paiva, J.M.; Kohlscheen, J.; Beake, B.D.; Veldhuis, S.C. An integrative approach to coating/carbide substrate design of CVD and PVD coated cutting tools during the machining of austenitic stainless steel. *Ceram. Int.* **2019**, *46*, 5149–5158. [CrossRef]
39. Mia, M.; Królczyk, G.; Maruda, R.; Wojciechowski, S. Intelligent optimization of hard-turning parameters using evolutionary algorithms for smart manufacturing. *Materials* **2019**, *12*, 879. [CrossRef]

40. Radha Krishnan, B.; Ramesh, M. Optimization of machining process parameters in CNC turning process of IS2062 E250 steel using coated carbide cutting tool. *Mater. Today Proc.* **2019**. [CrossRef]
41. Rao, V.D.P.; Ali, S.R.M.; Ali, S.M.S.; Geethika, V.N. Multi-objective optimization of cutting parameters in CNC turning of stainless steel 304 with TiAlN nano coated tool. *Mater. Today Proc.* **2018**, *5*, 25789–25797. [CrossRef]
42. Khan, A.; Maity, K. Influence of cutting speed and cooling method on the machinability of commercially pure titanium (CP-Ti) grade II. *J. Manuf. Process.* **2018**, *31*, 650–661. [CrossRef]
43. Harisha, S.K.; Rajkumar, G.R.; Pawar, V.; Keshav, M. Statistical investigation of tool geometry for minimization of cutting force in turning of hardened steel. *Mater. Today Proc.* **2018**, *5*, 11277–11282. [CrossRef]
44. Silva, F.J.G.; Casais, R.; Martinho, R.P.; Baptista, A. Mechanical and tribological characterization of TiB_2 thin films. *J. Nanosci. Nanotechnol.* **2012**, *12*, 9187–9194. [CrossRef] [PubMed]
45. Martinho, R.P.; Silva, F.J.; Alexandre, R.J.D.; Baptista, A.P.M. TiB_2 nanostructured coating for GFRP injection moulds. *J. Nanosci. Nanotechnol.* **2011**, *11*, 5374–5382. [CrossRef] [PubMed]
46. Gupta, K.M.; Ramdev, K.; Dharmateja, S.; Sivarajan, S. Cutting characteristics of PVD coated cutting tools. *Mater. Today Proc.* **2018**, *5*, 11260–11267. [CrossRef]
47. Koyilada, B.; Gangopadhyay, S.; Thakur, A. Comparative evaluation of machinability characteristics of Nimonic C-263 using CVD and PVD coated tools. *Measurement* **2016**, *85*, 152–163. [CrossRef]
48. Ginting, A.; Skein, R.; Cuaca, D.; Herdianto, P.; Masyithah, Z. The characteristics of CVD- and PVD-coated carbide tools in hard turning of AISI 4340. *Measurement* **2018**, *129*, 548–557. [CrossRef]
49. Ramanujam, R.; Vignesh, M.; Tamiloli, N.; Sharma, N.; Srivastava, S.; Patel, A. Comparative evaluation of performances of TiAlN, AlCrN, TiAlN/AlCrN coated carbide cutting tools and uncoated carbide cutting tools on turning Inconel 825 alloy using grey relational analysis. *Sens. Actuators A Phys.* **2018**, *279*, 331–342. [CrossRef]
50. Koseki, S.; Inoue, K.; Morito, S.; Ohba, T.; Usuki, H. Comparison of TiN-coated tools using CVD and PVD processes during continuous cutting of Ni-based superalloys. *Surf. Coat. Technol.* **2015**, *283*, 353–363. [CrossRef]
51. Mishra, S.K.; Ghosh, S.; Aravindan, S. Characterization and machining performance of laser-textured chevron shaped tools coated with AlTiN and AlCrN coatings. *Surf. Coat. Technol.* **2018**, *334*, 344–356. [CrossRef]
52. Tang, J.; Xiong, J.; Guo, Z.; Yang, T.; Liang, M.; Yang, W.; Liu, J.; Zheng, Q. Microstructure and properties of CVD coated on gradient cemented carbide with different WC grain size. *Int. J. Refract. Metals Hard Mater.* **2016**, *61*, 128–135. [CrossRef]
53. Krolczyk, G.M.; Maruda, R.W.; Krolczyk, J.B.; Wojciechowski, S.; Mia, M.; Nieslony, P.; Budzik, G. Ecological trends in machining as a key factor in sustainable production—A review. *J. Clean. Prod.* **2019**, *218*, 601–615. [CrossRef]
54. Thakur, A.; Gangopadhyay, S. Dry machining of nickel-based super alloy as a sustainable alternative using TiN/TiAlN coated tool. *J. Clean. Prod.* **2016**, *129*, 256–268. [CrossRef]
55. Helmersson, U.; Todorova, S.; Barnett, S.A.; Sundgren, J.-E.; Markert, L.C.; Greene, J.E. Growth of single-crystal TiN/VN strained-layer superlattices with extremely high mechanical hardness. *J. Appl. Phys.* **1987**, *62*, 481–484. [CrossRef]
56. Vereschaka, A.A.; Grigoriev, S.; Sitnikov, N.N.; Bublikov, J.I.; Batako, A.D.L. Effect produced by thickness of nanolayers of multilayer composite wear-resistant coating on tool life of metal-cutting tool in turning of steel AISI 321. *Procedia CIRP* **2018**, *77*, 549–552. [CrossRef]
57. Gassner, M.; Schalk, N.; Tkadletz, M.; Czettl, C.; Mitterer, C. Thermal crack network on CVD TiCN/α-Al_2O_3 coated cemented carbide cutting tools. *Int. J. Refract. Metals Hard Mater.* **2019**, *81*, 1–6. [CrossRef]
58. Bruni, C.; Forcellese, A.; Gabrielli, F.; Simoncini, M. Effect of the lubrication-cooling technique, insert technology and machine bed material on the workpart surface finish and tool wear in finish turning of AISI 420B. *Int. J. Mach. Tools Manuf.* **2006**, *46*, 1547–1554. [CrossRef]
59. Przestacki, D.; Chwalczuk, T.; Wojciechowski, S. The study on minimum uncut chip thickness and cutting forces during laser-assisted turning of WC/NiCr clad layers. *Int. J. Adv. Manuf. Technol.* **2017**, *91*, 3887–3898. [CrossRef]
60. Pervaiz, S.; Rashid, A.; Deiab, I.; Nicolescu, M. Influence of tool materials on machinability of titanium- and nickel-based alloys: A review. *Mater. Manuf. Process.* **2014**, *29*, 219–252. [CrossRef]

61. Aruna Prabha, K.; Srinivasa Prasad, B.; Srilatha, N. Comparative study of wear patterns of both coated and uncoated tool inserts in high speed turning of EN36 steel. *Mater. Today Proc.* **2018**, *5*, 4368–4375. [CrossRef]
62. Saketi, S.; Olsson, M. Influence of CVD and PVD coating micro topography on the initial material transfer of 316L stainless steel in sliding contacts—A laboratory study. *Wear* **2017**, *388–389*, 29–38. [CrossRef]
63. Faksa, L.; Daves, W.; Ecker, W.; Klünsner, T.; Tkadletz, M.; Czettl, C. Effect of shot peening on residual stresses and crack closure in CVD coated hard metal cutting inserts. *Int. J. Refract. Metals Hard Mater.* **2019**, *82*, 174–182. [CrossRef]
64. Skordaris, G.; Bouzakis, K.D.; Kotsanis, T.; Charalampous, P.; Bouzakis, E.; Breidenstein, B.; Bergmann, B.; Denkena, B. Effect of PVD film's residual stresses on their mechanical properties, brittleness, adhesion and cutting performance of coated tools. *CIRP J. Manuf. Sci. Technol.* **2017**, *18*, 145–151. [CrossRef]
65. Zheng, G.; Cheng, X.; Yang, X.; Xu, R.; Zhao, J.; Zhao, G. Self-organization wear characteristics of MTCVD-TiCN-Al 2 O 3 coated tool against 300 M steel. *Ceram. Int.* **2017**, *43*, 13214–13223. [CrossRef]
66. Zheng, G.; Xu, R.; Cheng, X.; Zhao, G.; Li, L.; Zhao, J. Effect of cutting parameters on wear behavior of coated tool and surface roughness in high-speed turning of 300 M. *Measurement* **2018**, *125*, 99–108. [CrossRef]
67. Liu, W.; Chu, Q.; Zeng, J.; He, R.; Wu, H.; Wu, Z.; Wu, S. PVD-CrAlN and TiAlN coated Si 3 N 4 ceramic cutting tools —1. Microstructure, turning performance and wear mechanism. *Ceram. Int.* **2017**, *43*, 8999–9004. [CrossRef]
68. Martinho, R.P.; Silva, F.J.G.; Baptista, A. Wear behaviour of uncoated and diamond coated Si3N4 tools under severe turning conditions. *Wear* **2007**, *263*, 1417–1422. [CrossRef]
69. Martinho, R.P.; Silva, F.J.G.; Baptista, A. Cutting forces and wear analysis of Si3N4 diamond coated tools in high speed machining. *Vacuum* **2008**, *82*, 1415–1420. [CrossRef]
70. Naskar, A.; Chattopadhyay, A.K. Investigation on flank wear mechanism of CVD and PVD hard coatings in high speed dry turning of low and high carbon steel. *Wear* **2018**, *396–397*, 98–106. [CrossRef]
71. Vereschaka, A.; Tabakov, V.; Grigoriev, S.; Sitnikov, N.; Milovich, F.; Andreev, N.; Bublikov, J. Investigation of wear mechanisms for the rake face of a cutting tool with a multilayer composite nanostructured Cr–CrN-(Ti,Cr,Al,Si)N coating in high-speed steel turning. *Wear* **2019**, *438–439*, 203069. [CrossRef]
72. Tamil Alagan, N.; Zeman, P.; Hoier, P.; Beno, T.; Klement, U. Investigation of micro-textured cutting tools used for face turning of alloy 718 with high-pressure cooling. *J. Manuf. Process.* **2019**, *37*, 606–616. [CrossRef]
73. Asha, P.B.; Rao, C.R.P.; Kiran, R.; Kumar, D.V.R. Effect of machining parameters on cutting tool temperature and tool life while turning EN24 and Hchcr grade alloy steel. *Mater. Today Proc.* **2018**, *5*, 11819–11826. [CrossRef]
74. Boing, D.; de Oliveira, A.J.; Schroeter, R.B. Limiting conditions for application of PVD (TiAlN) and CVD (TiCN/Al2O3/TiN) coated cemented carbide grades in the turning of hardened steels. *Wear* **2018**, *416*, 54–61. [CrossRef]
75. Vereschaka, A.; Grigoriev, S.; Sitnikov, N.; Oganyan, G.; Sotova, C. Influence of thickness of multilayer composite nano-structured coating Ti-TiN-(Ti,Al,Cr)N on tool life of metal-cutting tool. *Procedia CIRP* **2018**, *77*, 545–548. [CrossRef]
76. Ranjan Das, S.; Panda, A.; Dhupal, D. Hard turning of AISI 4340 steel using coated carbide insert: Surface roughness, tool wear, chip morphology and cost estimation. *Mater. Today Proc.* **2018**, *5*, 6560–6569. [CrossRef]
77. Zhao, J.; Liu, Z. Modelling for prediction of time-varying heat partition coefficient at coated tool-chip interface in continuous turning and interrupted milling. *Int. J. Mach. Tools Manuf.* **2019**, *147*, 103467. [CrossRef]
78. Zhang, X.; Zheng, G.; Cheng, X.; Li, Y.; Li, L.; Liu, H. 2D Fractal analysis of the cutting force and surface profile in turning of Iron-based superalloy. *Measurement* **2019**, *151*, 107125. [CrossRef]
79. Uhlmann, E.; Hinzmann, D.; Kropidlowksi, K.; Meier, P.; Prasol, L.; Woydt, M. Substitution of commercially coated tungsten carbide tools in dry cylindrical turning process by HiPIMS coated niobium carbide cutting inserts. *Surf. Coat. Technol.* **2018**, *354*, 112–118. [CrossRef]
80. Yazid, M.Z.A.; Cheharon, C.H.; Ghani, J.A.; Ibrahim, G.A.; Said, A.Y.M. Surface integrity of Inconel 718 when finish turning with PVD coated carbide tool under MQL. *Procedia Eng.* **2011**, *19*, 396–401. [CrossRef]
81. Maruda, R.W.; Krolczyk, G.M.; Wojciechowski, S.; Zak, K.; Habrat, W.; Nieslony, P. Effects of extreme pressure and anti-wear additives on surface topography and tool wear during MQCL turning of AISI 1045 steel. *J. Mech. Sci. Technol.* **2018**, *32*, 1585–1591. [CrossRef]

82. Maruda, R.W.; Krolczyk, G.M.; Michalski, M.; Nieslony, P.; Wojciechowski, M. Structural and microhardness changes after turning of the AISI 1045 steel for minimum quantity cooling lubrication. *J. Mater. Eng. Perform.* **2017**, *26*, 431–438. [CrossRef]
83. Marques, A.; Paipa Suarez, M.; Falco Sales, W.; Rocha Machado, Á. Turning of Inconel 718 with whisker-reinforced ceramic tools applying vegetable-based cutting fluid mixed with solid lubricants by MQL. *J. Mater. Process. Technol.* **2018**, *266*, 530–543. [CrossRef]
84. Elmunafi, M.H.S.; Noordin, M.Y.; Kurniawan, D. Tool life of coated carbide cutting tool when turning hardened stainless steel under minimum quantity lubricant using castor oil. *Procedia Manuf.* **2015**, *2*, 563–567. [CrossRef]
85. Behera, B.C.; Ghosh, S.; Rao, P.V. Wear behavior of PVD TiN coated carbide inserts during machining of Nimonic 90 and Ti6Al4V superalloys under dry and MQL conditions. *Ceram. Int.* **2016**, *42*, 14873–14885. [CrossRef]
86. Kumar, U.; Senthil, P. Performance of cryogenic treated multi-layer coated WC insert in terms of machinability on titanium alloys Ti-6Al-4V in dry turning. *Mater. Today Proc.* **2019**. [CrossRef]
87. Dhananchezian, M.; Priyan, M.R.; Rajashekar, G.; Narayanan, S.S. Study The effect of cryogenic cooling on machinability characteristics during turning duplex stainless steel 2205. *Mater. Today Proc.* **2018**, *5*, 12062–12070. [CrossRef]
88. Khare, S.K.; Agarwal, S. Optimization of machining parameters in turning of AISI 4340 steel under cryogenic condition using taguchi technique. *Procedia CIRP* **2017**, *63*, 610–614. [CrossRef]
89. Sivaiah, P.; Chakradhar, D. Performance improvement of cryogenic turning process during machining of 17-4 PH stainless steel using multi objective optimization techniques. *Measurement* **2018**, *136*, 326–336. [CrossRef]
90. Gutnichenko, O.; Bushlya, V.; Bihagen, S.; Ståhl, J.-E. Influence of GnP additive to vegetable oil on machining performance when MQL-assisted turning alloy 718. *Procedia Manuf.* **2018**, *25*, 330–337. [CrossRef]

Article

Sustainable Milling of Ti-6Al-4V: Investigating the Effects of Milling Orientation, Cutter's Helix Angle, and Type of Cryogenic Coolant

Asif Iqbal [1,*], Hazwani Suhaimi [1], Wei Zhao [2], Muhammad Jamil [2], Malik M Nauman [1], Ning He [2] and Juliana Zaini [1]

1 Faculty of Integrated Technologies, Universiti Brunei Darussalam, Jalan Tungku Link, Gadong BE 1410, Brunei; hazwani.suhaimi@ubd.edu.bn (H.S.); malik.nauman@ubd.edu.bn (M.M.N.); juliana.zaini@ubd.edu.bn (J.Z.)

2 College of Mechanical & Electrical Engineering, Nanjing University of Aeronautics & Astronautics, 29-Yu Dao Street, Nanjing 210016, China; nuaazw@nuaa.edu.cn (W.Z.); engr.jamil@nuaa.edu.cn (M.J.); drnhe@nuaa.edu.cn (N.H.)

* Correspondence: asif.asifiqbal@gmail.com; Tel.: +673-892-9497

Received: 9 December 2019; Accepted: 21 January 2020; Published: 17 February 2020

Abstract: Ti-6Al-4V, the most commonly used alloy of titanium, possesses excellent mechanical properties and corrosion resistance, which is the prime reason for the continual rise in its industrial demand worldwide. The extraordinary mechanical properties of the alloy are viewed as a hindrance when it comes to its shaping processes, and the process of milling is no exception to it. The generation of intense heat flux around the cutting zones is an established reason of poor machinability of the alloy and unacceptably low sustainability of its machining. The work presented in this paper attempts to enhance sustainability of milling Ti-6Al-4V by investigating the effects of milling orientation, cutter's helix angle, cutting speed, and the type of cryogenic coolant and lubricant on the sustainability measures, such as tool damage, cutting energy consumption, process cost, milling forces, and work surface roughness. It was found that micro-lubrication is more effective than the two commonly used cryogenic coolants (carbon dioxide snow and liquid nitrogen) in reducing tool wear, work surface roughness, process cost, and energy consumption. Furthermore, down-milling enormously outperformed up-milling with respect to tool wear, work surface quality, and process cost. Likewise, the high levels of cutter's helix angle and cutting speed also proved to be beneficial for milling sustainability.

Keywords: cutting energy; tool damage; machining; liquid nitrogen; carbon dioxide snow

1. Introduction

Titanium alloys gained an unprecedented rise in their demand from various engineering sectors due to their excellent mechanical properties and corrosion resistance. The same properties considered as excellent during the "use" phase are termed as unfavorable during the "manufacturing" phase of their life cycle. With regard to the machining domain, the same unfavorable properties are responsible for their low machinability, which render the cutting process unsustainable. The low machinability is attributed mainly to high strength, chemical affinity with the tool materials, and a short chip-rake contact length [1]. A titanium work is, thus, machined with formation of an intense heat flux around the cutting edge and consumption of exceedingly high cutting energy, leading to acceleration of the temperature-dependent modes of tool wear [2]. The high tool wear rates, resulting in frequent changes of cutting tools or edges, leave the machining process highly unsustainable, economically as well as environmentally [3]. The problem is generally negotiated either by lowering the material removal rates or by applying emulsion-based coolants, neither of which actually offers a sustainable solution.

With regard to quashing the intense heat flux around the cutting areas, the application of cryogenic fluids, especially liquid nitrogen (LN_2), fares very well. The fluids offer a viable solution because of extremely low operational temperatures, no waste generation, and controllable flow rates. Additionally, application of micro-lubrication, also known as minimum quantity of lubrication (MQL), also proved to be very beneficial in enhancing tool life and improving surface quality in machining of titanium alloys conducted at medium cutting speeds [4,5]. MQL is a near-dry machining method in which a miniscule quantity of lubricating oil is pulverized into a stream of air, and the resulting aerosol is applied onto the cutting areas [6].

Although steady progress is being made regarding quantification of the effects of cryogenic coolants with regard to the continuous machining processes, not much effort is being put up concerning interrupted machining processes, such as milling. Cryogenic milling is distinct from continuous machining processes performed under cryogenic environment in the sense that the cutting teeth of a milling tool periodically and rapidly engage and disengage with the work material, leading to cyclic heating (caused by thermal energy released by work material's plastic deformation) and cooling (caused by interaction with the incoming cryogenic fluid) of the cutting edges. Such a course is expected to induce thermal shocks in the teeth, leading to cracking, chipping, and more catastrophic forms of tool damage. Furthermore, a comparative analysis regarding the cooling effects of an evaporative cryogenic fluid, such as LN_2, and a throttling-based cryogenic fluid, such as CO_2 snow, is also required. With regard to the milling process, two distinct parameters, milling orientation and cutter's helix angle, are also expected to have effects on the process's sustainability measures, such as tool damage rate, cutting energy consumption, work surface roughness, process cost, and cutting forces.

Literature Review

This sub-section presents a review of the published work concerning the issues of milling titanium alloys, application of cryogenic coolants in machining (especially milling) of titanium, potentials of using micro-lubrication in milling, and machining sustainability measures.

It is reported that an increase in the flow rate of liquid nitrogen can prolong the tool life in machining of titanium alloys [7]. Furthermore, the surface integrity would be greatly improved when the pressure and flow rate of the coolant are increased. Milling of Ti-6Al-4V with liquid carbon dioxide can greatly reduce the lateral crack propagation and chipping [8]. Therefore, it can be used to prolong tool life as compared to emulsion-based cooling. The effects of tool life criterion, work material's temper state, cutting parameters, and micro-lubrication on the sustainability measures of a milling process were studied [9]. It was found that the material's temper state and the option of using MQL were the most influential parameters with respect to the sustainability measures, such as specific cutting energy, tool life, and process cost. Sartori et al. reported that an MQL system amalgamated with a CO_2 and LN_2 distribution system could optimize the lubrication and cooling effect, leading to a significant reduction in crater wear [10]. In another study, it was reported that machining under flood coolant does not reduce surface roughness [11]. The authors also reported that LN_2 hybridized with oil-based MQL can yield the lowest cutting forces of all the tested coolants. Isakson et al. reported that the cooling methods utilizing LN_2 and an emulsion-based coolant yielded similar effects on surface integrity [8]. Furthermore, the authors also managed to reduce consumption of the cryogenic coolant to provide a good surface quality without conceding any negative environmental or economic impact. It was found that milling of Ti-6Al-4V with liquid CO_2 could greatly reduce chipping and lateral crack propagation; therefore, it can be used to significantly prolong tool life in comparison with emulsion-based cooling [12]. The effects of using CO_2 snow as a coolant and its merger with MQL were investigated in continuous machining of a high-strength β-titanium alloy [13]. It was found that the usage of CO_2 snow and the location of its application was highly influential with respect to the sustainability measures. In another work, it was reported that the cryogenic cooling with LN_2 could considerably reduce tool wear and, thus, lead to an increase in material removal rate [14]. It was concluded that cryogenic machining operating at a given cutting speed can cause much lower energy consumption than machining with a flood coolant. Mia et al. reported that the use of dual jets of LN_2 is an excellent way to reduce energy consumption and working temperature, as well as to improve work surface quality [15]. In another experimental study, it was reported that the use of liquid CO_2 at a temperature of −79.5 °C in cutting of a nickel-chromium alloy could reduce average surface roughness by 42–47%, 24–27%, and 16–21% over dry, wet, and MQL cutting, respectively [16]. Furthermore, the cryogenic cooling was also found to increase the compressive stresses on the surface and decrease the flank wear. Li et al. presented optimization of milling Ti-6Al-4V alloy with a graphene-dispersed vegetable-oil-based cutting fluid [17]. The results showed significant improvements in the milling performance measures including milling force, temperature, surface micro-hardness, and work surface roughness. Dry and MQL-based milling processes were compared for machining of Inconel 718 [18]. MQL was found to improve the tool life, as well as the work surface finish. An experimental study focused on modeling the effects of tool wear rate on economic sustainability of milling a titanium alloy [19]. In total, 47.5% and 47.59% less electricity consumption cost and machine operational cost, respectively, were achieved for the cryogenic cooling approach in comparison with dry machining. In another experimental study concerning end-milling of Ti–6Al–4V titanium alloy, the effects of cryogenic cooling on work's surface integrity were compared with those under dry and flood cooling environments [20]. The authors reported that cryogenic cooling resulted in up to 31% and 39% lower surface roughness when compared to flood cooling and dry approaches, respectively. A significant reduction in microscopic surface defects under the cryogenic environment was also reported. Dawood et al. studied the effects of the three cutting parameters on machining performance under flood cooling and sustainable dry environments [21]. Dry machining was found to yield better surface finish, but it also sustained more severe adhesion wear, crater wear, and formation of built-up edge. An experimental study evaluated the effects of applying cryogenic cooling with MQL lubrication in contour milling of Inconel 718 [22]. The authors claimed superiority of the proposed CroMQL method over the other lubri-cooling techniques. Pusavec et al. presented an experimental study on sustainable machining of Inconel 718 under various lubri-cooling environments such as dry, MQL, cryogenic, and cryo-lubrication [23]. Based on the statistical analyses of the results, the authors concluded that the cooling/lubrication condition had significant effects on the sustainability measures including tool life, cutting forces, and power consumption.

Milling orientation (up- and down-milling) and cutter's helix angle are amongst the milling parameters, which do not receive much attention with respect to their influence on the process's sustainability measures. It was reported that both the parameters possessed significant effects on tool life and work surface roughness in high-speed milling of hardened steels using carbide cutters [24]. Milling cutters with 45° helix angle yielded significantly longer tool life and marginally better surface finish than the 30° cutters. Moreover, down-milling was found to provide much better surface finish, but equal tool life as compared to up-milling. A tool orientation optimization model was presented that includes the effect of deflection error caused by milling forces to achieve better machining precision controlling in five-axis surface milling [25]. The effects of up- and down-milling were compared in peripheral milling of a high-alloy steel [26]. The up-milling approach was found to generate compressive residual stresses in the work surface, but with a poor surface finish in comparison with down-milling. An experimental study was performed to compare the effects of up- and down-milling in end-milling of Inconel 718 [27]. It was found that the down-milling approach yielded better results in terms of tool wear as compared to up-milling. Furthermore, the chips formed in up-milling were segmented and continuous as compared to discontinuous ones produced in down-milling. Wan et al. presented an analytical model to quantify the influence of tool's helix angle on peak cutting force [28]. The authors found that the peak value of cutting forces decreased with an increase in helix angle for a single engaged cutting edge and that the optimal helix angle corresponding to the minimum peak cutting force was a function of the number of flutes, axial depth of cut, and cutter diameter. Another work focused on quantifying the effect of helix angle on performance of coated carbide end mills for dry side-milling of 304L stainless steel [29]. It was found that the number of axis contact points and effective cutting length increased with increasing helix angle, leading to reduced tool wear and thinner chips, but with higher cutting temperature. It was concluded that the TiAlN-coated end mill with a high helix angle of 60° yielded the best surface finish with an acceptably long tool life. Another study presented a mathematical model for predicting surface topography and various surface roughness metrics by considering the effects of cutter's helix angle, feed rate, and tool's eccentricity [30].

The literature review reveals the following gap between the state of the art and the objectives of this work: (1) a very limited amount of work is done so far regarding application of cryogenic fluids to milling of titanium alloys as most of the published investigations have focused on turning process only; (2) no searchable work was found regarding comparison of cooling effectiveness between throttling and evaporation based cryogenic coolants applied to intermittent cutting processes; (3) a limited amount of investigative work is available that quantifies the effects of milling orientation and tool's helix angle on the sustainability measures of titanium machining.

In perspective of the abovementioned research gap, the presented work aims to quantify and improve the sustainability measures in respect of side- and end-milling of an $\alpha + \beta$ titanium alloy (Ti-6Al-4V) while employing three kinds of cutting fluids (an evaporative cryogenic coolant, a throttling-based cryogenic coolant, and micro-lubrication), two modes of milling orientation, and two levels each of cutting speed and milling cutter's helix angle. The sustainability measures of the milling process to be evaluated are cutting energy consumption, tool damage, work surface roughness, machining forces, and process cost.

2. Experimental Work

This section presents the details regarding work material and tooling, predictor variables, responses (sustainability measures), design of experiments, fixed parameters, equipment, and measuring instruments.

2.1. Work Material and Tooling

The work material used in the study is Ti-6Al-4V, a commonly used $\alpha + \beta$ alloy of titanium. The annealed form of the material is used in the form of a plate having dimensions 75 mm × 200 mm × 19 mm. The heat treatment was done by soaking the work pieces at a temperature between 778 and 782 °C for about 70 min, followed by air cooling. The work material, after carrying out the heat treatment process, possessed ultimate tensile strength, yield strength (0.2% proof stress), and elongation of 1003.5 MPa, 927.3 MPa, and 15%, respectively.

The milling cutters used in this study were FIRE-coated cemented tungsten carbide flat end mills from Guhring Inc., Berlin, Germany, having diameter of 8 mm and number of cutting flutes equal to four. FIRE is a multi-layer TiN + TiAlN ceramic coating system that provides extreme wear- and heat-resistant properties to the tool. The hardness of the coating was 3300 HV, and the coefficient of friction was 0.6. The cutters with a helix angle of 30° had total and cutting lengths of 68 mm and 22 mm, respectively, while those with a helix angle of 42° had 63 mm and 19 mm, respectively. A new end mill cutter was used for each experimental run. Figure 1 presents the two kinds of milling cutters used in the experiments.

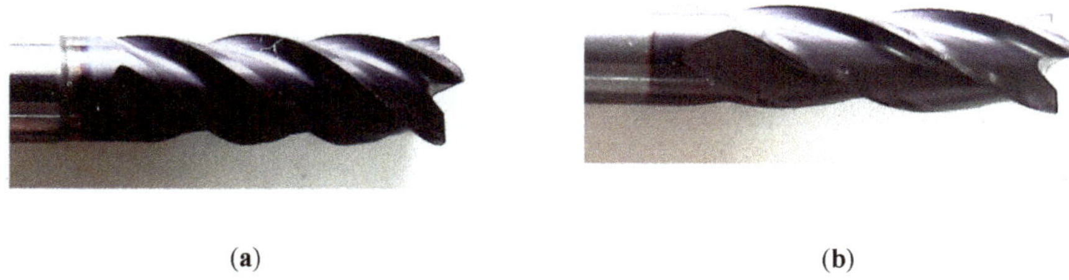

(a) (b)

Figure 1. Ceramic-coated tungsten carbide side- and end-mill cutters having helix angles of (**a**) 42° and (**b**) 30°.

2.2. Predictors, Responses, and Design of Experiments

The following four predictor variables were controlled in the milling experiments:

(1) Cutting fluid. The following three levels were tested with respect to this predictor: (a) evaporative cryogenic fluid (liquid nitrogen); (b) throttle cryogenic fluid (CO_2 snow); (c) micro-lubrication (minimum quantity of lubrication).
(2) Cutter's helix angle, λ (degrees). The two levels chosen for this predictor were 30° and 42°.
(3) Milling orientation. The two options for this binary predictor were up-milling and down-milling.
(4) Cutting speed, v_c (m/min). The two levels chosen for this predictor were 100 m/min and 175 m/min.

The aforementioned levels of the four predictor variables yielded a total of 24 (= 3 × 2 × 2 × 2) experimental runs for the sake of executing a full-factorial design of experiments. Table 1 presents the details regarding the levels of the four predictors controlled in the experiments. It is to be noted that the first and the third predictors were categorical while the other two were numerical. The two levels of the cutting speed (100 and 175 m/min) were selected on the basis of the preliminary tests. Cutting speeds in excess of 200 m/min at the given feed rate and radial depth of cut resulted in the tools getting red-hot toward the end of the cuts and the cutting edges getting covered with thick adhesions. This observation led to fixation of the upper level of the cutting speed equal to 175 m/min. The lower level, thereupon, was decided as a value between 50% and 60% of the upper one.

Table 1. Levels of the four predictor variables controlled in the experiments. MQL—minimum quantity of lubrication.

Predictor	Units	Level 1	Level 2	Level 3
Cutting fluid	-	LN_2	CO_2	MQL
Cutter's helix angle	°	30	42	-
Milling orientation	-	up	down	-
Cutting speed	m/min	100	175	-

Each experimental run involved removing 600 mm^3 of volume of the work material under the following dimensions: 0.5 mm (radial depth of cut) × 8 mm (axial depth of cut) × 150 mm (length of cut). The 150-mm length of cut was completed in two passes of equal length. Figure 2 presents the pictorial description regarding the length of cut and the two depths of cut. A new side- and end-mill cutter was used for each experimental run. Up-milling is a cutting approach in which a cutting tooth of the milling cutter enters the work surface with zero chip thickness and exits with a maximum. On the contrary, the tooth enters and exits the work surface with a maximum and zero chip thickness in the down-milling approach.

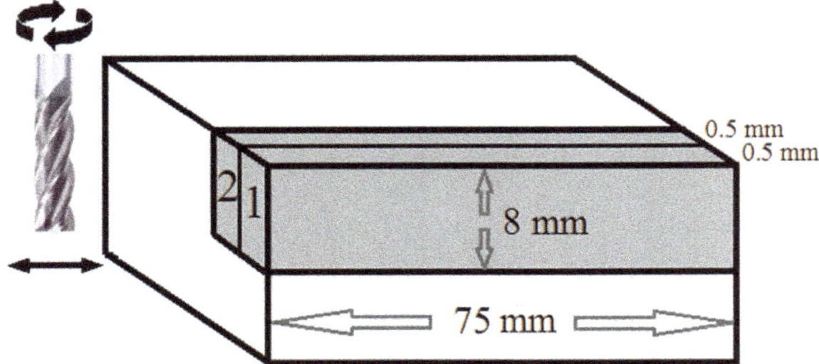

Figure 2. Cutting schematic for each experimental run showing length of cut = 2 × 75 mm^2 (not to scale), axial depth of cut = 8 mm, and radial depth of cut = 0.5 mm. The numbers 1 and 2 show the order of the slice removal. The total volume (600 mm^3) of material to be removed in each run is shaded gray.

The following responses were evaluated for each of the 24 experimental runs:

(1) Average width of flank wear land on the four cutting teeth of the milling cutter, *VB* (mm), to be determined after removing the given volume (600 mm^3) of the work material.
(2) Average arithmetic roughness of the milled surface, *R*a (μm).
(3) Specific cutting energy, *SCE* (J/mm^3), averaged for the entire length of cut.
(4) Machining force components, F_x, F_y, and F_z (N), averaged for the entire length of cut. F_x and F_z are aligned with the feed direction and the cutter's axis, respectively, while F_y is the third orthogonal component that falls perpendicular to the milled side surface of the work.
(5) Process cost, *PC* (Brunei Dollars (BND)/mm^3 of work material removed, converted to 100-scale comparative cost values).

Additionally, the other sustainability measures of the milling process, such as waste generation, operator's safety, and health are discussed in a qualitative way.

2.3. Experimental Set-Up and Measurements

All experiments were performed on Mikron UCP 710, a five-axis, vertical machining center (Mikron Holding, Biel, Switzerland) having maximum rotational speed, feed rate, and power of 18,000 rpm, 20 m/min, and 16 kW, respectively. Milling was performed in a straight line, cutting through the 75-mm side twice during each run. Figure 3 presents the experimental set-up.

The throttling-based cryogenic cooling equipment consisted of a storage bottle containing CO_2 gas compressed at a pressure of 5.5 MPa. The compressed gas was transported from the bottle to the exit nozzle through a copper tube. The mass flow rate of the CO_2 gas at the exit of the 2-mm-diameter nozzle was measured to be 0.5 kg/min against the storage pressure of 5.5 MPa. The exit of the nozzle was located very close to the machining area such that the CO_2 gas impacted directly on to the cutter's teeth. The gas on exit expanded and absorbed heat from its surroundings due to the Joule-Thomson effect [31]. The throttling gas, consequently, cooled down to a temperature of about −72 °C, which caused it to convert to a semi-solid state (CO_2 snow) adhering to the tooling system and the work's surface. The evaporation-based cryogenic cooling equipment consisted of a storage dewar containing nitrogen, which was cooled down to a liquid state. The jet of LN_2 impinged on to the milling cutter under a flow rate and pressure of 0.5 L/min and 0.1 MPa, respectively, through a 6-mm-diameter nozzle. The temperature of the LN_2 jet, measured at the nozzle's exit, was −197 °C. The direction of the nozzle was adjusted such that the maximum mass of the fluid directly impacted that portion of the cutting teeth which periodically engaged and disengaged with the work material. The micro-lubrication, in the form of minimum quantity of lubrication, was supplied by mixing a vegetable-based oil at a rate of 25 mL/h in the flow of air compressed to a pressure of 0.6 MPa. The resulting aerosol was applied to the milling cutter at the region adjacent to the work surface.

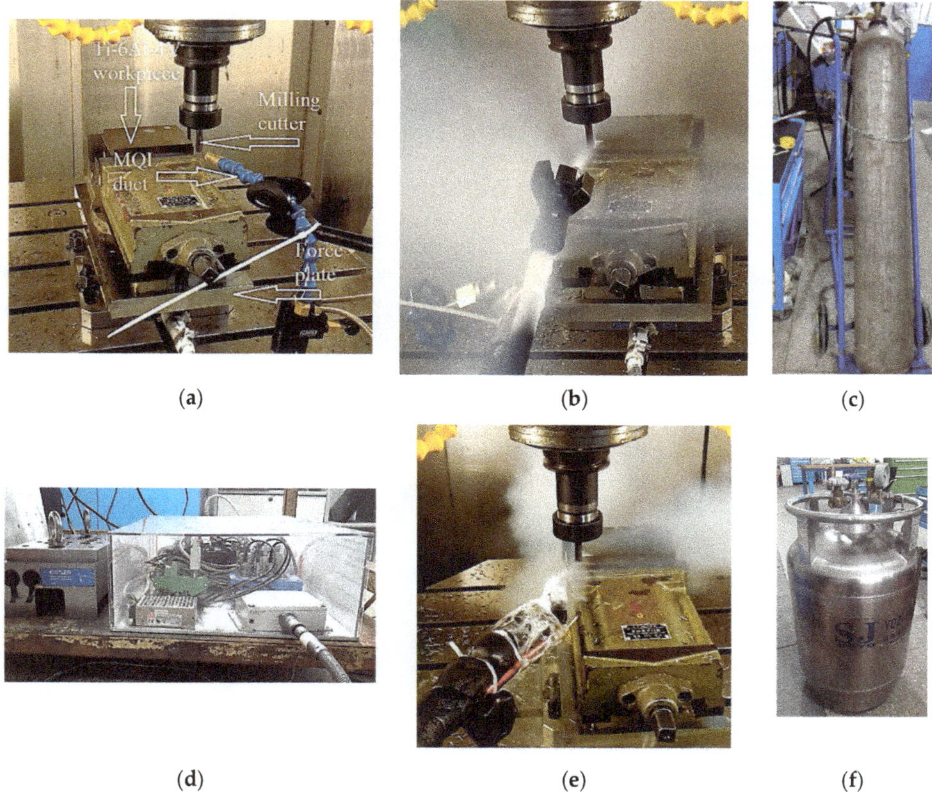

Figure 3. Experimental set-up: (**a**) Ti-6Al-4V work, milling cutter, and MQL duct; (**b**) application of CO_2 snow; (**c**) CO_2 gas storage bottle; (**d**) force data acquisition system; (**e**) application of LN_2; (**f**) storage dewar for LN_2.

Flank wear land of the used cutters' teeth was measured using a camera-fitted optical microscope ARTCAM 130-MT-WOM (Tokyo, Japan). The captured images of the flank faces were processed according to the scale to determine average width of flank wear land on each of the four cutting teeth. *VB* was then calculated out by taking average of the flank wears of the four teeth. The roughness of the side surface, after finishing two passes, was measured using Mahr MarSurf M 300 C (Mahr GmbH, Göttingen, Germany) a mobile roughness measuring instrument. The instrument used a 2-µm contact stylus to find the arithmetical mean height (*R*a) of the milled surface according to ISO 11562 standard. The sampling length for each measurement was 4 mm, and Gaussian Profile Filter was used to get the roughness values. Four measurements were taken for each experimental run at the distances of 15, 30, 45, and 60 mm from the starting edge of the milled side surface. The *R*a was then obtained by taking the average of the four measurements. Cutting forces were measured using a Kistler piezoelectric dynamometer 9265B (Kistler AG, Bern, Switzerland), utilizing a force plate 9443B. The dynamometer possessed a measuring range of 0–15 kN in the *x*- and *y*-directions and 0–30 kN in the *z*-direction.

Process cost, *PC*, comprised five components. Firstly, tooling cost was quantified by multiplying *VB*(mm)/(0.3 mm × 600 mm^3) by the current market price (BND) of the milling cutter, where *VB* is the average flank wear of a milling cutter measured after removing 600 mm^3 of the work material's volume.

The number "0.3" in mm represents the commonly adopted tool life criterion for the machining tools. As the used tools can be resold after grinding and recoating, the average resale price of the cutter was subtracted from the purchase cost of the new tools. Clearly, a larger *VB* results in a higher tooling cost. Secondly, the direct electricity consumption cost, for each experimental run, was equal to the product of the commercial electricity tariff (BND/kWh) and the specific energy (kWh/mm^3) taken by the CNC machine tool and other associated equipment (LN_2/MQL) during the actual cutting process. Thirdly, the overhead cost included wage cost of one skilled operator (BND/h) and costs of using lighting and heating, ventilation, and air conditioning (HVAC) in a small room containing the milling machine. The latter was obtained as a product of the electricity tariff and the total wattage of the lights and the HVAC system (BND/kWh × kW = BND/h). The overhead cost was calculated for the actual duration of the cutting process. The cost in BND/h was divided by the respective *MRR* to obtain the overhead cost in BND/mm^3. Fourthly, the equipment's depreciation cost was calculated using the actual purchase cost of each equipment (CNC milling machine, MQL, LN_2 equipment, and CO_2 storage bottle), proportioned for the actual time spent removing 1 mm^3 of work material's volume. The useful life and the salvage value of each of the four equipment were taken as 10 years and zero, respectively. Machining times required for removing 600 mm^3 of work material with respect to the runs employing the cutting speeds of 100 and 175 m/min were 5.65 s and 3.23 s, respectively. The numbers of working days in a year and working hours in a day were taken as 250 and eight, respectively. The straight-line depreciation model was used for calculating the depreciation cost. Lastly, the cutting fluids' consumption cost was obtained by multiplying the mass flow rate (kg/min) of the fluid (LN_2, vegetable oil, or CO_2) by its per mass unit purchase cost (BND/kg) and dividing by the material removal rate (*MRR*) (mm^3/s) of the run. Fine details of the employed costing approach can be read in Reference [9]. Consequently, the *PC*, for each run, was obtained in BND/mm^3 of the work material removed. It was quantified using the following equation:

$$PC\left(\mathrm{BND/mm^3}\right) = \frac{VB \times (A_1 - A_2)}{180} + \frac{SE \times B}{3.6e6} + \frac{(C + D \times B)}{MRR} + \frac{F \times t_m}{10 \times 7.2e6 \times 600} + \frac{\dot{m} \times G}{MRR}, \quad (1)$$

where A_1 = purchase cost of a new end mill cutter (BND), A_2 = average resale price of the cutter (BND), *SE* = specific energy taken in by the machine tool and the associated cooling/lubricating equipment (J/mm^3), *B* = commercial electricity tariff (BND/kWh), *C* = hourly wage of a skilled machine operator (BND/h), *D* = total wattage of the lights and HVAC in kW, *F* = procurement cost of all the relevant equipment; *G* = purchase cost of the cutting fluid (BND/kg), t_m = actual machining time (5.65 s and 3.23 s), and $\dot{m}$ = mass flow rate of the cutting fluid. The numbers 3.6e6 and 7.2e6 represent the factors for converting kWh to Joules and years to seconds, respectively.

Considering the ever-present inflation, it is more meaningful to present the process cost in terms of a 100-scale comparative cost structure. In such an arrangement, the costliest result is presented as 100, the most economical as 0, and all others as proportionately determined numbers lying between 0 and 100.

The consumption of specific cutting energy for each experimental run was determined as follows:

I. Three current clamp meters, Hantek CC 65, were applied to the three phases of the AC supply of the CNC machine tool to measure the electric power drawn during the machining process (P_{total}). The input voltage (*V*) and the power factor (*PF*) were measured as 220 V and 0.85, respectively. The total power was calculated using the following formula:

$$P_{total} = \frac{\sqrt{3}.PF.V(I_1 + I_2 + I_3)}{3}, \quad (2)$$

where I1, I2, and I3 represent the current, in amperes, as measured by the three clamp meters.

II. The non-cutting power ($P_{non\text{-}cut}$) was determined by rotating and linearly moving the milling cutter in the direction of feed at the given combination of rotational speed and feed speed.

The cutter was moved linearly at the feed speeds of 1592 and 2785 mm/min for the runs employing the cutting speeds of 100 and 175 m/min, respectively.

III. The $P_{non\text{-}cut}$ for each run was lessened from the relevant average total power consumed by the CNC machine during the milling process to obtain the average cutting power (P_{cut}).

IV. The P_{cut} (J/sec) for each run was divided by the relevant material removal rate (*MRR*) to get the specific cutting energy, *SCE* (J/mm^3). The *MRR* values for the runs conducted at the cutting speeds of 100 and 175 m/min were 106.2 and 185.8 mm^3/s, respectively.

Analysis of variance (ANOVA) was also performed on the experimental data in order to have a better understanding of the individual effects of the controlled predictors, as well as their interactive effects on the measured responses. ANOVA not only helps to isolate the effect of each individual predictor; it also reveals the strength of the effect.

2.4. Fixed Parameters

The values of feed per tooth, axial depth of cut, and radial depth of cut were fixed to 0.1 mm/z, 8 mm, and 0.5 mm, respectively, for all the experimental runs. Each cutter was held in a collet at a distance of 28 mm from the cutting end. All the runs were performed by carrying out the milling cuts twice in a straight line. The supply of the relevant cutting fluid (LN_2/CO_2/MQL) was started 20 s prior to first engagement of the tool with the work and was not stopped during the transition between the two passes. The supply was shut down immediately after completion of the second pass.

3. Experimental Results

The section provides details on the experimental results regarding tool wear, surface roughness, specific cutting energy consumption, cutting forces, and process cost.

3.1. Tool Wear

Figure 4 presents the experimental results of the 24 runs regarding *VB* obtained after removing 600 mm^3 of the work material. The results are grouped into three plots based on the type of cutting fluid used. A striking result is evident that, in general, micro-lubrication yielded lower levels of tool wear as compared to the two cryogenic cooling options. Of the two cryogenic cooling options, it is evident that the evaporation-based coolant (LN_2) fared better. Furthermore, the smallest *VB* (0.04 mm) of all the runs was also yielded by LN_2. Furthermore, it is clearly observable that down-milling generated remarkably smaller wear than up-milling when the applied cutting fluid was LN_2.

With regard to cutter's helix angle, the larger angle (42°) performed better than the smaller angle (30°) with respect to tool wear. Unfortunately, the dependence of helix angle on any other predictor with respect to *VB* was not clear from the plots. Although a higher helix angle favored higher material removal rates (realized by high levels of speeds and feeds), the resulting smaller flute spacing caused a problem in chip evacuation at these rates, especially for a sticky material such as Ti-6Al-4V. This is why no interactive effect between helix angle and cutting fluid was visible on tool wear. Down-milling (also known as climb milling) generally yielded lower tool wear than up-milling (conventional milling). This effect was especially prominent in the milling conducted under LN_2. The gradual decrease in chip thickness, as the cut proceeded in down-milling, reaching zero at the end, prevented the cutting edge and adjacent flank face from rubbing and burnishing against the work surface. The lower levels of flank wear observed in down-milling were attributed to this mode of frottage-free cutting. With respect to cutting speed, its higher level was generally found to yield higher levels of tool wear, although the effect was quite insignificant. A high level of cutting speed instigated or accelerated the temperature-dependent modes of tool damage because of an enhancement in the rate of heat generation and led to intensification of tool wear. As a result, the cutting edge and the adjacent faces incurred higher magnitudes of wear per unit volume of work material removed.

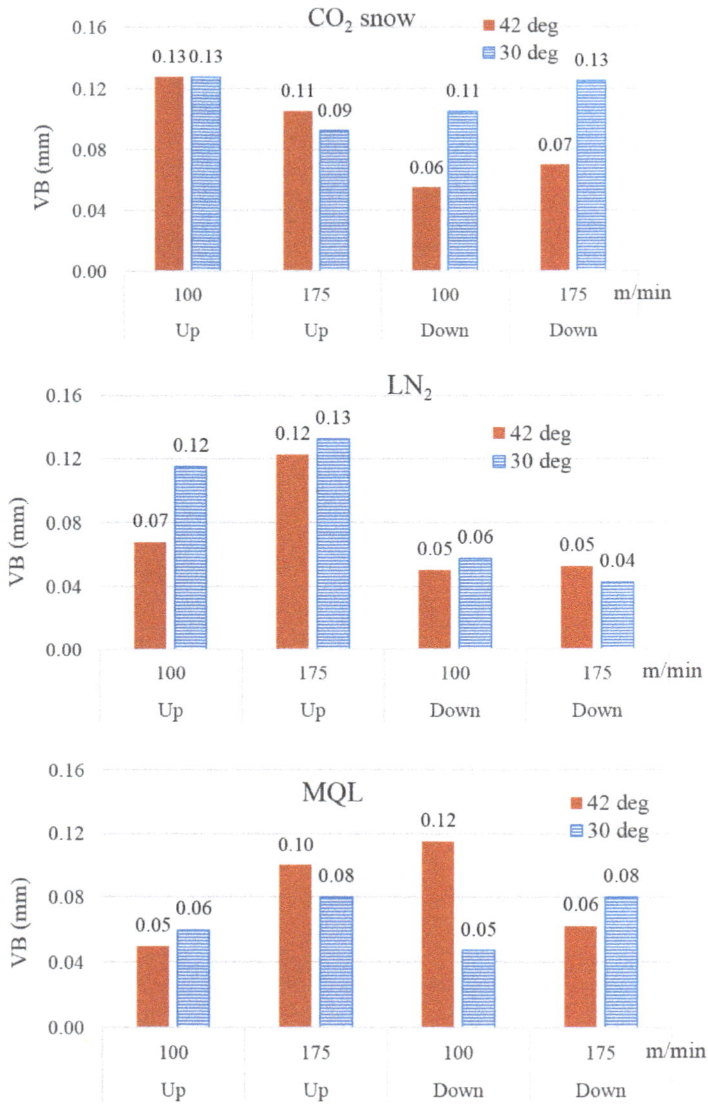

Figure 4. Bar graphs present the measurements of the average width of flank wear land (*VB*) for the 24 experimental runs.

ANOVA was performed on the data shown in Figure 4. The analysis revealed milling orientation as the most influential factor regarding *VB*, followed by the choice of cutting fluid. The interaction between milling orientation and cutting fluid stood third in terms of statistical significance, followed by the solitary effects of cutting speed and cutter's helix angle. The strong interactive effect between milling orientation and cutting fluid on tool wear is evident from the three plots of Figure 4 in the manifestation that down-milling yielded vastly lower tool wear than up-milling only when the applied cutting fluid was liquid nitrogen. Shokrani et al. claimed that the coated carbide cutter used for cryogenic (LN_2) milling of Ti-6Al-4V, operated at 200 m/min cutting speed and 0.03 mm/tooth feed rate, showed the minimum level of flank wear [14]. The current work, on the other hand, found MQL to be a better cutting fluid than the cryogenic fluids with respect to tool damage. The finding was based on milling performed at a slightly lower cutting speed but an enormously larger feed rate.

Figure 5 presents the microscopic images of the selected used cutters. The six cutters were selected so as to ensure a maximum level of diversity regarding the predictor variables. The microscopic visual analysis revealed occurrence of progressive mechanical wear and adhesion of work material as the major modes of tool damage, while chipping of cutting edge was also observed in a few tools. As can be seen from Figure 5a,d, all the cutters used in the runs involving up-milling and either of the two cryogenic fluids experienced thick adhesion of work material at their cutting edges. The adhered material, in all these runs, was in the form of minute flakes. The observation tallies with the findings of the previous works. It was reported that adhesion wear was the main tool damage mechanism in LN_2-assisted machining of Ti-6Al-4V [7]. In addition, up-milling was also reported to instigate more severe tool wear in the peripheral milling of a hardened steel [24].

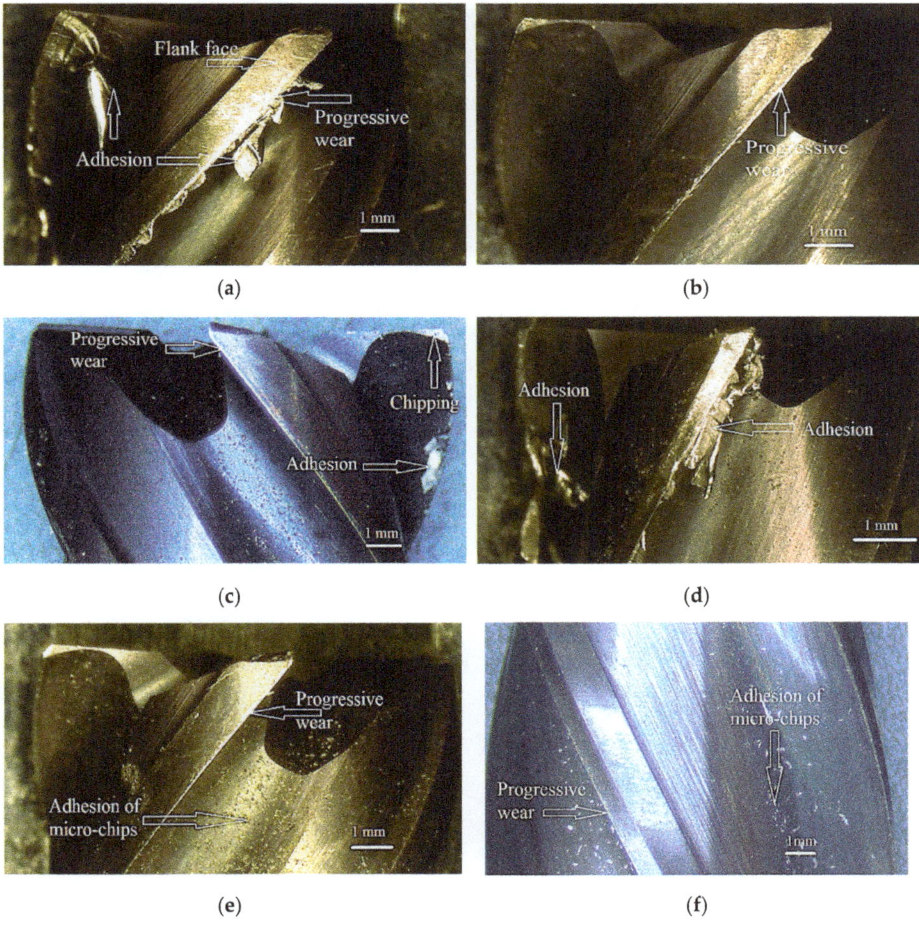

Figure 5. Micrographs showing the modes of damage incurred by the teeth of the milling cutters used in the following runs: (**a**) CO_2, 42°, up, 175 m/min; (**b**) CO_2, 42°, down, 175 m/min; (**c**) LN_2, 42°, down, 175 m/min; (**d**) LN_2, 30°, up, 100 m/min; (**e**) MQL, 42°, up, 100 m/min; (**f**) MQL, 30°, down, 100 m/min.

As a cutting edge, in up-milling, is known to enter the work with a minimum thickness and exit with a maximum, it sees a tendency of the removed chip to remain adhered to it under the influence of high temperature. Rapid cooling offered by the incoming cryogenic fluid strengthens the attachment of the chip, or part of it, to the edge and the adjacent faces. On the other hand, as is visible from Figure 5e,f, all the runs involving micro-lubrication yielded adhesion of micro-chips on the tool surface located inside of the flutes and away from the cutting edges. Micro-sized particles of the removed material were vulnerable to be caught up by the oily surface of the flutes and remained clung even against the high centrifugal forces of the rotating tool. Progressive mechanical wear (abrasion), in the form of a bright line existing close to the cutting edge, is evident in all the six images. Not much information can be deduced from the images regarding dependence of its severity on the various predictors controlled in the experiments. It was reported that abrasion was not the wear-determining mechanism in cryogenic or wet milling of Ti-6Al-4V using coated and uncoated tools irrespective of the coolant's choice [12]. Chipping was observed in no more than three milling cutters. One of them is shown in Figure 5c. All of them were involved in milling at the high level of cutting speed (175 m/min). Furthermore, the chipping, in all three instances, occurred at the corner of the relevant tooth. In a previous work, the application of CO_2 was reported to slow down the chipping process [12]. Thermal cracking of the cutting teeth, normally caused by rapid heating and cooling cycles of cryogenic milling, was not evident, probably, due to a considerably short length of cut employed in the experimental runs.

3.2. Work Surface Roughness

Figure 6 presents the measurements of *Ra* for the 24 runs, categorized by the type of cutting fluid used. The error bars present the standard deviation of the measured data for each experimental run. Quite a few conclusions can be drawn directly from the plots. As far as the type of cutting fluid is concerned, it is evident that micro-lubrication yielded better surface finish than the cryogenic fluids. Furthermore, down-milling fared better than up-milling. The effects of other two predictors were not clear from the graphs. ANOVA was performed on the *Ra* data to get further insights. The analysis revealed milling orientation to be the most influential predictor, followed by cutting speed and cutting fluid. The effect of cutter's helix angle was found to be statistically insignificant. With regard to milling of a hardened tool steel, the cutters with a 45° helix angle were reported to yield significantly lower surface roughness than the 30° cutters [24]. Up-milling is clearly not the better choice for the sake of good surface finish. As described in the previous sub-section, the cutting teeth, while following the up-milling approach, remove the work material with a lot of work material adhesion. The adhesion is believed to compromise the true geometry and sharpness of the cutting edge, thus leading to generation of a rougher surface. The superiority of down-milling regarding work surface finish was also reported for the milling of a tool steel [26].

It is quite surprising to see micro-lubrication yielding better surface finish than the two cryogenic coolants. As far as a continuous machining process is concerned, many papers reported strikingly improved work surface finish caused by the use of a cryogenic coolant. On the contrary, cryogenic cooling, in this study, yielded a poor work surface quality regarding the milling process. A plausible explanation for this observation is that milling is an interrupted machining process in which cutting teeth periodically engage and disengage with the work. Each tooth engagement causes an increase in temperature of the tooth, as well as the newly generated work surface. The surface tends to expand due to the resulting increase in temperature but immediately gets impinged upon with the flow of a cryogenic fluid as the cutter clears the area following its feed. The impact of the super-cool fluid immediately lowers its temperature, pushing the surface into the contraction mode. Such an intense thermal effect instigated upon the newly generated surface, sending it into abrupt modes of expansion and contraction, causes deterioration of the surface quality. A 25% reduction in work surface roughness was reported when graphene-dispersed vegetable-oil-based micro-lubrication was applied in the milling of Ti-6Al-4V [17]. An increase in cutting speed was found to reduce the surface roughness when the milling orientation was down-milling. Its influence on *Ra* with up-milling is not clear. An increase in cutting speed was also reported to reduce the work surface roughness in milling of a hardened steel (AISI D2) [9].

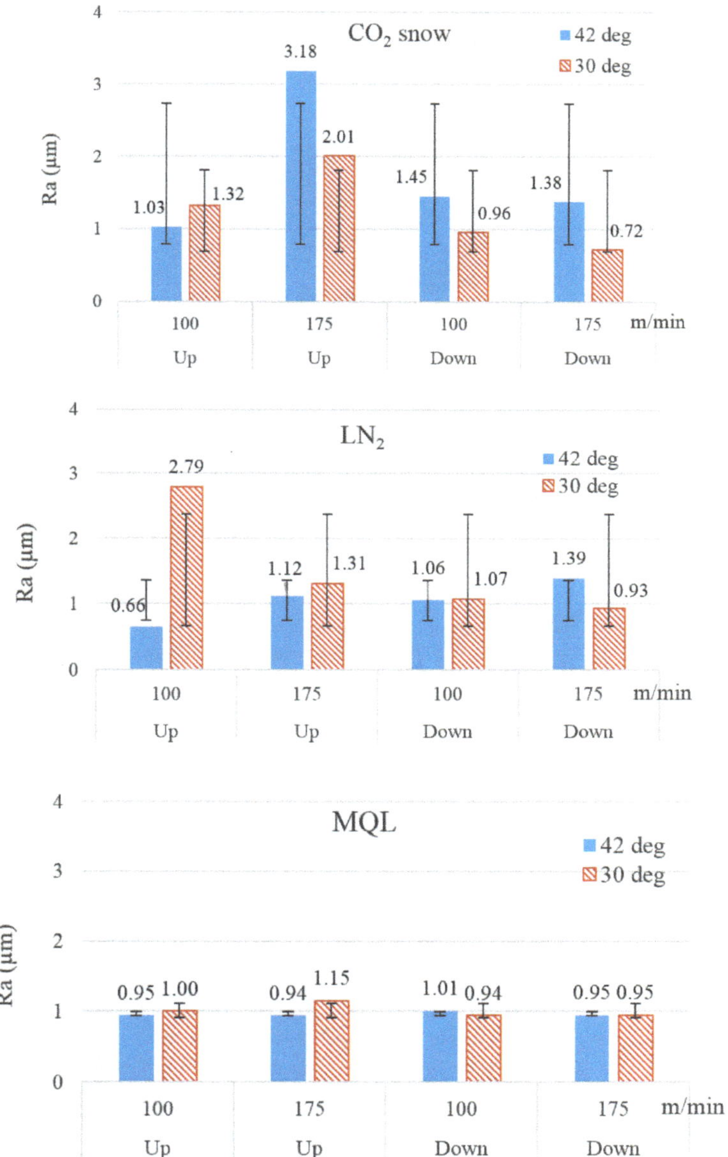

Figure 6. Bar graphs present the experimental results regarding arithmetic average surface roughness for the 24 runs.

Figure 7 presents the textures of the four milled surfaces obtained after completion of the selected experimental runs. Clearly, the surfaces produced by up-milling (Figure 7a,b) were marred by the adhesion of micro-chips. The adhesion caused serious deterioration of the work surface. Figure 7a presents the roughest of all the surfaces generated in the 24 runs. All the runs employing the combination of CO_2 snow and up-milling yielded severe adhesion of micro-chips on the milled surfaces.

Figure 7c,d do not show any sign of adhesion, as both of them were associated with the runs employing the down-milling orientation. It is to be noted that the pattern of vertical lines on the texture was the same in all the images. This is because all the runs were carried out at a fixed value of feed per tooth. Figure 7b shows a white horizontal line running through the generated surface. The line was generated by the chipped portions of the four flutes of the cutter, which might have encountered a hard phase during the first pass of the run.

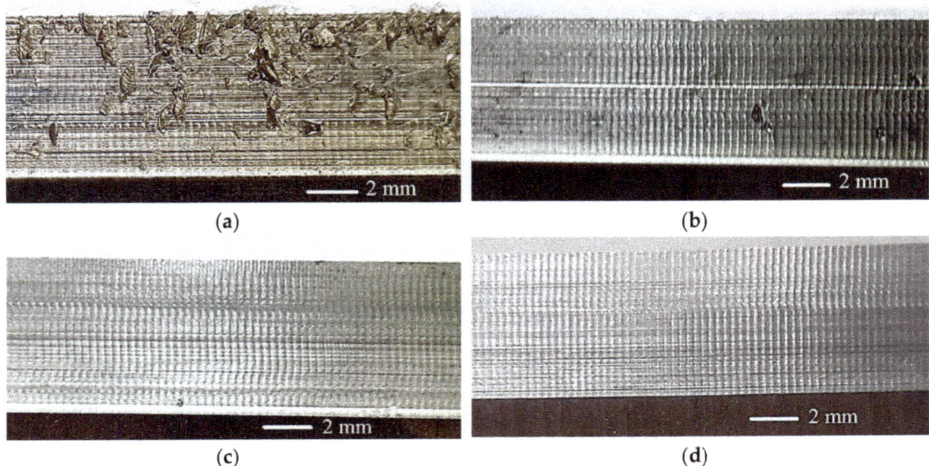

Figure 7. Work surface texture obtained after completion of the following runs: (**a**) CO_2, 42°, up, 175 m/min; (**b**) LN_2, 30°, up, 175 m/min; (**c**) LN_2, 42°, down, 100 m/min; (**d**) MQL, 30°, down, 100 m/min.

3.3. *Specific Cutting Energy and Machining Forces*

Table 2 presents the data regarding measurement of total electric power and cutting and non-cutting powers and evaluation of the *SCE*s for the 24 experiments. The procedure for measuring the electric powers was already detailed in sub-Section 2.3. Figure 8 presents the experimental results of the 24 runs regarding consumption of specific cutting energy. Two conclusions can be drawn directly from the plots. Firstly, the cutters with a 42° helix angle consumed less energy than those with a 30° helix angle. Secondly, the higher level of cutting speed caused a decline in energy consumption. ANOVA applied to the *SCE* data revealed extreme statistical influences of cutting speed and cutter's helix angle. The effect of cutting fluid was found to be marginally significant, and that of milling orientation was found to be insignificant.

The resulting diminution in specific cutting energy caused by cutting speed was attributed to an increase in material removal rate to a larger extent and an increase in cutting power consumption to a smaller extent. As *SCE* is the ratio of cutting power to material removal rate, an increase in cutting speed resulted in lessening of its magnitude. The cutting power (and energy) is required at the cutting edge to plastically deform the work material in shear and convert it into a chip. The formation of the chip is realized along with the generation of process heat. A higher cutting speed results in a higher heat generation rate. The process heat was reported to reduce the flow stress of the work material causing reduction in the cutting energy required to remove the same volume of material [32]. This phenomenon is termed thermal softening, and it was the reason for a lesser than expected increase in cutting power with increase in cutting speed. Hence, the specific cutting energy was found to decline with a step up in cutting speed. An increase in cutting speed, in side- and end-milling of hardened tool steel, was also reported to significantly reduce the specific energy consumption [9].

Table 2. Measurements and calculations regarding total power consumed by the machine tool, non-cutting power, cutting power, and specific cutting energy. MRR—material removal rate.

Test No.	Coolant	λ (°)	Milling Orientation	v_c (m/min)	P_{total} (W)	$P_{non\text{-}cut}$ (W)	P_{cut} (W)	MRR (mm³/s)	SCE (J/mm³)
1	CO_2	42	Up	100	3449.8	3033	416.8	106.2	3.92
2	CO_2	42	Up	175	3838.0	3268	570.0	185.8	3.07
3	CO_2	42	Down	100	3438.4	3033	405.4	106.2	3.82
4	CO_2	42	Down	175	3878.2	3268	610.2	185.8	3.28
5	CO_2	30	Up	100	3497.9	3033	464.9	106.2	4.38
6	CO_2	30	Up	175	3982.0	3268	714.0	185.8	3.84
7	CO_2	30	Down	100	3491.1	3033	458.1	106.2	4.31
8	CO_2	30	Down	175	4060.3	3268	792.3	185.8	4.26
9	LN_2	42	Up	100	3405.8	3033	372.8	106.2	3.51
10	LN_2	42	Up	175	3829.0	3268	561.0	185.8	3.02
11	LN_2	42	Down	100	3495.7	3033	462.7	106.2	4.36
12	LN_2	42	Down	175	3865.7	3268	597.7	185.8	3.22
13	LN_2	30	Up	100	3596.6	3033	563.6	106.2	5.31
14	LN_2	30	Up	175	3813.5	3268	545.5	185.8	2.94
15	LN_2	30	Down	100	3560.2	3033	527.2	106.2	4.96
16	LN_2	30	Down	175	3967.1	3268	699.1	185.8	3.76
17	MQL	42	Up	100	3522.9	3033	489.9	106.2	4.61
18	MQL	42	Up	175	3846.4	3268	578.4	185.8	3.11
19	MQL	42	Down	100	3472.0	3033	439.0	106.2	4.13
20	MQL	42	Down	175	3845.0	3268	577.0	185.8	3.11
21	MQL	30	Up	100	3617.2	3033	584.2	106.2	5.50
22	MQL	30	Up	175	4030.4	3268	762.4	185.8	4.10
23	MQL	30	Down	100	3508.8	3033	475.8	106.2	4.48
24	MQL	30	Down	175	4018.1	3268	750.1	185.8	4.04

A cutter with a larger helix angle cuts the work material with a larger ratio of the axial force component to the radial. The radial component of the machining force is responsible for causing vibrations in the tool and stirring instability in the milling process. Such a phenomenon becomes more prominent in machining a difficult-to-cut material, such as Ti-6Al-4V. Stronger vibrations and higher instability means a higher consumption of energy for removing the same volume of material. As such, a larger helix angle causes a reduction in the proportion of the radial force component, thus stabilizing the milling process and causing a diminution in cutting energy.

Among the three cutting fluids, the throttling-based cryogenic coolant (CO_2 snow) produced the best results regarding specific cutting energy, although not by a momentous difference, followed by the evaporative cryogenic coolant (LN_2). Both the cryogenic coolants reduced cutting energy consumption by curbing the tool wear progress. The super-cool fluid put a check on the temperature-dependent tool wear modes, which helped to maintain the cutting-edge geometry for a longer period of time. Consequently, the maintained sharpness of the cutting edge needed a smaller cutting force and lesser energy to cut the given volume of work material. In another work, cryogenic milling of Ti-6Al-4V was also found to be hugely energy conservative in comparison with wet milling due to avoidance of additional power drawn in by the coolant pump [14].

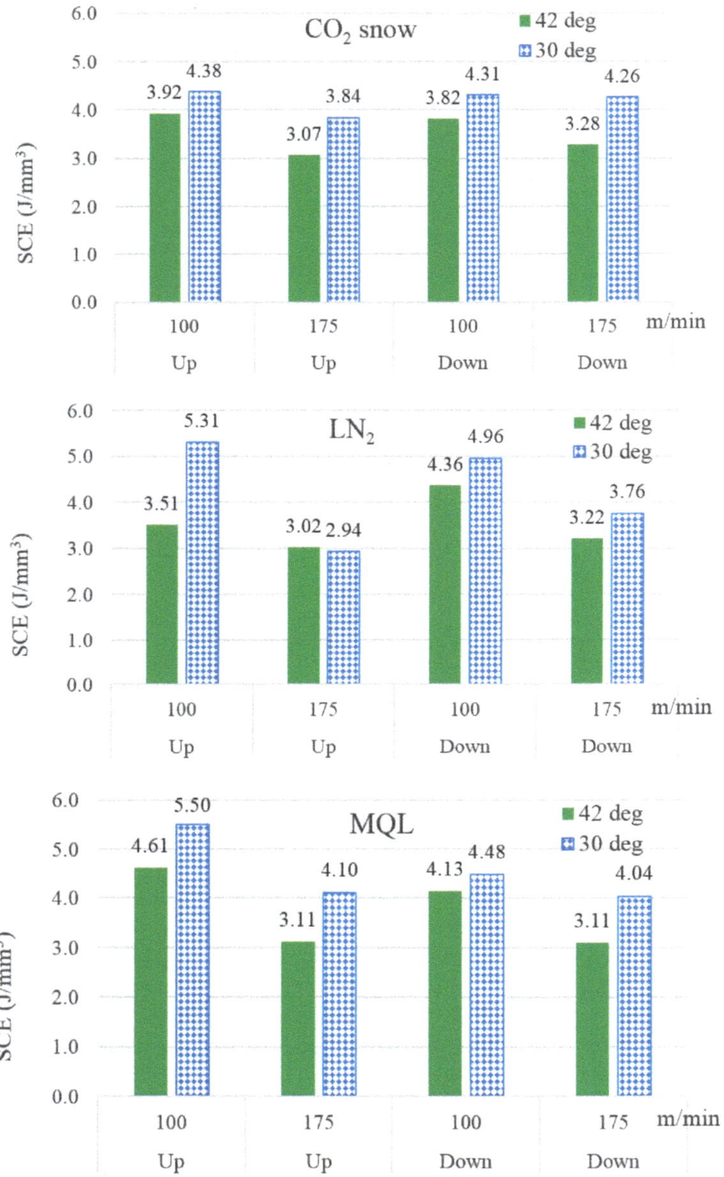

Figure 8. Bar graphs present the experimental results regarding the *SCE* for the 24 runs.

From the perspective of machining sustainability, it is often urged to include the relevant energies required to liquefy, compress, and deliver per unit mass of the cutting fluids in the *SCE* calculations. The authors would like to present some quantifications in this regard. For this study, the measured volumetric flow rate of LN_2 as a coolant was 0.5 L/min. Furthermore, the density of nitrogen in the liquid state was equal to 804 kg/m^3 at a temperature of −195.8 °C. European Industrial Gases Association (EIGA) reported that separating nitrogen at 0.1013 MPa and liquefying it from a temperature of 285 K requires consumption of electrical energy of about 549 kWh per ton of the gas [33]. Converting this figure into the power requirement against the mass flow rate of 0.402 kg/min gives 13.24 kW. Moreover, the storage dewar used in the experimental work consumes additional 50 watts of power to pump the LN_2 at the given flow rate. It was reported that capture and compression of CO_2 gas consumes electrical energy in the range of 250–300 kWh/ton of the gas [34]. For the average value of the reported range, 275 Wh/kg, and the mass flow rate of 0.5 kg/min, the average electric power required to supply and maintain the given flow rate of the CO_2 gas is 8.25 kW. Lastly, the average power required to drive the employed MQL system at the air pressure, oil mixing rate, and air flow rate of 0.6 MPa, 30 mL/h, and 0.4 L/s, respectively, is 418 Watts. The details regarding the calculations can be seen in Reference [35]. By adding the above calculated values of power requirements, the *SCE* values ranged from 57.6 to 82 J/mm^3, 74.5 to 131 J/mm^3, and 5.2 to 8.3 J/mm^3, respectively, for CO_2 snow, LN_2, and micro-lubrication. Clearly, the modified values of *SCE* rendered the comparisons among the three cutting fluids trivial, with LN_2 topping the list of the tested fluids by an unprecedented margin.

Figure 9 presents the experimental data regarding the three components of the machining forces obtained from the 24 experimental runs. As described before, F_x and F_z are aligned with the feed direction and the cutter's axis, respectively, while F_y is perpendicular to the side surface of the work. Thus, it can be said that the axial component of the milling force is F_z, while the radial component is the resultant of the other two.

A few inferences can be drawn directly from the three plots. Firstly, the cutter with the larger helix angle experienced lower magnitudes of all the force components. Secondly, F_x rose unusually for all the three cutting fluids when a 30° helix cutter was used to mill at the high level of cutting speed using the down-milling orientation. Thirdly, with respect to the choice of cutting fluid, LN_2 yielded the lowest magnitudes of the milling force components, followed by CO_2 snow. ANOVA was applied to get in-depth results. For all the three force components, cutter's helix angle was found to be a highly and the most influential predictor. The effects of milling orientation and cutting fluid were found to be marginally significant, whereas that of cutting speed was found to be insignificant. Furthermore, an interactive effect between cutting fluid and milling orientation was found to be significant on the radial components only.

A milling cutter with a large helix angle is ideal for dynamic stability in milling, machining with small depths of cut, and gaining a good work surface quality. On the other hand, a cutter with a small helix angle is good for rough machining and removing work material at high removal rates. Dynamic stability of milling is reflected by small magnitudes of machining forces. The results in this study show that a 42° helix angle yielded better dynamic stability than a 30° angle, as reflected by the results regarding the milling force components. The observed effectiveness of LN_2, in curbing machining forces, can be attributed to its more effective heat dissipation capability. An extremely low operational temperature of the fluid is believed to equip it with such a capability. Better heat dissipation mechanism helps keeping a check on tool wear progress, which, in turn, helps maintaining cutting edge's geometry and suppressing the machining forces. In other works, hybrid cryogenic cooling/lubrication was reported to yield the lowest machining forces compared to both emulsion and cryogenic conditions in machining of titanium alloys [11,13]. The effect of milling orientation was neither very significant nor clear. Down-milling yields lower magnitudes of force components for CO_2 snow and MQL and higher magnitudes for the third fluid. Moreover, up-milling was found to be the better orientation to be used at higher cutting speeds. Li et al. reported an 18% reduction in milling forces when graphene-dispersed vegetable-oil-based MQL was applied in the milling of Ti-6Al-4V [17].

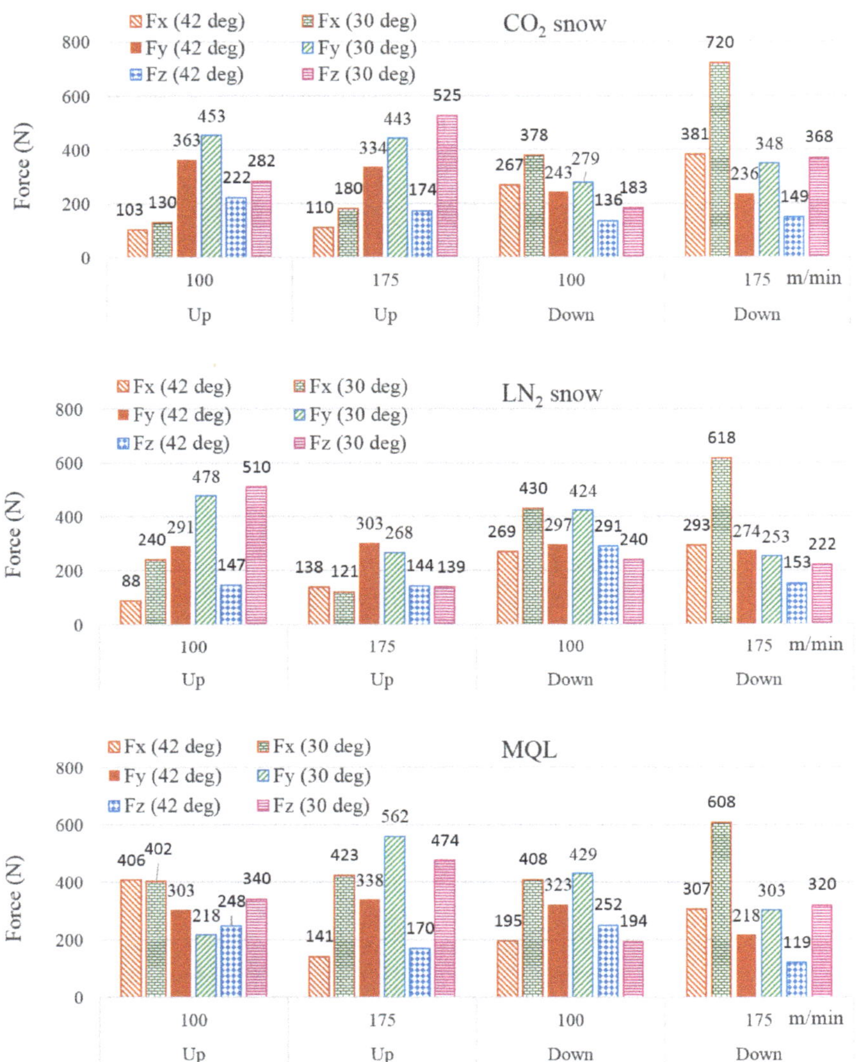

Figure 9. Bar graphs present the experimental results regarding the three components of the machining forces (F_x, F_y, and F_z) for the 24 runs.

Figure 10 presents the progress of the three force components with time for the selected experimental runs. Two concentration zones are visible in each of the three graphs representing the two cutting passes per experimental run. The two zones in Figure 10a are thinner than those in the others because they represent the cutting passes conducted at the higher speed level. Based on the feed directions, the radial force components (F_x and F_y) were found in the positive and negative regions for down- and up-milling, respectively. The axial component (F_z) was located in the negative region irrespective of the milling orientation.

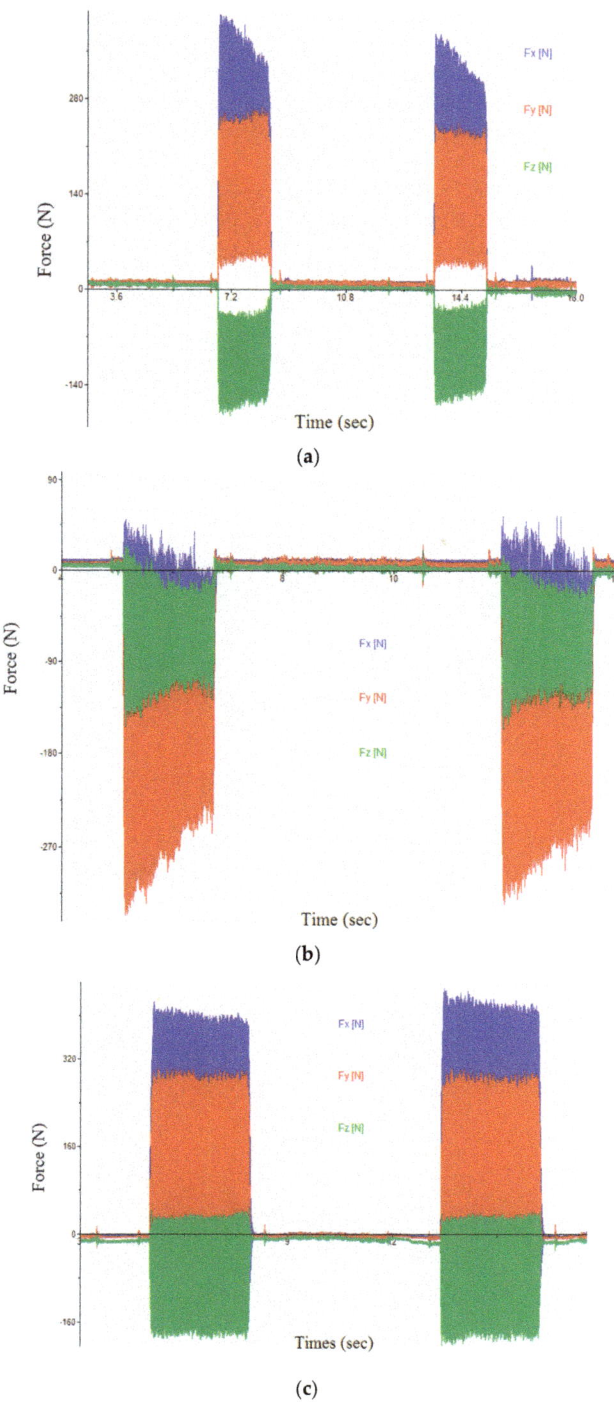

(c)

Figure 10. Progress of the milling force components with time for the following experimental runs: (**a**) CO_2, 42°, down, 175 m/min; (**b**) LN_2, 30°, up, 100 m/min; (**c**) MQL, 30°, down, 100 m/min.

The graphs highlight a phenomenon that sets apart the cryogenic fluids from micro-lubrication. It can be seen from Figure 10a,b that the instantaneous values of the force components sharply decreased as the cutting pass progressed. The magnitudes remained fairly constant in the third graph. This is attributed to the temperature dependent variation in yield strength/flow stress of the work material. As described before, the supply of a cryogenic fluid was opened 20 s prior to the start of the first milling pass. As the cryogenic fluid was also received by the work, in addition to the tool, the surface temperature fell sharply, raising the material's yield strength. Thereafter, as the cutting process started, the process heat raised the temperature of the work material, thereby reducing its flow stress. The fall in the work material's flow stress is reflected by the diminution in the force components as the cutting process proceeded.

The cutting power required at the cutting edge for plastically deforming the work material into a chip is presented as the product of cutting force and cutting speed ($P_{cut} = v_c \times F_c$), where F_c, in peripheral milling, is a function of the three orthogonal force components, tool geometry, and shear plane angle [36,37]. This means that the cutting power should have positive correlations with the measured orthogonal machining force components. The correlation coefficient between P_{cut} and the resultant of the two force components acting perpendicular to the cutter's axis ($= [F_x{}^2 + F_y{}^2]^{1/2}$) was found to possess a value of 0.66. Likewise, the correlation coefficient between P_{cut} and the resultant of all the three force components ($= [F_x{}^2 + F_y{}^2 + F_z{}^2]^{1/2}$) was 0.68. The two correlations suggest that strong uphill relationships exist between the cutting power and the machining forces. It can, thus, be safely stated that machining force data can estimate cutting energy consumption fairly accurately.

3.4. Process Cost

While evaluating *PC* for the 24 experimental runs, it was found that tooling cost was the most weighted contributing factor, followed, in descending order of weightage, by acquisition cost of cutting fluids, overhead cost, direct energy consumption cost, and equipment depreciation cost. The process cost's estimation is shown in Figure 11. The figure clearly shows that, of the three cutting fluids tested, micro-lubrication yielded the most economical results, which is attributed to the low levels of *VB* and low unit cost and consumption rate of the associated cutting fluid. Of the two cryogenic coolants, CO_2 snow yielded better results when milling orientation was up, while LN_2 was clearly more economical for down-milling. Furthermore, a high helix angle produced better results, especially under the cryogenic cooling environments, due to the associated low levels of tool damage incurred. Lastly, the effect of cutting speed on process cost was not very clear, as this predictor asserted its effects on all five constituents of *PC* in different ways. With regard to a continuous cutting process, it was reported that high-speed machining of Ti-6Al-4V yielded better surface finish and caused lower consumption of specific energy than conventional machining, but fared poorly with respect to the sustainability measures of tool life and process cost [35].

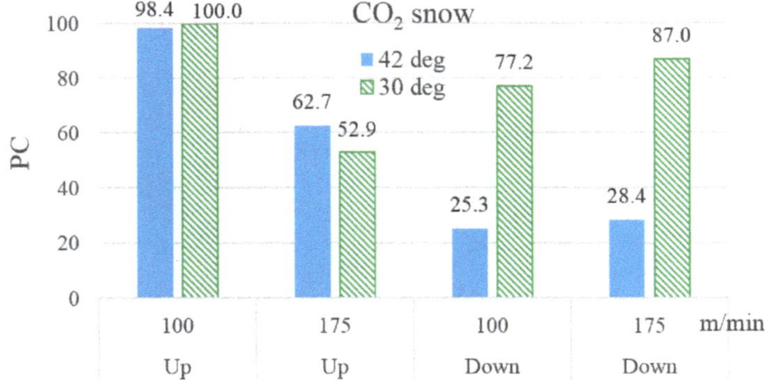

Figure 11. *Cont.*

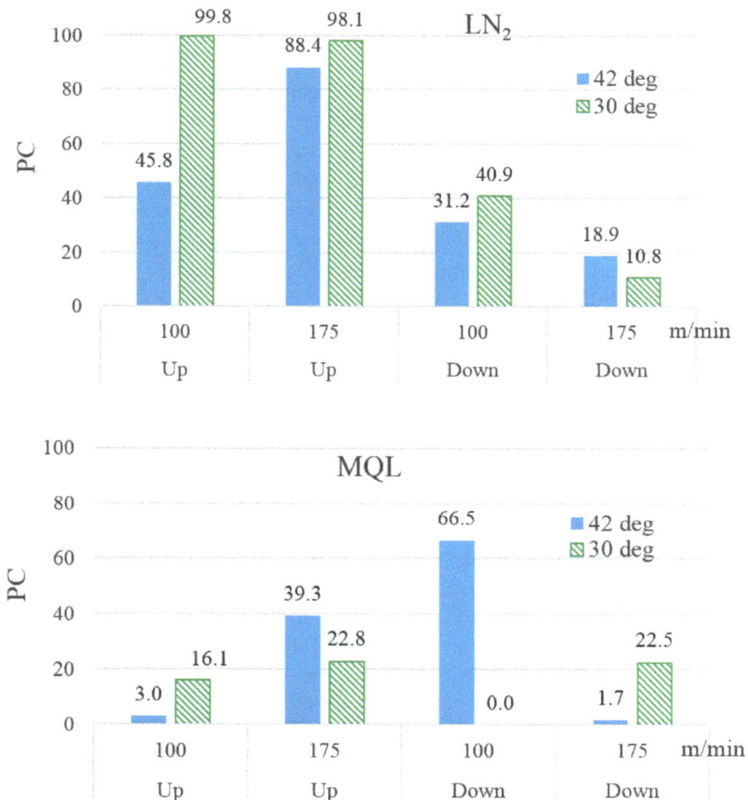

Figure 11. Bar graphs present the process costs for the 24 experimental runs.

4. Discussion on Milling Sustainability

The previous section presented comprehensive quantifications and analyses with respect to the vital sustainability metrics regarding milling of the most commonly used titanium alloy. The current section discusses the implications of the experimental results in the perspective of the three pillars of sustainability: economic, social, and environmental.

Process cost is the most important performance measure related to the economic aspect of sustainability. As detailed in the previous section, tooling cost and cutting fluid's acquisition cost are the most important contributors. Tooling cost is governed by tool wear (*VB*); a larger value results in a higher cost. On the other hand, *VB* is also influenced in a positive way by the application of cutting fluids. The experimental results regarding *PC* suggest that micro-lubrication, combined with the down-milling approach, ensured the most economical milling of Ti-6Al-4V, realized in the form of slightest tool wear. Work surface roughness also possesses significance with respect to the economic aspect. An acceptable level of surface finish avoids financial losses due to the following factors: (1) additional processing cost caused by rework, and (2) additional material and processing cost incurred by rejection of the work. In this regard, once again, micro-lubrication and down-milling arose as the ideal combination, as explained in Section 3.2.

The social dimension of machining sustainability is highly qualitative. The factors covered under this pillar are workers' safety, health protection, ergonomics, payment equity, and work pressure and intensification. Cutting speed is expected to affect ergonomics and work pressure as its low levels may increase working hours of the machine operators and build-up on production slack, leading to enhancement of work pressure. Choice of cutting fluid is another influential parameter with respect to the social pillar. All three levels of cutting fluid tested in this work demand extra space and workload for setting up and operating the fluid supply system. A minor cleaning effort is also required after the milling process utilizing micro-lubrication and cryogenic fluids is over. The oil delivered to the machining area in micro-lubrication is of minute quantity, which requires an easy cleaning effort. Moreover, the commonly used oils are vegetable-based and are, thus, not hazardous. Application of LN_2 to the milling area immediately turns the fluid into its gaseous form, which simply escapes into the environment without affecting the operators. Likewise, compressed CO_2 gas, after throttling, converts into snow, which gradually sublimates to a gaseous form and escapes. A small flow rate (0.5 kg/min) of the gas does not pose any asphyxiation risk.

The environmental pillar of milling sustainability is estimated by specific cutting energy, tool damage, and waste generation. The experimental results regarding *SCE* suggest that Ti-6Al-4V should be milled at a high level of cutting speed using a cutter of a medium-to-high helix angle. Moreover, the application of micro-lubrication is much more energy-conservative than the cryogenic fluids. A high *VB* causes more frequent tool replacements, leading to a higher number of tools required to remove a given volume of work material. As a result, more energy is consumed for remanufacturing/recycling the worn-out tools or, in a worse case, more landfilling of the end-of-life tooling and extraction of the raw material for making new tools is required. Such a situation quickens depletion of natural resources and causes a harmful effect on the environment. The analysis carried out on tool wear data suggests that the titanium alloy should be milled with the down-milling orientation and under the effect of micro-lubrication. Regarding the issue of machining waste generation, the tested levels of milling orientation, cutter's helix angle, and cutting speed did not make any difference. The total mass of chips generated was the same for any combination of these three predictors. On the other hand, all three options of cutting fluid tested in the work were better than the conventional approach of emulsion-based flood cooling. Emulsion coolant is known to create swarf, leading to the need for cleaning the chips before the commencement of their recycling process. Furthermore, the emulsion fluid requires filtration and pumping for its repeated use until it becomes toxic due to contamination from swarf, sump, and exposure to air. Milling under cryogenic and MQL environments does not create any swarf; thus, the chips remain clean. Furthermore, no greenhouse gases are emitted and recycling of the used fluid is not required. Regarding the option of LN_2, nitrogen gas is harmless to human health and is also environmentally benign. Moreover, the use of CO_2 gas, as a cryogenic coolant, does not increase carbon footprint because a minute quantity is used for cooling. A viable cutting speed of up to 200 m/min was claimed in machining titanium alloys using a natural diamond tool and under the effect of flood coolant [2]. On the other hand, the current work presented sustainable machining of the titanium alloy up to a speed of 175 m/min, but with more economical tooling and environmentally friendly coolants.

In the context of the discussion provided above, the authors would conclude that, with respect to side- and end-milling of Ti-6Al-4V, the application of micro-lubrication is more favorable than the two cryogenic fluids tested. It fared significantly better in terms of all the sustainability measures, namely, tool damage, process cost, work surface quality, and energy consumption (with inclusion of the fluids' energies). Not all the previously published reports claimed superiority of the cryogenic fluids over conventional approaches. Isakson et al. reported development of tool wear at a faster rate under LN_2 than under emulsion-based coolant in the machining of titanium alloy, although the limited supply of coolant was blamed for its below-par performance [8]. Additionally, in the current work, the down-milling orientation clearly outperformed up-milling with respect to tool damage, process cost, and work surface quality. With regard to the other two predictors, the high levels of cutter's helix angle and cutting speed generally yielded better results, but the observed effects were not emphatic.

5. Conclusions

The presented work aimed to quantify and enhance sustainability of milling a commonly used titanium alloy under the environments of micro-lubrication and cryogenic cooling. The cryogenic cooling environment was set up by the application of two kinds of coolants, throttle and evaporative, realized by expansion of compressed CO_2 gas and evaporation of liquid nitrogen, respectively. Additionally, the effects of employing two milling orientations and two levels each of tool's helix angle and cutting speed on the sustainability measures were also quantified.

The most prominent finding of the work is that application of micro-lubrication in milling of Ti-6Al-4V is more sustainable than that of a cryogenic coolant. Minimum quantity of lubrication outperformed both the cryogenic coolants in terms of tool wear, work surface quality, process cost, and energy consumption. Dissipation of process heat using a cryogenic coolant is not as viable in milling as it is in a continuous machining process such as turning. The interrupted nature of cutting caused by periodic engagement and disengagement of cutting teeth in the work under the action of a super-cool fluid renders the cryogenic coolant far less effective. Among the other predictors, the effects of milling orientation on the sustainability measures in milling of the titanium alloy were highly significant. Down-milling was found to be enormously better than up-milling with respect to tool wear, work surface quality, and process cost. Thus, it is highly recommended to use the combination of micro-lubrication and down-milling for sustainable milling of Ti-6Al-4V.

The effects of tool's helix angle and cutting speed were different for different measures of sustainability, and they were not highly assertive. The high level of helix angle was excellent for reducing specific cutting energy and also performed well for the sake of reduction in process cost, tool wear, and milling forces. Likewise, high cutting speed was good for reducing work surface roughness and specific cutting energy consumption.

Author Contributions: A.I. conceived and designed the experiments, analyzed the results, and partially wrote the paper; H.S. arranged the tooling and work material; W.Z. and M.J. performed the experiments; M.M.N. determined the responses from the measurements; N.H. arranged the measuring instruments and performed the sustainability analysis; and J.Z. developed the graphs and partially wrote the paper. All authors have read and agreed to the published version of the manuscript.

Funding: The work presented in the article is financially supported by Universiti Brunei Darussalam, Brunei through its University Research Grant scheme (grant number: UBD/RSCH/URC/RG(b)/2018/003).

Acknowledgments: The authors of the article are grateful to the management of the Laboratory of Advanced Cutting Technology at Nanjing University of Aeronautics and Astronautics, China, for providing access to the CNC milling machine, cutting fluid equipment, and measuring instruments.

Conflicts of Interest: The authors declare no conflict of interest.

References

1. Ezugwu, E.O.; Wang, Z.M. Titanium alloys and their machinability—A review. *J. Mater. Proc. Technol.* **1997**, *68*, 262–274. [CrossRef]
2. Narutaki, N.; Murakoshi, A.; Motonishi, S.; Takeyama, H. Study on machining of titanium alloys. *CIRP Ann.* **1983**, *32*, 65–69. [CrossRef]
3. Pusavec, F.; Krajnik, P.; Kopac, J. Transitioning to sustainable production—Part I: Application on machining technologies. *J. Clean. Prod.* **2010**, *18*, 174–184. [CrossRef]
4. Dhar, N.R.; Kamruzzaman, M.; Ahmed, M. Effect of minimum quantity lubrication (MQL) on tool wear and surface roughness in turning AISI-4340 steel. *J. Mater. Proc. Technol.* **2006**, *172*, 299–304. [CrossRef]
5. Boubekri, N.; Shaikh, V. Minimum quantity lubrication (MQL) in machining: Benefits and drawbacks. *J. Ind. Intel. Infor.* **2015**, *3*, 205–209. [CrossRef]
6. Braga, D.U.; Diniz, A.E.; Miranda, G.W.; Coppini, N.L. Using a minimum quantity of lubricant (MQL) and a diamond coated tool in the drilling of aluminum–silicon alloys. *J. Mater. Proc. Technol.* **2002**, *122*, 127–138. [CrossRef]

7. Ayed, Y.; Germain, G.; Melsio, A.P.; Kowalewski, P.; Locufier, D. Impact of supply conditions of liquid nitrogen on tool wear and surface integrity when machining the Ti-6Al-4V titanium alloy. *Int. J. Adv. Manuf. Technol.* **2017**, *93*, 1199–1206. [CrossRef]
8. Isakson, S.; Sadik, M.I.; Malakizadi, A.; Krajnik, P. Effect of cryogenic cooling and tool wear on surface integrity of turned Ti-6Al-4V. *Procedia CIRP* **2018**, *71*, 254–259. [CrossRef]
9. Iqbal, A.; Al-Ghamdi, K.A.; Hussain, G. Effects of tool life criterion on sustainability of milling. *J. Clean. Prod.* **2016**, *139*, 1105–1117. [CrossRef]
10. Sartori, S.; Ghiotti, A.; Bruschi, S. Hybrid lubricating/cooling strategies to reduce the tool wear in finishing turning of difficult-to-cut alloys. *Wear* **2017**, *376*, 107–114. [CrossRef]
11. Schoop, J.; Sales, W.F.; Jawahir, I.S. High speed cryogenic finish machining of Ti-6Al4V with polycrystalline diamond tools. *J. Mater. Proc. Technol.* **2017**, *250*, 1–8. [CrossRef]
12. Sadik, M.I.; Isakson, S. The role of PVD coating and coolant nature in wear development and tool performance in cryogenic and wet milling of Ti-6Al-4V. *Wear* **2017**, *386*, 204–210. [CrossRef]
13. Iqbal, A.; Biermann, D.; Abbas, H.; Al-Ghamdi, K.A.; Metzger, M. Machining β-titanium alloy under carbon dioxide snow and micro-lubrication: A study on tool deflection, energy consumption, and tool damage. *Int. J. Adv. Manuf. Technol.* **2018**, *97*, 4195–4208. [CrossRef]
14. Shokrani, A.; Dhokia, V.; Newman, S.T. Energy conscious cryogenic machining of Ti-6Al-4V titanium alloy. *Proc. Inst. Mech. Eng. Part B J. Eng. Manuf.* **2018**, *232*, 1690–1706. [CrossRef]
15. Mia, M.; Gupta, M.K.; Lozano, J.A.; Carou, D.; Pimenov, D.Y.; Królczyk, G.; Khan, A.M.; Dhar, N.R. Multi-objective optimization and life cycle assessment of eco-friendly cryogenic N_2 assisted turning of Ti-6Al-4V. *J. Clean. Prod.* **2019**, *210*, 121–133. [CrossRef]
16. Ross, K.N.S.; Manimaran, G. Effect of cryogenic coolant on machinability of difficult-to-machine Ni–Cr alloy using PVD-TiAlN coated WC tool. *J. Braz. Soc. Mech. Sci. Eng.* **2019**, *41*, 44. [CrossRef]
17. Li, M.; Yu, T.; Yang, L.; Li, H.; Zhang, R.; Wang, W. Parameter optimization during minimum quantity lubrication milling of TC4 alloy with graphene-dispersed vegetable-oil-based cutting fluid. *J. Clean. Prod.* **2019**, *209*, 1508–1522. [CrossRef]
18. Aslantas, K.; Çiçek, A. The effects of cooling/lubrication techniques on cutting performance in micro-milling of Inconel 718 superalloy. *Procedia CIRP* **2018**, *77*, 70–73. [CrossRef]
19. Masood, I.; Jahanzaib, M.; Haider, A. Tool wear and cost evaluation of face milling grade 5 titanium alloy for sustainable machining. *Adv. Prod. Eng. Man.* **2016**, *11*, 239–250. [CrossRef]
20. Shokrani, A.; Dhokia, V.; Newman, S.T. Investigation of the effects of cryogenic machining on surface integrity in CNC end milling of Ti–6Al–4V titanium alloy. *J. Manuf. Proc.* **2016**, *21*, 172–179. [CrossRef]
21. Dawood, A.; Arbuckle, G.K.; Jahan, M.P. A comparative study on the machinability of Ti-6Al-4V using conventional flood coolant and sustainable dry machining. *Int. J Mach. Mach. Mater.* **2015**, *17*, 507–528. [CrossRef]
22. Pereira, O.; Urbikain, G.; Rodríguez, A.; Fernández-Valdivielso, A.; Calleja, A.; Ayesta, I.; de Lacalle, L.L. Internal cryolubrication approach for Inconel 718 milling. *Proc. Manuf.* **2017**, *13*, 89–93. [CrossRef]
23. Pusavec, F.; Deshpande, A.; Yang, S.; M'Saoubi, R.; Kopac, J.; Dillon, O.W., Jr.; Jawahir, I.S. Sustainable machining of high temperature Nickel alloy–Inconel 718: Part 1–predictive performance models. *J. Clean. Prod.* **2014**, *81*, 255–269. [CrossRef]
24. Iqbal, A.; He, N.; Li, L.; Xia, Y. Influence of cutter's helix angle, workpiece hardness, milling orientation, and MQL in high-speed side milling of AISI D2. *Mat. Sci. Forum* **2006**, *532*, 45–48. [CrossRef]
25. Duan, X.; Peng, F.; Zhu, K.; Jiang, G. Tool orientation optimization considering cutter deflection error caused by cutting force for multi-axis sculptured surface milling. *Int. J. Adv. Manuf. Technol.* **2019**, *103*, 1925–1934. [CrossRef]
26. Laamouri, A.; Ghanem, F.; Braham, C.; Sidhom, H. Influences of up-milling and down-milling on surface integrity and fatigue strength of X160CrMoV12 steel. *Int. J. Adv. Manuf. Technol.* **2019**. [CrossRef]
27. Hadi, M.A.; Ghani, J.A.; Haron, C.C.; Kasim, M.S. Investigation on wear behavior and chip formation during up-milling and down-milling operations for Inconel 718. *Jurnal Teknol.* **2014**, *66*, 15–21. [CrossRef]
28. Wan, M.; Feng, J.; Zhang, W.H.; Yang, Y.; Ma, Y.C. Working mechanism of helix angle on peak cutting forces together with its design theory for peripheral milling tools. *J. Mater. Proc. Technol.* **2017**, *249*, 570–580. [CrossRef]

29. Baowan, P.; Saikaew, C.; Wisitsoraat, A. Influence of helix angle on tool performances of TiAlN-and DLC-coated carbide end mills for dry side milling of stainless steel. *Int. J. Adv. Manuf. Technol.* **2017**, *90*, 3085–3097. [CrossRef]
30. Buj-Corral, I.; Vivancos-Calvet, J.; González-Rojas, H. Influence of feed, eccentricity and helix angle on topography obtained in side milling processes. *Int. J. Mach. Tools Manuf.* **2011**, *51*, 889–897. [CrossRef]
31. Roebuck, J.R.; Murrell, T.A.; Miller, E.E. The Joule-Thomson effect in carbon dioxide. *J. Am. Chem. Soc.* **1942**, *64*, 400–411. [CrossRef]
32. Sun, J.; Guo, Y.B. Material flow stress and failure in multiscale machining titanium alloy Ti-6Al-4V. *Int. J. Adv. Manuf. Technol.* **2009**, *41*, 651–659. [CrossRef]
33. Indirect CO_2 Emissions Compensation: Benchmark proposal for Air Separation Plants. In *Position Paper by EIGA*; 33/19 (Revalidation of PP 33 dated December 2010); European Industrial Gases Association: Belgium, Brussels, 2019.
34. Jackson, S.; Brodal, E. A comparison of the energy consumption for CO_2 compression process alternatives. In *IOP Conference Series: Earth and Environmental Science*; IOP Publishing: Bristol, UK, 2018; Volume 167, p. 012031.
35. Al-Ghamdi, K.A.; Iqbal, A. A sustainability comparison between conventional and high-speed machining. *J. Clean. Prod.* **2015**, *108*, 192–206. [CrossRef]
36. Lee, P.; Altintaş, Y. Prediction of ball-end milling forces from orthogonal cutting data. *Int. J. Mach. Tools Manuf.* **1996**, *36*, 1059–1072. [CrossRef]
37. Yellowley, I. Observations on the mean values of forces, torque and specific power in the peripheral milling process. *Int. J. Mach. Tools Des. Res.* **1985**, *25*, 337–346. [CrossRef]

Article

Sustainable High-Speed Finishing Turning of Haynes 282 Using Carbide Tools in Dry Conditions

Antonio Díaz-Álvarez *, José Díaz-Álvarez, José Luis Cantero and Henar Miguélez

Department of Mechanical Engineering, University Carlos III of Madrid, 30 Avida. de la Universidad, 28911 Leganés, Madrid, Spain

* Correspondence: andiaza@ing.uc3m.es

Received: 30 July 2019; Accepted: 3 September 2019; Published: 6 September 2019

Abstract: Nickel-based superalloys exhibit an exceptional combination of corrosion resistance, enhanced mechanical properties at high temperatures, and thermal stability. The mechanical behavior of nickel-based superalloys depends on the grain size and the precipitation state after aging. Haynes 282 was developed in order to improve the creep behavior, formability, and strain-age cracking of the other commonly used nickel-based superalloys. Nevertheless, taking into account the interest of the industry in the machinability of Haynes 282 because of its great mechanical properties, which is not found in other superalloys like Inconel 718 or Waspaloy, more research on this alloy is necessary. Cutting tools suffer extreme thermomechanical loading because of the high pressure and temperature localized in the cutting zone. The consequence is material adhesion during machining and strong abrasion due to the hard carbides included in the material. The main recommendations for finishing turning in Haynes 282 include the use of carbide tools, low cutting speeds, low depth of pass, and the use of cutting fluids. However, because of the growing interest in sustainable processes and cost reduction, dry machining is considered to be one of the best techniques for material removal. During the machining of Haynes 282, at both the finishing and roughing turning, cemented carbide inserts are most commonly used and are recommended all over the industry. This paper deals with the machining of Haynes 282 by means of coated carbide tools cutting fluids (dry condition). Different cutting speeds and feeds were tested to quantify the cutting forces, quality of surface, wear progression, and end of tool life. Tool life values similar to those obtained with a lubricant under similar conditions in other studies have been obtained for the most favorable conditions in dry environments.

Keywords: dry; carbide tool; Haynes 282; finishing turning

---

1. Introduction

Turbine components suffer the extreme conditions of thermomechanical loading during their service life. Significant tensile stresses in rotative elements induce fatigue phenomena [1]. The development of new advanced materials and the continuous improvement of the processing routes are required in order to improve the performance of the turbine components [2]. Nickel-based superalloys are widely used in turbine elements because of their excellent mechanical properties at high temperatures and their resistance to corrosion [3]. Being about 50 wt. % of the materials used in these applications [4], Ni alloys are also used in other applications such as pressure vessels, marine equipment, different elements of aircraft engines, and petrochemical plants [5,6]. The excellent mechanical properties of this family of superalloys also include low formability, with different problems during component processing that could affect its service life. New-generation alloys are developed in order to solve these problems. For example, Haynes 282 focuses on the improvement of the weldability and fabricability with a similar creep strength. These combinations of properties are of great interest for critical steam applications. This superalloy has already been adopted for hot section parts in gas turbines for aircraft and power generation, and it can be a baseline for the further improvement of superalloys [7].

Haynes 282 is highlighted by a high percentage of molybdenum (>6 wt. %), which develops carbide particles at temperatures ranging from 815 to 870 °C in a complex cubic structure, with it being more stable at high temperatures [8]. Haynes 282 was developed at the beginning of the 21st century. It is strengthened by the precipitation of the γ' phase, which is the $L1_2$ ordered structure Ni_3 (Al,Ti), and has a coherent relationship with the γ matrix [9]. These γ' precipitates, characterized by their size, distribution, morphology, and composition, influence the mechanical properties of the alloy [10].

Despite the renewed interest in Haynes 282, there is a lack of information concerning its machinability. The machining of nickel-based superalloys presents great challenges, mainly because of the high work hardening tendency, structure stability, low thermal conductivity, adhesion of materials in the tool, and carbide particles in its structure [11]. All of the above features result in hard loads and temperatures (up to 1200 °C [12]) at the chip–tool interface, resulting in the rapid wear of the tool [13,14], which influences the surface integrity of the piece, generating residual stresses and increments in roughness [15,16]. Elevated temperatures combined with a high chemical affinity between the workpiece and the materials used for the cutting tools promote oxidation and diffusion wear, as well as the adhesion of the work material at the cutting tool area (mostly related to the damage on the tool–rake face) [3]. Moreover, the adhesion and abrasion on the clearance surface normally induce flank wear, chipping, and catastrophic failures [17].

Thus, the selection of the tool is critical during the machining of nickel-based superalloys, requiring elevated wear resistance and hardness, high strength, and chemical stability at elevated temperatures [18]. The industry recommendation for the turning of nickel-based alloys involves the use of ceramic and carbide tools, the latter being used in finishing the turning [4].

Concerning the tool coating physical vapor deposition (PVD), TiAlN, ALTiN, or AlCrTiN are widely used for carbide tools in the turning of nickel-based alloys for improving the competitiveness of carbides as opposed to ceramic tools, because of their lower cost [19]. The TiAlN coating in comparison to the TiN coating decreases the machining forces, whereas it improves tool resistance to flank wear because of its chemical inertness, adhesion resistance, high hardness at elevated temperatures (up to 1000 °C), and high oxidation resistance [20,21]. Coated carbide tools are recommended for a medium cutting speed, ranging between 30–70 m/min [22] for the turning of nickel-based alloys, because of its thermomechanical instability [14].

Traditionally, cutting fluids have been used to lubricate (helping to reduce the friction in the area of contact between the chip and the tool), eliminate the chip from the cutting area, and, above all, to eliminate the heat produced during the process of machining, cooling the tool and the workpiece at the interface. Thus, the use of cutting fluids during the machining process (generally between 10–100 L per minute [23]) has a huge impact on the temperature of the tool, its wear evolution, and life, as well as on the surface finishing of the workpiece (roughness of the surface, generation of residual stresses, etc.). However, the impact of cutting fluids on the environment is significant. Therefore, industrial activities are encouraging manufacturers to implement new green techniques, replacing the use of traditional cutting fluids. Moreover, the use of cutting fluids is both harmful to the environment and very expensive, not only for its acquisition, but also for the costs associated with its recovery and disposal management [24].

Sustainability requirements are leading to the use of new vegetable-based cutting fluids that are sustainable, environmentally friendly, biodegradable, and less toxic, and they are becoming a real alternative to petrol-based cutting fluids [25]. Moreover, cutting fluids are normally applied with flood coolant systems (FC), systems that can account for up to 17% of the total production costs [26] and that sometimes do not reach the area of machining because of the obstruction of the chips. Alternatives for the application of cutting fluids have been developed, such as near-dry-machining (NDM) systems, also known as minimum quantity lubrication (MQL) [27] or minimum quantity cutting fluids (MQCF) [28]. However, dry machining that avoids the use of cutting fluid would be the best technique, if possible. Cantero et al. [3,29] analyzed the performance of the carbide and PCBN tools in the dry finishing turning of Inconel 718, obtaining a tool life of 29 min and 2 min, respectively, for competitive cutting conditions, confirming the industrial viability of the carbide inserts but not of the PCBNs in the dry finishing of the Inconel 718.

Few papers are available on the topic of Haynes 282 machinability. Suarez et al. [30] carried out an experimental investigation focusing on the effect of lubricant pressure and material heat treatment on the turning of this alloy. A negligible effect for the high-pressure cooling was observed, while the solution annealing large grain solution (LGS) state presented enhanced machinability when compared to the precipitation hardened large grain aged (LGA) state in terms of force levels and tool wear. Díaz-Álvarez et al. [11] studied the performance of a coated carbide tool during the finishing turning of Haynes 282 with a cutting fluid at the conventional pressure, observing that, for all of the cutting conditions, the tool broke because of the fragile fracture of the cutting edge.

There is a lack of research focusing on the machining of Haynes 282. Moreover, the tool wear analysis of the carbide inserts when machining Haynes 282 in a dry environment has not been studied. The present work deals with the finishing turning of the Haynes 282 alloy in dry conditions. Dry machining tests using coated carbide tools were performed under different cutting conditions in order to evaluate the viability of the cutting fluid removal in finishing turning of Haynes 282 with carbide tools. Roughness, cutting forces, and tool wear were quantified in each test. Although the industrial dry machining of Haynes 282 has not yet been applied, in this study, tool life values similar to those obtained with lubricants under similar conditions in other studies have been obtained for the most favorable conditions in dry conditions.

2. Experimental Setup

2.1. Material Properties and Cutting Tools.

A Haynes 282 alloy was tested in a round bar with a 90 mm diameter shape, which was manufactured following the AMS5951 specification. The Haynes 282 workpiece was annealed at 1135 °C (in the typical range 1121–1149 °C) and age hardened according to the following stages: It was heated up to 1283 °C, maintained at this temperature for 2 h, and then cooled in air. Afterwards, it was heated up to 1061 °C, maintained at this temperature for 8 h, and then cooled in air. The hardness of each specimen tested was quantified at different points, obtaining values that varied between 42.2 and 43.5 HRc. Each element percentage of the Haynes 282 that was tested in the present paper is summarized in Table 1.

Table 1. Haynes 282 chemical composition [11].

Element (%)	Ni	Cr	Fe	Nb	Mo	Ti	Al	Co	Si	Cu	Mn	C
Haynes 282	57	19.42	0.87	<0.01	8.52	2.22	1.41	10.2	<0.05	<0.01	0.06	0.062

A carbide tool (CW, TS200 grade) with a multilayer coating of TiAl/TiAlN, provided by SECO (SECO tools, Fagersta, Sweden), were used for turning tests. These coated carbide tools are especially recommended for finishing turning of Nickel superalloy. Insert presents a tip and honing radius of 0.4 mm and 25 μm respectively, tip angle equal to 80°, rake angle of 16° and a relief angle of 7°. The cutting tool with the code CCMT 09T304F1 was fixed in a tool holder type SCLCR 2525M09JET provided by SECO.

2.2. Experimental Setup and Instrumentation

Haynes 282 turning tests were carried out in a lathe Pinacho Smart turn 6/165 (Pinacho, Castejón del Puente, Spain) equipped with a Kistler 9257B dynamometer (Kistler, Winterthur, Switzerland) for the cutting force measurement (Figure 1).

During the development of the turning tests and at the end of each pass, a rounded surface remained because of the effect of the tool tip radius. Therefore, because of the consequent increase of material needing to be removed in that zone in the next pass, which did not allow for a continuous cut, a sudden increment of undeformed chip cross-section was caused [29]. The finishing operation was characterized by small cutting depths, so this increase in material as a result of the tool tip radius at the end of the pass led to a significant increase in the cutting forces, hence influencing the tool wear. To avoid this phenomenon, a second tool was attached in the tool holder in the lathe (see Figure 1) in order to remove this zone once the cutting force had been stabilized and measured using the tested tool.

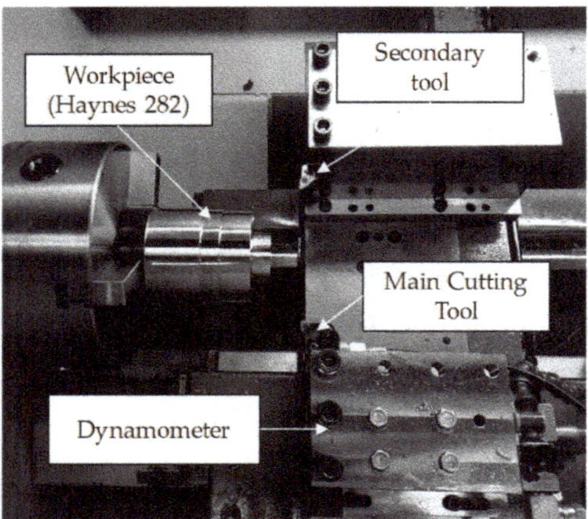

Figure 1. Instrumentation and setup.

The tool wear level was periodically evaluated during the turning tests for each cutting condition, tested by means of obtaining images from a stereo microscope Optika SZR (Optika, Ponteranica BG, Italy). Also, a scanning electron microscopy (SEM) Philips XL-30 (Philips, Eindhoven, Netherlands) with an EDSDX4i system was used to analyze the wear evolution. At the same time, the surface finish of the workpiece was evaluated by means of the surface roughness through a Mitutoyo model SJ-201 (Mitutoyo, Kawasaki, Japan) rugosimeter, obtaining the mean of nine measurements as the representative roughness value.

All of the cutting tests in this study were carried out without any type of coolant by analyzing the finishing turning of Haynes 282 under dry conditions.

As knowledge of the machining of the Haynes 282 alloy at an industrial level is poor, tool manufacturers do not include the relevant information for the selection of the cutting parameters for its process. Nevertheless, in the bibliography, there are general recommendations establishing the ranges for the cutting speed (30–35 m/min), feed rate (0.1–0.18 mm/rev), and depth (1 mm) [31]. Moreover, Díaz-Álvarez et al. [11] investigated the machining of Haynes 282 with carbide tools under a conventional pressure coolant using cutting speeds between 50–90 m/min, feeds between 0.1–0.15 mm/rev, and a depth of pass of 0.25 mm, obtaining a maximum tool life of 33 min. Thus, the cutting parameters selected for the present study are summarized in Table 2.

Table 2. Cutting parameters for the turning tests.

Cutting Speed (m/min)	Feed (rev/min)	Pass Depth (mm)
50	0.1	0.25
	0.15	
70	0.1	
	0.15	
90	0.1	
	0.15	

3. Results and Discussion

3.1. Cutting Forces Analysis

The evolution of the cutting forces—cutting force (F_c), feed force (F_f), and back force (F_p)—were recorded for each preformed test using a frequency of acquisition of 100 Hz. To guarantee the repeatability of the results, each test was performed twice, obtaining deviations lower than 5% with respect to the mean value. Thus, the average values have been used for the subsequent analyses. For the sake of simplicity, in the following analysis, the specific force components (k_c, k_f, and k_p) have been defined as the each of the cutting forces over the undeformed cross section of the chip. In the subsequent points, the results of each component are compared with the observed tool wear damage (notch, chipping, flank, and built up edge).

3.1.1. Fresh Tools Results for the Specific Cutting Force

In Figure 2, the obtained results for each component plus the resultant specific cutting force (k_r) quantified at the first stages of each of the tests through fresh tools are represented. For the series of cutting parameters that were studied, the results of the specific cutting force (k_c) ranged from 3580 N/mm^2 (case: V_c = 90 m/min and feed = 0.15 mm/rev) to 4200 N/mm^2 (case: V_c = 50 m/min and feed = 0.1 mm/rev). The values of the resultant cutting forces that take into account all of the cutting forces components range from 4330 N/mm^2 (case: V_c = 90 m/min and feed = 0.15 mm/rev) to 5700 N/mm^2 (case: V_c = 90 m/min and feed = 0.1 mm/rev).

Cutting Speed vs. Specific Cutting Forces

- For the lowest feed (0.1 mm/rev) used, the specific cutting force (k_c) was not significantly affected by the cutting speed for the studied range. However, the rest of the components increased by up to 26% for the specific feed force (k_f), and up to 100% for the specific back force (k_p) when the cutting speed was increased from 50 m/min to 90 m/min. This behavior was not observed for the feed equal to 0.15 mm/rev, whereas the cutting speed was increased from 50 m/min to 90 m/min, the values of the specific cutting forces were decreased by up to 57%, 70%, and 30% for the specific cutting force (k_c), the specific feed force (k_f), and the specific back force (k_p), respectively. By increasing the cutting speed, the temperature of the material to be cut rose, so that it softened, thus requiring lower cutting forces. At the same time, increasing the cutting speed also increased the strain rate by increasing the resistance of the material to be cut. For a feed of 0.1 mm/rev, it was observed that, because of the higher proportion of chip sections with high levels of deformation, when increasing the cutting speed, the specific cutting forces increased because of the strain hardening effect; however, for the feed value of 0.15 mm/rev, the softening effect of the material, because of the increment of the cutting speed, was the predominant effect.

Feed vs. Specific Cutting Forces

- Regarding the specific cutting force component induced by the feed, reductions of up to 13% for the specific cutting force (k_c), up to 39% for the specific feed force (k_f), and up to 38% for the specific back force (k_p) were recorded with increments on the feed from 0.1 mm/rev to 0.15 mm/rev. The specific cutting force component induced by the feed was as expected. For the lowest feed, as the proportion of material subjected to large deformation (along the cutting edge) was higher, the specific cutting force results were also higher; this tendency can also be verified through the resultant specific cutting force (k_r).

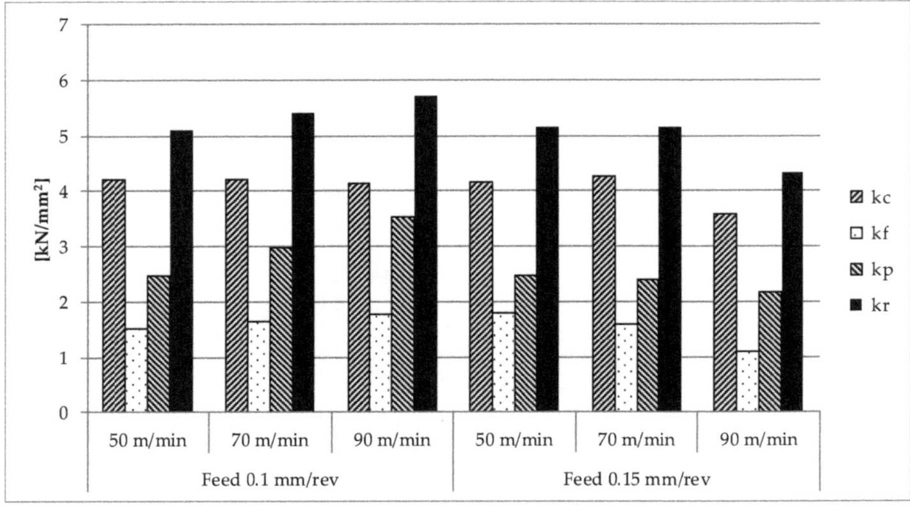

Figure 2. Specific cutting forces for fresh carbide tools at the beginning of the tests.

3.1.2. Specific Cutting Forces Evolution During Haynes 282 Turning

The specific force progression and the resultant specific force for the different components with the cutting time are represented in Figure 3. For all of the cutting conditions, all of the specific cutting force components increased with the time of use of the tool. However, k_c presented an increasing linear trend for the tool life, while the growth of the k_f- and k_p-specific forces showed other trends in all of the cases being highlighted in two regions, as follows: the first one with a linear growth and the second one with a more pronounced increment.

Cutting Speed vs. Specific Cutting Forces

- Cutting speed 50 m/min: As mentioned in the previous paragraph, a slight linear increase in the cutting forces was recorded in the first region. In the subsequent region, the k_p component underwent a drastic increase, contrasting the k_f components with a lower increase. These two regions, clearly identifiable during the tests, exhibited a close relationship with the different wear modes observed. Thus, while for the first region of the force evolution a moderate chipping combined with a progressive erosion of the tool flank was observed, for the second region, the k_f and k_p components of the specific force through the loss of the cutting edge integrity were affected by a more aggressive chipping combined with a rapid progression of the notch. Therefore, a clear trend can be observed, according to which, as the cutting speed increases, both the tool wear rate and consequently the specific forces increase.
- Cutting speed 70 m/min: The evolution of the specific cutting force showed a similar trend to those obtained for the 50 m/min cutting speed. Thus, for k_p, it could be clearly seen in the two

regions, with a drastic increase in the second one, whereas for the k_f component, this increase was not so evident. Therefore, all of the components of the cutting speed exhibited a slight increase during the first region, with the flank wear progression moderated by means of a light chipping. However, the components of force k_f and k_p suffered suspected growth during the second region because of a great deterioration of the cutting edge through the notch and more intense chipping in this final stage.

- Cutting speed of 90 m/min: As in the previous cases, there were two clearly differentiated regions of specific forces of growth, the main difference between them being the use time of the tool, in which the trend change appeared much smaller than for the lower speeds. These two growth zones were also related by a moderate growth of flank wear together with chipping, until the chipping was dominant, progressing in quick increment of the specific force components k_f and k_p.

Feed vs. Specific Cutting Forces

- For all of the conditions analyzed, a remarkable influence of the feed on the evolution of the components of the specific cutting force were not found during the turning tests.

Near the end of the tool life, values up to 10 times of those obtained with a fresh tool were obtained for the specific back force, whereas values up to 7 and 2.5 times were obtained for the feed and specific cutting forces respectively, when compared with the ones obtained for the fresh tool. Therefore, especially for lower cutting speeds, the evolution of the specific back force could be a suitable indicator of tool wear progression. The value of the resultant specific cutting force included in Figure 3 can be used as a more stable variable to evaluate the wear state of the tool.

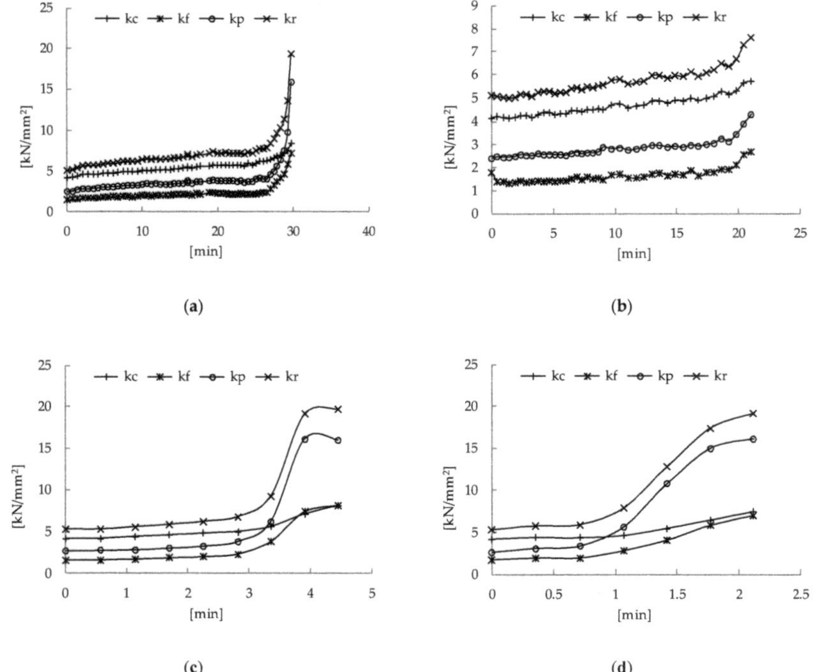

Figure 3. *Cont.*

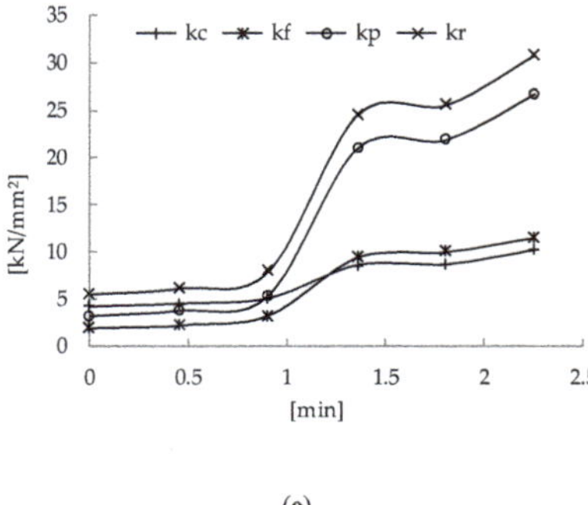

(e)

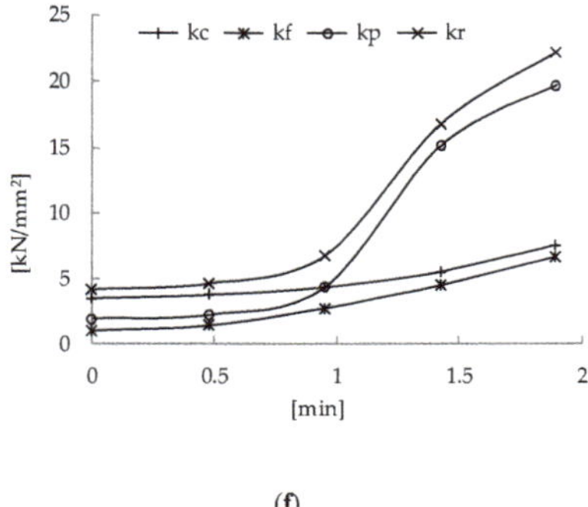

(f)

Figure 3. Specific cutting force evolutions with the times for the different cutting conditions tested. (**a**) V_c = 50 m/min and f = 0.1 mm/rev; (**b**) V_c = 50 m/min and f = 0.15 mm/rev; (**c**) V_c = 70 m/min and f = 0.1 mm/rev; (**d**) V_c = 70 m/min and f = 0.15 mm/rev; (**e**) V_c = 90 m/min and f = 0.1 mm/rev; (**f**) V_c = 90 m/min and f = 0.15 mm/rev.

3.2. Analysis of Wear and Tool Life

During each experiment, the test tool wear progression was checked, with the main wear modes identified being the notch, chipping, flank, and built-up edge (BUE). In order to quantify the wear for all of the cutting conditions analyzed, the tool wear was periodically studied within each test. Tools reached the end of tool life by means of the breakage of the cutting edge, or through the end of tool life criterion, established by means of a notch or flank wear larger than 0.4 mm; however, only one cutting condition reached a value of flank wear close to 0.4 mm. The value of 0.4 mm for the notch or flank wear was established by attending to the behavior of the tools, and for values of flank wear close to 0.4 mm, high increments of the cutting forces and a rapid growth of the chipping wear leading to the catastrophic failure of the cutting edge were observed (see Figure 3, V_c = 50 m/min and a feed of 0.1 mm/rev to check the rapid increments of the cutting forces when the value of the flank wear reached values close to 0.4 mm).

Although both BUE and the adhesion of the material were observed for all of the cutting conditions analyzed, as can be seen in Figure 4, they have not supposed a significant influence on the tool life.

Chipping, together with notch wear, were predominant along the entire cutting edge of the tools at the beginning of the performed tests. The wear progression was similar, regardless of the cutting conditions. Chipping became larger, being filled by material (BUE). Furthermore, the area of the notch that grew throughout the tests was clearly differentiated, and, at the same time, the flank wear progressed. It was found that for higher cutting speeds, the chipping progressed more rapidly, exposing more of the flank surface (causing the flank wear to grow much faster than with lower cutting speeds, where chipping was not so aggressive).

The progression of chipping wear is enhanced with the increment of the cutting speed, and, to a lesser extent, by increasing the feed, causing a reduction in the tool life and leading to the final catastrophic breakage of the cutting edge for all of the cases analyzed. Increasing the feed results in obtaining higher forces and a more unstable cut because of the increase of material that is to be removed in each pass, thus, favoring the appearance of fragile breaks in the cutting edge. Increasing the cutting speed is related to an increase in the temperature at the cutting area [12]. This increase in temperature favors the adhesion of materials in the tool (and the consequent chipping) through a reduction of the tool material strength.

In Table 3, both the tool life values by means of cutting time, the machined surface per time ($S_{mach.t}$) and the machined surface per cutting edge (S_{edge}), quantified through Equations (1) and (2), respectively, have been summarized for all of the cases analyzed [11].

$$S_{mach.t} = V_c \cdot f \cdot 1000/60 \quad (1)$$

$$S_{edge} = S_{mach.t}\ T \cdot 60 \quad (2)$$

where $S_{mach.t}$ is the machined surface per unit time (mm^2/s), S_{edge} is the machined surface per edge (mm^2), V_c is the cutting speed (m/min), f is the feed (mm/rev), and T is the tool life (min).

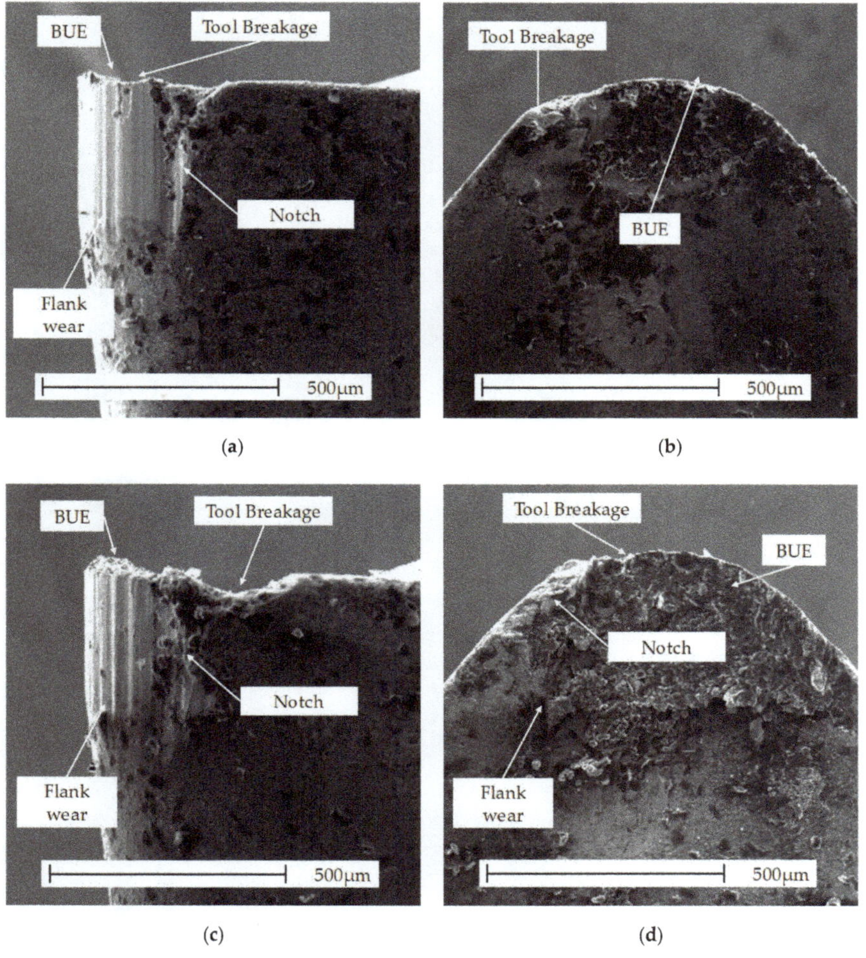

Figure 4. *Cont.*

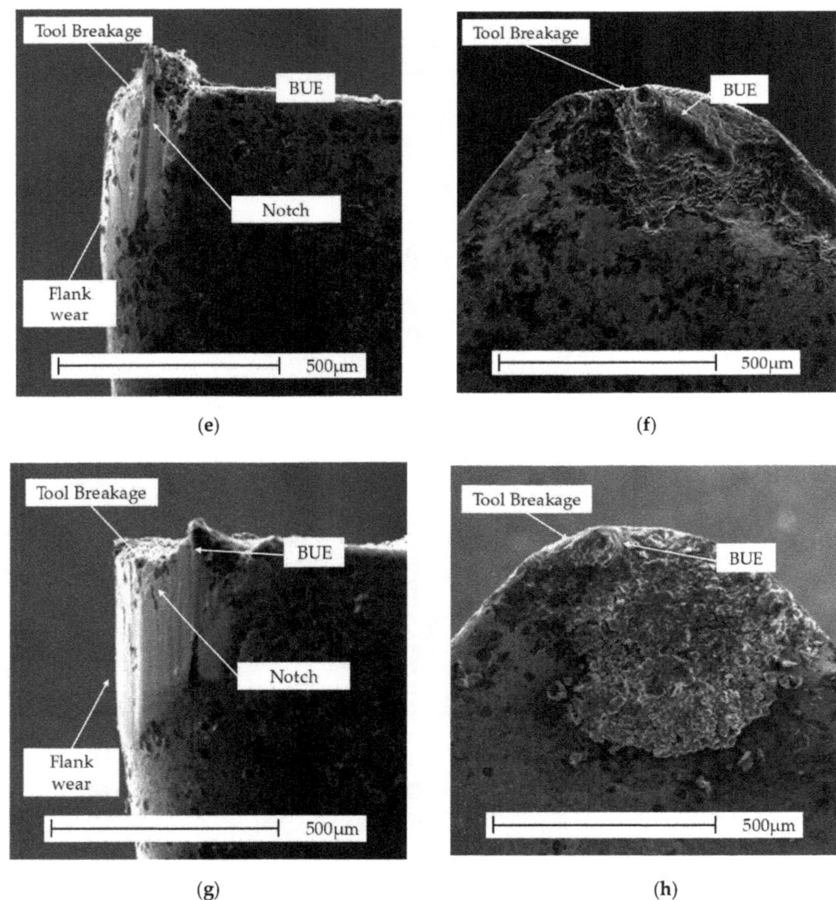

(e) (f)

(g) (h)

Figure 4. Scanning electron microscopy (SEM) images at the end of the tool life for the different conditions tested. V_c = 50 m/min and f = 0.1 mm/rev: (**a**) relief and (**b**) rake surface view. V_c = 50 m/min and f = 0.15 mm/rev: (**c**) relief and (**d**) rake surface view. V_c = 90 m/min and f = 0.1 mm/rev: (**e**) relief and (**f**) rake surface view. V_c = 90 m/min f = 0.15 mm/rev: (**g**) relief and (**h**) rake surface view. BUE—built up edge.

Table 3. Tool life, machined surface per unit time (mm^2/s), and machined surface per edge (mm^2) for the different cutting conditions tested.

Tool	Cutting Speed (m/min)	Feed (mm/rev)	Depth (mm)	Life (min)	Machined Surface per Unit Time (mm^2/s)	Machined Surface per Cutting Edge (mm^2)
Carbide (TS2000, Seco)	50	0.1	0.25	30.1	83.3	150,520
		0.15		21.0	125.0	157,629
	70	0.1	0.25	4.5	117	31,196
		0.15		2.1	175.0	22,266
	90	0.1	0.25	2.3	150.0	20,263
		0.15		1.9	225.0	25,525

For the lowest cutting speed (50 m/min), values of 30.1 and 21 min of tool life for 0.1 mm/rev and 0.15 mm/rev feeds, respectively, were obtained when reaching the point of the highest level of flank extension (almost 0.4 mm), which is near to the end of tool life criterion that has been established (Figure 4a–d). The end of tool life by means of cutting-edge breakage was reached because of predominant chipping.

Wear due to chipping appears during tests at cutting speeds of 70 m/min, compared with the cutting speed of 50 m/min, reaching the end of tool life through a breakage of the cutting edge. Thus, tool life values of 4.5 and 2.1 min for feeds of 0.1 and 0.15 mm/rev, respectively, were obtained during turning tests at 70 m/min.

In the turning tests at 90 m/min cutting speed, wear due to chipping appeared at the first stages, with a fast progression up to cutting edge breakage at the end of tool life (Figure 4e–h). As expected, the lowest tool life values were reported at 90 m/min, obtaining 2.3 and 1.9 min for feeds of 0.15 and 0.1 mm/rev, respectively.

As shown in Table 3, tool life values of 1.9 (V_c = 90 m/min) up to 30.1 min (V_c = 50 m/min) were obtained during the dry turning tests on Haynes 282. As mentioned above, increasing the cutting speed increased the adhesion of the material in the tool and therefore the chipping wear. Thus, the best results in terms of tool life time were those obtained for both low cutting speeds and feeds. The tool life obtained for the cutting speed of 50 m/min and 0.1 mm/rev feed was very close to those obtained by the authors in analogous tests, where a conventional pressure coolant was used [11]. However, a great influence of the cutting speed in the tool life has been found, decreasing its life up to 85% when the cutting speed increases from 50 to 70 for a feed of 0.1 mm/rev, and up to 90% for a feed of 0.15 mm/rev.

The machined surface per cutting edge (known as an indicator for tool industrial performance) at 50 m/min cutting speed was similar for both feeds (0.1 and 0.15 mm/rev), whereas, because of the short tool life derived from increasing the cutting speed from 70 m/min to 90 m/min, there was no significant variation in the machined surface per cutting edge. It is necessary to highlight the important result obtained in terms of the mechanized surface and tool life for the less aggressive tool parameters (50 m/min and 0.1 mm/rev) in dry conditions, these being very similar to those obtained at the conventional coolant pressure [11]. This result makes the use of this type of tool suitable for the finishing the machining of Haynes 282 under dry conditions, which, until today, was done with cutting fluid.

3.3. Analysis of Surface Quality

The surface roughness progression was evaluated at different stages during the development of the tests. The surface quality was measured three times at three different zones over the machined surfaces in terms of the average roughness (R_a). Thus, the maximum value of these measured values for each stage were taken as the value of the roughness for each condition tested (Figure 5).

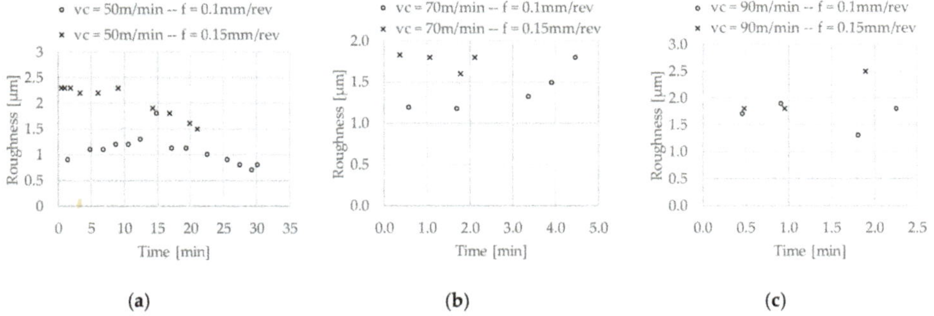

Figure 5. Roughness evolution at the machined surface for all of the cutting conditions tested: (**a**) V_c = 50 m/min, (**b**) V_c = 70 m/min, (**c**) V_c = 90 m/min.

During the first stage of tests, with fresh tools and no significant wear, the values of roughness within the range of 0.7 and 2.5 μm were obtained. It should be noted that the R_a values were reduced with the wear of the tool for a cutting speed of 50 m/min, which is related to the type of wear found, with the flank for this cutting speed evolving progressively, reaching values close to 0.4 mm, causing an artificial increase in the tip radius, resulting in lower values of R_a. However, for higher cutting speeds, the chipping was dominant, as the beginning caused the original honing of the cutting edge, which was not so defined. The authors obtained similar results during the finishing turning of Inconel 718 [32].

The best roughness values were those obtained for the 50 m/min cutting speed and 0.1 mm/rev of the feed. This phenomenon is related to the lower chipping obtained at the beginning for the lowest cutting speed, because a more linear trend was observed in the roughness progression.

On the contrary, for cutting speeds of 70 and 90 m/min, where chipping wear affects the tool more severely from the first moments of the test, it has not been possible to establish a clear trend in the roughness progression.

The feed shows a clear influence on the roughness values obtained, with it being generally greater for higher feeds, regardless of the cutting speed. This result agrees with that which is theoretically expected from the application of Equation (3) [33], namely,

$$R_a = 0.0321 \cdot f / r_c \tag{3}$$

where f is the feed (mm/tooth), and r_c is the tool nose radius (mm).

4. Conclusions

This work dealt with the sustainable finishing turning of Haynes 282 by means of coated carbide tools without cutting fluids (dry condition). Different cutting speeds and feeds for Ni-based alloys were tested in order to quantify the cutting forces, the quality of surface, the wear progression, and the end of tool life. The main contributions of the analysis are summarized below.

- The less aggressive conditions for the tool when working on dry conditions are low cutting speeds, 50 m/min in this study. The effect of the feed is not so significant in terms of the mechanized surface, whereas, on the contrary, it is significant in terms of the tool life, with shorter lives having the greatest feed. This effect must be taken into account for productivity purposes.
- It should be noted that life values similar to those obtained in other studies, with lubricant under similar conditions, have been obtained for the most favorable conditions in dry conditions. This is a significant result, demonstrating the suitability of implementing dry turning in an industrial environment.
- The great influence of the cutting speed in the tool life was demonstrated by decreasing it by 85% when going from 50 to 70 m/min for a feed of 0.1 mm/rev and 90% for a feed of 0.15 mm/rev.
- Great increases in cutting forces have been appreciated for all of the tested conditions, obtaining values for the specific back force at the end of tool life, of 10 times the value obtained when the tool was fresh. It has been possible to observe a clear relationship between this force and the tool life.
- Regarding the wear, since the beginning of the trials, chipping, built up edge (BUE), and notch wear were found. The catastrophic failure of the cutting edge has been found at the end of tool life in all of the cutting conditions tested because of the chipping progression.
- For all of the tests, the R_a values were low, regardless of the time of use of the tool (between 0.7 and 2.5 μm). However, the homogeneity in terms of R_a for the test with a 50 cutting speed and 0.1 mm/rev of feed stands out, which gives one an idea of the stability of this condition in relation to the other cutting parameters, where large variations in this value were found.

Author Contributions: Conceptualization, J.D.-Á. and J.L.C.; data curation, J.D.-Á. and A.D.-Á.; formal analysis, J.D.-Á., A.D.-Á., and J.L.C.; funding acquisition, H.M. and J.L.C.; investigation, J.D.-Á. and A.D.-Á.; project administration, H.M. and J.L.C.; resources, J.L.C.; supervision, H.M. and J.L.C.; validation, J.D.-Á., A.D.-Á., H.M., and J.L.C.; visualization, A.D.-Á.; writing (original draft), J.D.-Á. and A.D.-Á.; writing (review and editing), A.D.-Á., J.D.-Á., and H.M.

Funding: This research was funded by the Ministry of Economy, Industry, and Competitiveness, and the FEDER program, grant number DPI2017-89197-C2-1-R.

Conflicts of Interest: The authors declare no conflict of interest.

References

1. Hu, D.; Wang, T.; Ma, Q.; Liu, X.; Shang, L.; Li, D.; Pan, J.; Wang, R. Effect of inclusions on low cycle fatigue lifetime in a powder metallurgy nickel-based superalloy FGH96. *Int. J. Fatigue* **2019**, *118*, 237–248. [CrossRef]
2. Zhu, D.; Zhang, X.; Ding, H. Tool wear characteristics in machining of nickel-based superalloys. *Int. J. Mach. Tools Manuf.* **2013**, *64*, 60–77. [CrossRef]
3. Cantero, J.L.; Díaz-Álvarez, J.; Miguélez, M.H.; Marín, N.C. Analysis of tool wear patterns in finishing turning of Inconel 718. *Wear* **2013**, *297*, 885–894. [CrossRef]
4. Ezugwu, E.O.; Bonney, J.; Yamane, Y. An overview of the machinability of aeroengine alloys. *J. Mater. Process. Technol.* **2003**, *134*, 233–253. [CrossRef]
5. Wang, F.; Cheng, X.; Liu, Y.; Yang, X.; Meng, F. Micromilling Simulation for the Hard-to-cut Material. *Procedia Eng.* **2017**, *174*, 693–699. [CrossRef]
6. Thakur, A.; Gangopadhyay, S. State-of-the-art in surface integrity in machining of nickel-based super alloys. *Int. J. Mach. Tools Manuf.* **2016**, *100*, 25–54. [CrossRef]
7. Kruger, K.L. *HAYNES 282 Alloy*; Elsevier Ltd.: Amsterdam, The Netherlands, 2016; ISBN 9780081005583.
8. Donachie, M.J. *Superalloys: A Technical Guide*, 2nd ed.; ASM International: New York, NY, USA, 2002; pp. 1–409.
9. Shin, K.Y.; Kim, J.H.; Terner, M.; Kong, B.O.; Hong, H.U. Effects of heat treatment on the microstructure evolution and the high-temperature tensile properties of Haynes 282 superalloy. *Mater. Sci. Eng. A* **2019**, *751*, 311–322. [CrossRef]
10. Rodríguez-Millán, M.; Díaz-Álvarez, J.; Bernier, R.; Cantero, J.; Rusinek, A.; Miguelez, M. Thermo-Viscoplastic Behavior of Ni-Based Superalloy Haynes 282 and Its Application to Machining Simulation. *Metals* **2017**, *7*, 561. [CrossRef]
11. Díaz-Álvarez, J.; Díaz-Álvarez, A.; Miguélez, H.; Cantero, J. Finishing Turning of Ni Superalloy Haynes 282. *Metals* **2018**, *8*, 843. [CrossRef]
12. Díaz-Álvarez, J.; Tapetado, A.; Vázquez, C.; Miguélez, H. Temperature measurement and numerical prediction in machining inconel 718. *Sensors* **2017**, *17*, 1531. [CrossRef]
13. Ezugwu, E.O. Key improvements in the machining of difficult-to-cut aerospace superalloys. *Int. J. Mach. Tools Manuf.* **2005**, *45*, 1353–1367. [CrossRef]
14. Choudhury, I.; El-Baradie, M. Machinability of nickel-base super alloys: A general review. *J. Mater. Process. Technol.* **1998**, *77*, 278–284. [CrossRef]
15. Muñoz-Sánchez, A.; Canteli, J.A.; Cantero, J.L.; Miguélez, M.H. Numerical analysis of the tool wear effect in the machining induced residual stresses. *Simul. Model. Pract. Theory* **2011**, *19*, 872–886. [CrossRef]
16. Urbikain, G.; de Lacalle, L.N.L. Modelling of surface roughness in inclined milling operations with circle-segment end mills. *Simul. Model. Pract. Theory* **2018**, *84*, 161–176. [CrossRef]
17. Hoier, P.; Malakizadi, A.; Stuppa, P.; Cedergren, S.; Klement, U. Microstructural characteristics of Alloy 718 and Waspaloy and their influence on flank wear during turning. *Wear* **2018**, *400–401*, 184–193. [CrossRef]
18. Thellaputta, G.R.; Chandra, P.S.; Rao, C.S.P. Machinability of Nickel Based Superalloys: A Review. *Mater. Today Proc.* **2017**, *4*, 3712–3721. [CrossRef]
19. Grzesik, W.; Niesłony, P.; Habrat, W.; Sieniawski, J.; Laskowski, P. Investigation of tool wear in the turning of Inconel 718 superalloy in terms of process performance and productivity enhancement. *Tribol. Int.* **2018**, *118*, 337–346. [CrossRef]
20. Thakur, D.G.; Ramamoorthy, B.; Vijayaraghavan, L. Study on the machinability characteristics of superalloy Inconel 718 during high speed turning. *Mater. Des.* **2009**, *30*, 1718–1725. [CrossRef]

21. Bouzakis, K.D.; Michailidis, N.; Skordaris, G.; Bouzakis, E.; Biermann, D.; M'Saoubi, R. Cutting with coated tools: Coating technologies, characterization methods and performance optimization. *CIRP Ann. Manuf. Technol.* **2012**, *61*, 703–723. [CrossRef]
22. Olovsjö, S.; Nyborg, L. Influence of microstructure on wear behaviour of uncoated WC tools in turning of Alloy 718 and Waspaloy. *Wear* **2012**, *282–283*, 12–21. [CrossRef]
23. Goindi, G.S.; Sarkar, P. Dry machining: A step towards sustainable machining—Challenges and future directions. *J. Clean. Prod.* **2017**, *165*, 1557–1571. [CrossRef]
24. Devillez, A.; Schneider, F.; Dominiak, S.; Dudzinski, D.; Larrouquere, D. Cutting forces and wear in dry machining of Inconel 718 with coated carbide tools. *Wear* **2007**, *262*, 931–942. [CrossRef]
25. Gajrani, K.K.; Suvin, P.S.; Kailas, S.V.; Sankar, M.R. Hard machining performance of indigenously developed green cutting fluid using flood cooling and minimum quantity cutting fluid. *J. Clean. Prod.* **2019**, *206*, 108–123. [CrossRef]
26. Amiril, S.A.S.; Rahim, E.A.; Syahrullail, S. A review on ionic liquids as sustainable lubricants in manufacturing and engineering: Recent research, performance, and applications. *J. Clean. Prod.* **2017**, *168*, 1571–1589. [CrossRef]
27. Gutnichenko, O.; Bushlya, V.; Bihagen, S.; Ståhl, J.-E. Influence of GnP additive to vegetable oil on machining performance when MQL-assisted turning Alloy 718. *Procedia Manuf.* **2018**, *25*, 330–337. [CrossRef]
28. Gajrani, K.K.; Ram, D.; Ravi Sankar, M. Biodegradation and hard machining performance comparison of eco-friendly cutting fluid and mineral oil using flood cooling and minimum quantity cutting fluid techniques. *J. Clean. Prod.* **2017**, *165*, 1420–1435. [CrossRef]
29. Cantero, J.; Díaz-Álvarez, J.; Infante-García, D.; Rodríguez, M.; Criado, V. High Speed Finish Turning of Inconel 718 Using PCBN Tools under Dry Conditions. *Metals* **2018**, *8*, 192. [CrossRef]
30. Suárez, A.; Veiga, F.; de Lacalle, L.N.L.; Polvorosa, R.; Wretland, A. An investigation of cutting forces and tool wear in turning of Haynes 282. *J. Manuf. Process.* **2019**, *37*, 529–540. [CrossRef]
31. Pike, L.M.; Weatherhill, A.E. *HAYNES® 282TM Alloy: A New Wrought Superalloy Designed for Improved Creep Strength and Fabrication Ability*; ASME: New York, NY, USA, 2006; pp. 1031–1039.
32. Díaz-Álvarez, J.; Criado, V.; Miguélez, H.; Cantero, J. PCBN Performance in High Speed Finishing Turning of Inconel 718. *Metals* **2018**, *8*, 582. [CrossRef]
33. Asiltürk, I.; Çunkas, M. Expert Systems with Applications Modeling and prediction of surface roughness in turning operations using artificial neural network and multiple regression method. *Expert Syst. Appl.* **2011**, *38*, 5826–5832. [CrossRef]

Article

On the Surface Quality of CFRTP/Steel Hybrid Structures Machined by AWJM

Fermin Bañon [1,*], Bartolome Simonet [2], Alejandro Sambruno [1], Moises Batista [1] and Jorge Salguero [1]

1 Faculty of Engineering, Mechanical Engineering and Industrial Design Department, University of Cadiz, Av. Universidad de Cadiz 10, Puerto Real, 11519 Cadiz, Spain; alejandro.sambruno@uca.es (A.S.); moises.batista@uca.es (M.B.); jorge.salguero@uca.es (J.S.)

2 Nanotures SL, C. Inteligencia 19, Tecnoparque Agroalimentario, Jerez de la Frontera, 11591 Cadiz, Spain; bartolome.simonet@nanotures.com

* Correspondence: fermin.banon@uca.es; Tel.: +34-956-48-32-91

Received: 23 June 2020; Accepted: 18 July 2020; Published: 21 July 2020

Abstract: The joining of dissimilar materials in a hybrid structure is a line of research of great interest at present. Nevertheless, the machining of materials with different machinability requires specific processes capable of minimizing defectology in both materials and achieving a correct surface finish in terms of functional performance. In this article, abrasive water jet machining of a hybrid carbon fiber-reinforced thermoplastics (CFRTP)/Steel structure and the generated surface finish are studied. A parametric study in two stacking configurations (CFRTP/Steel and Steel/CFRTP) has been established in order to determine the range of cutting parameters that generates the lowest values in terms of arithmetic mean roughness (*Ra*) and maximum profile height (*Rz*). The percentage contribution of each cutting parameter has been identified through an ANOVA analysis for each material and stacking configuration. A combination of 420 MPa hydraulic pressure with an abrasive mass flow of 385 g/min and a travel speed of 50 mm/min offers the lowest *Ra* and *Rz* values in the CFRTP/Steel configuration. The stacking order is a determining factor, obtaining a better surface quality in a CFRTP/Steel stack. Finally, a series of contour diagrams relating surface quality to machining conditions have been obtained.

Keywords: AWJM (abrasive water jet machining); CFRTP (carbon fiber-reinforced thermoplastics); hybrid structure; surface quality; Ra; Rz; C/TPU (carbon/thermoplastic polyurethane)

---

1. Introduction

Carbon fiber-reinforced thermoplastics (CFRTP) composites have an excellent weight-to-mechanical property ratio and high impact and corrosion resistance [1]. These are very interesting materials due to their ability to be remolded after curing, adopting new geometries and being of great interest for mass production [2]. Applications such as chassis in the automotive sector or the development of a lighter fuselage with better fatigue resistance developed by the company STELIA are an example of the current interest in these materials. In addition, in comparison with thermosets (CFRP), they have shorter production times and the possibility of storing the matrix at room temperature which reduces the final costs [3]. Within the wide range of thermoplastic polymers, thermoplastic polyurethane (TPU) can achieve high performance in service. The shaping of this matrix together with carbon fibers results in a flexible compound that can be adapted to various uses [4].

In order to increase the performance of these materials, current research is focused on their bonding with metal alloys in the form of hybrid structures through laser welding or friction stir welding processes [5]. These materials are essential elements in structural applications in the industry nowadays due to their mechanical properties, lightness, and corrosion resistance. Nevertheless, they must be joined to metallic elements to obtain a more robust structure that combines the performance of both materials in the form of a hybrid design [6]. Furthermore, in terms of production, the manufacture and subsequent machining of both materials at the same time means a reduction in operating times. A kind of hybrid structure of great interest is the union of CFRTP with a steel in order to obtain a structural element of high performance. This turns them into elements of great interest for the automotive sector where weight reduction is required and, at the same time, their ability to produce them in mass [2,7].

Their combination with a structural steel allows to obtain a hybrid structure of excellent performance and lightness minimizing energy consumption and CO_2 emissions. Nevertheless, the quality of the interlayer of these materials on applying an adhesive or thermal bonding process has been studied due to the formation of thermal defects or the formation of bubbles due to poor surface preparation of the steel [8,9]. In addition, according to the selected process, the thermoplastic matrix of the CFRTP can be used as an integrating element of the hybrid structure affecting the final thickness [10].

In addition, to achieve a final geometry, specific machining processes are required due to the anisotropic behavior of these materials, as well as the low glass transition temperature of some thermoplastics [11]. Inside conventional processes, such as drilling or milling, wear caused to the cutting edges increases the final costs and reduces the efficiency of these processes. Processes such as milling generate a smooth and clean surface with Rz values close to 9 μm and Ra values close to 2 μm [12,13]. This is in line with the results obtained in the conventional machining of thermoset composite materials, where results below 3.2 μm are required due to aeronautical tolerances [14].

Nevertheless, although very low values are obtained, the machining temperatures deteriorate the thermoplastic matrix and cause delaminations in regions in which the reinforcement is left unprotected [12]. In addition, the fact of machining two materials of different machinability requires the use of specific cutting geometries and complex machining strategies with change of cutting parameters in the interlayer. This means an increase in operating time and costs.

On the contrary, within nonconventional technologies, abrasive water jet machining (AWJM) has proved to be a very effective technology for machining this type of material [13,15,16]. It is a flexible process, capable of achieving high material removal rates and low cutting forces and machining different materials at the same time. In addition, due to the nature of the cutting process, the temperatures reached are very low, which minimizes thermal defects [17]. Furthermore, it is a clean and environmentally friendly technology, a fundamental aspect within the field of "Green machining," and does not generate suspended particles that could affect the health of the operators. This technology offers advantages such as the recovery of abrasive particles after machining, which can be reused after treatment, and no harmful gases are generated [18]. Another important point is the retention of particles of the machined material in the pool pit, especially in composite materials, preventing them from remaining in suspension, avoiding the exposure of the operators to a harmful atmosphere. The cutting tool is water, which can be reused after machining, and abrasive particles, which can be recovered later and treated to be reused.

Nevertheless, there is little literature on abrasive water jet machining of dissimilar material stacks. Most focus on machining FMLs or CFRP/Titanium stacks due to their relevance within the aeronautical industry [19–23]. These studies are focused on the influence of the parameters that govern abrasive water jet machining on the surface quality generated, as well as on the defectology associated with this process such as the taper angle. Although there is literature on water jet machining of thermoplastic composites [24] and steel alloy [25], little information exists on machining these materials in the form of a hybrid structure.

A crucial aspect in the machining processes is the surface quality obtained. The divergence of the water jet during machining generates a reduction in the kinetic energy that results in regions of different surface quality. Thus, in the initial moments of machining, the overlapping of the abrasive particles generates a highly eroded zone known as IDR (initial damage region) [26]. Subsequently, the convergence of the jet generates a stable zone in which a homogeneous surface is obtained, known as the SCR (smooth cutting region) [27]. Finally, due to the reduction of the kinetic energy and cutting capacity of the water jet, the surface generated has a region with very high irregularities in the form of grooves known as RCR (rough cutting region) [28,29].

In this sense, the stacking order of the hybrid structure has a fundamental role to play in the development of these regions. Depending on which material receives the first impact of the water jet, the final quality obtained in the second material will be conditioned by the difference in its machinability [23].

For this reason, minimizing water jet divergence during machining is a key aspect for achieving an acceptable surface quality. Thus, the correct selection of cutting parameters according to the order of stacking is essential to achieve this objective [30]. In studies carried out, the difference in results obtained between the composite material and the metal alloy when they are machined is also highlighted. This is due to the fact that the composite material is anisotropic and the water jet is able to eliminate the matrix generating greater irregularities or the formation of delaminations. In contrast, the isotropy of the metal alloy allows a more stable and smoother cut, but is more affected by the effect of the abrasive particles in the initial zone (IDR).

Within the cutting parameters, traverse speed and hydraulic pressure seem to be the most relevant [31]. Increases in traverse speed lead to increased water jet divergence, especially in the second material, significantly raising the roughness in the material [25]. Pahuja et al. [30] also explains the importance of traverse speed in water jet machining of a hybrid CFRP/Ti structure. Here, by increasing the speed from 1 to 10 mm/min, the Ra values increase by 14% for the titanium alloy and 260% for the composite material. On the other hand, an increase in hydraulic pressure increases the jet's machining capacity, allowing for an improvement in surface quality and obtaining *Ra* values that are very close to each other [22].

Nevertheless, the combination of the different machinability between materials, their stacking order, and the lack of knowledge about a range of cutting parameters capable of machining this type of hybrid structures require further studies. Due to this, this article proposes a parametric study in abrasive water jet machining of a hybrid CFRTP/Steel structure. The surface quality in terms of *Ra* and *Rz* has been evaluated in each material for the two stacking configurations (CFRTP/Steel and Steel/CFRTP). The difference in results between the two configurations has been evaluated, as well as a range of cutting parameters that improve surface quality has been determined. The machining of dissimilar materials in the form of a hybrid structure generates a difference in surface quality between the two elements that must be minimized or eliminated. Nevertheless, the study of the surface quality in both materials separately has been the main objective of this work. Thus, a range of cutting parameters that improves the surface quality has been determined. In order to obtain a homogeneous cut, a ratio between the results obtained for the composite material and the metal alloy has been established to identify which variation in cutting parameters generates the greatest difference in surface quality. Finally, an ANOVA analysis and a set of contour diagrams using predictive mathematical models have been obtained for the most relevant roughness parameter.

2. Methodology

2.1. Materials

This article focuses on the machining of a hybrid structure in order to evaluate the surface quality obtained. The materials selected to obtain this structure were a thermoplastic composite material reinforced with carbon fiber (Twill 200 g/m^2) and a steel alloy S275. The main characteristics of the composite material are shown in Table 1. Reinforced thermoplastic laminates were produced by hot compression molding.

The CFRTP used has a thermoplastic polyurethane matrix with a melting temperature of 145° and the final thickness of 2.1 mm and is composed of 7 layers in 0° and 90° orientations.

Table 1. Mechanical properties of thermoplastic composite material (carbon fiber-reinforced thermoplastics, CFRTP).

Tensile Strength (MPa)	Tensile Modulus (GPa)	Flexural Strength (MPa)	Flexural Modulus (GPa)	Compression Strength (MPa)	ILSS (MPa)
749	27.8	640	8.4	136	9.8

On the other hand, the thickness of the steel was 3 mm, obtaining a final thickness of the hybrid structure of 5.1 mm. This carbon steel is a structural type with wide applications in the industrial sector due to its mechanical properties, and it is of great interest to combine it with a composite material in order to obtain a hybrid structure. Its main characteristics are shown in Table 2.

Table 2. Characteristics of S275 steel.

%C	%Fe	%Mn	%P	%S	%Si	Yield Strength (MPa)	Tensile Strength (MPa)
0.25	98.01	1.60	0.04	0.05	0.05	275	450

The bonding between these materials was carried out by thermoforming in a hot plate press with the aim of obtaining a continuous and quality bond to avoid the formation of delaminations in the interlayer and to relate possible defects to the machining conditions. The characteristics of the thermoplastic matrix allow the matrix itself to be used as an integrating element between both materials by changing from a solid to a liquid state when its glass transition temperature is exceeded. Subsequently, the matrix expands and impregnates the surface of the steel alloy to generate a constant bond after it has cooled down. To ensure a quality bond, the steel surface is modified by sand-blasting, using a pressure of 5 bar, 630 μm corundum particles, and an impact distance of 100 mm. Because of it, a surface free energy value of 50 mJ/m^2 [32] was obtained.

2.2. Abrasive Water Jet Machining (AWJM)

The equipment used consisted of a water jet cutting machine (TCI Cutting, BP-C 3020, Valencia, Spain). The nozzle of the machine had a diameter of 0.8 mm, an orifice diameter of 0.3 mm, and a nozzle length of 94.7 mm. The AWJM machine was equipped with an ultrahigh capacity pump (KMT, 158 Streamline PRO-2 60, Bad Nauheim, Germany). All trials were carried out by 120 mesh Indian Garnet abrasive particles.

Three cutting parameters were modified according to the literature consulted. Three levels of hydraulic pressure (P), traverse speed (TS), and abrasive mass flow (AMF) were established (Table 3). At the same time, in order to obtain a greater robustness and repeatability in the results obtained, each combination of cutting parameters were carried out twice, obtaining a total of 54 tests. Due to its importance in the conservation of kinetic energy of the water jet, the jet-piece distance was set at 3 mm.

Table 3. Cutting parameters for abrasive water jet texturing.

Hydraulic Pressure—P (MPa)	Abrasive Mass Flow—AMF (g/min)	Traverse Speed—TS (mm/min)
250, 340, 420	225, 340, 385	50, 100, 300

The tests consisted of slots with a machining length of 30 mm and a gap between cuts of 6 mm. In order to guarantee a constant flow and traverse speed, the cuts were started 15 mm before the beginning of the material (Figure 1).

On the other hand, the order of the materials during machining is a key parameter in the final quality. Because of this, each test has been performed in two stacking configurations, CFRTP/Steel, and Steel/CFRTP.

Figure 1. Abrasive water jet machining of a hybrid structure Steel/carbon fiber-reinforced thermoplastics (CFRTP).

2.3. Surface Quality Evaluation

The surface finish after machining processes is a key parameter in determining the functional performance of the geometry obtained. The importance of the surface of the parts in the functional behavior of the latter is relevant when considering that it is through their surfaces that contact is established between them, being the main basis of most mechanical functions [33].

Surface quality in abrasive water jet machining is a key factor. The loss of kinetic energy of the water jet in combination with poor selection of cutting parameters produces large variations. This results in a first region that is highly affected by abrasive particles and a final zone of high roughness due to the formation of grooves. This in combination with the fact that dissimilar materials are being machined at the same time makes it an essential parameter to study.

Due to this, the surface quality has been evaluated in terms of arithmetic mean roughness (*Ra*) and maximum profile height (*Rz*). The anisotropy of the composite material and the possible formation of defectology associated with the loss of thermoplastic matrix requires the study of several parameters that provide more complete and real information about the surface obtained [34]. Three measurements were made at three height levels in each material (Figure 2). Each roughness profile was measured at three different levels, i.e., at 25%, 50%, and 75% of the thickness of each material. The goal was to determine the presence of the three characteristic regions in abrasive water jet machining in terms of surface quality: IDR (initial damage region), SCR (smooth cutting region), and RCR (rough cutting region). The measurements were made in a perpendicular direction to the grooves generated by the water jet.

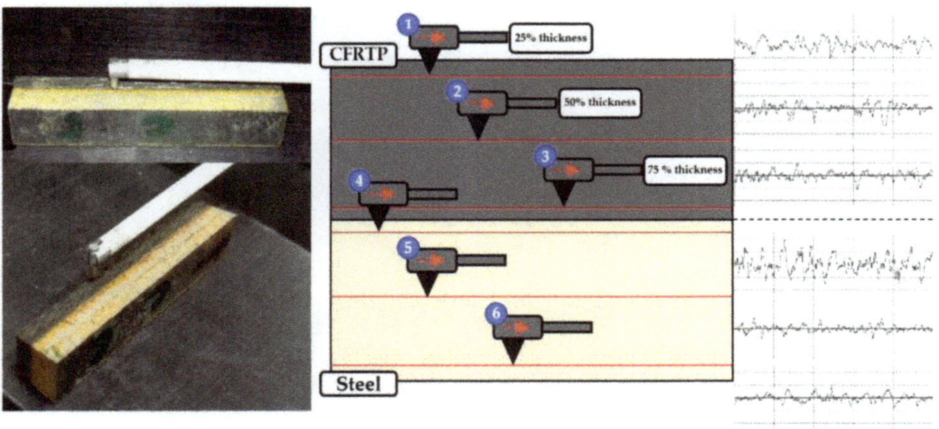

Figure 2. Graphical representation of surface quality evaluation.

A roughness-meter (Mahr Perthometer PGK 120, Göttingen, Germany) was used. The surface quality evaluation was carried out following ISO 4288:1999 standard. A cut-off of 2.5 mm was established for a total evaluation length of 12.25 mm. Stylus with 2 μm tip radius and 90° tip angle was used for the measurements, reference M-250 from Mahr.

Finally, the surface generated after the surface modification was evaluated by visual inspection using a scanning electron microscope (Hitachi, VP-SEM SU1510, Schaumburg, IL, USA).

3. Results and Discussion

3.1. CFRTP/Steel

This section shows the results obtained in the first CFRTP/Steel configuration in order to determine the influence of the cutting parameters. The surface quality obtained in the thermoplastic composite material in terms of Ra is shown in Figure 3. For pressures of 250 MPa, two tendencies are observed when increasing the abrasive flow. When the traverse speed is 50 mm/min, high values of abrasive increase the roughness. This may be due to an excess amount of particles impacting the surface. In combination with reduced pressure, the jet does not have enough energy to obtain a clean cut (Figure 4). Thus, the intercollisions of the abrasive particles reduce its cutting capacity, producing a more eroded zone [35].

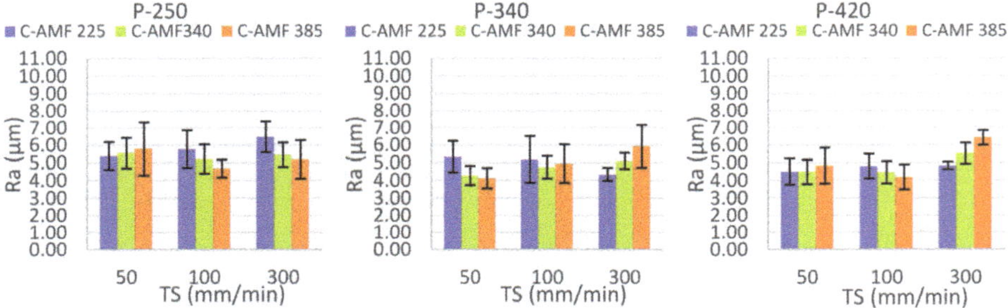

Figure 3. CFRTP surface quality results in terms of arithmetic mean roughness (*Ra*) as a function of the cut-off parameters set for the CFRTP/Steel configuration.

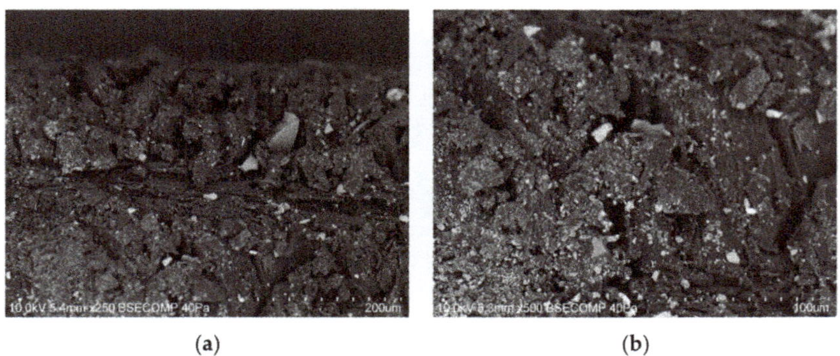

(**a**) (**b**)

Figure 4. Initial damaged region (IDR) due to the erosive effect of the abrasive particles (hydraulic pressure (P) 250 MPa, traverse speed (TS) 300 mm/min, and abrasive mass flow (AMF) 385 g/min) at: (**a**) 250× and (**b**) 500×.

On the other hand, when the traverse speed increases its value, this trend for a hydraulic pressure of 250 MPa is totally opposite due to the increase in the kinetic energy of the water jet. This produces a rougher RCR zone due to the curvature of the water jet due to the loss of kinetic energy [36].

Nevertheless, as hydraulic pressure increases, the surface quality is directly influenced by the traverse speed, especially for pressures of 420 MPa.

Thus, an increase in pressure and abrasive flow at reduced traverse speeds produces a stable water jet capable of homogeneously machining all composite material [37].

When the traverse speed is maximum, the increase in pressure has a varying effect depending on the amount of abrasive particles applied in the machining. With a minimum flow rate of 225 g/min, there is a significant increase in the machining capacity of the water jet, obtaining a more constant material removal and reducing the *Ra* values from 7 to almost 5 μm.

The increase of the abrasive particles improves the cutting capacity of the jet allowing a smoother region. This can be seen in the pressure of 250 MPa. However, a combination of high values of both hydraulic pressure and abrasive flow can be excessive resulting in a deterioration in surface quality. This may be due mainly to an excess of abrasive particles that intercollide, minimizing their erosive effect. In turn, this produces an increase in the IDR region increasing the abrasive particles adhered in the initial moments of machining and increasing the average roughness [15]. This in combination with the divergence between the inlet and outlet zones of the water jet due to a very high travel speed results in these variations when hydraulic pressure is increased.

Also, an increased pressure leads to a greater offset between the inlet area of the water jet and the outlet area. During this time, the kinetic energy of the water jet allowing the material to be machined is reduced. Thus, the impacts generated on the surface are more abrupt, generating irregularly shaped machined areas [26]. This could produce an area of higher roughness and poorer surface quality (Figure 5). These roughness profiles were made at three different levels as shown in Figure 2. The first profile at a distance equivalent to 25% of the thickness, the second profile at 50% of the thickness, and the third profile at 75% of the thickness of the material were obtained.

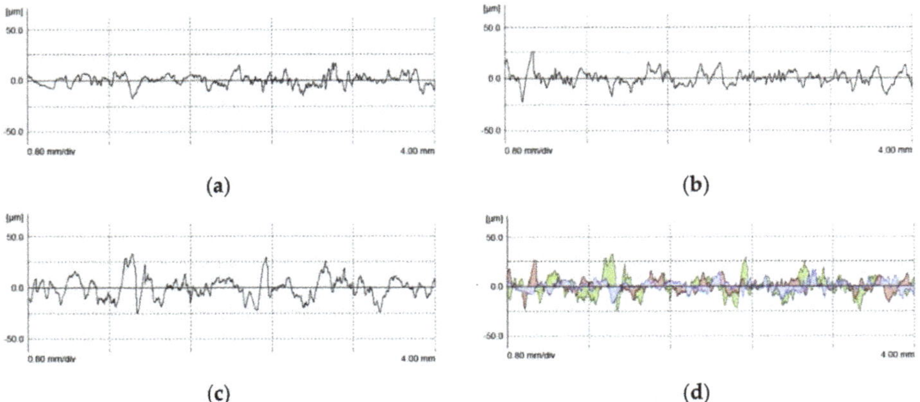

Figure 5. Roughness profiles for the combination of *P* of 420 MPa, *TS* of 300 mm/min, and *AMF* of 225 g/min for: (**a**) IDR, (**b**) smooth cutting region (SCR), (**c**) rough cutting region (RCR) region, and (**d**) overlapping roughness profiles.

This is corroborated by the results obtained for the parameter *Rz* (Figure 6). The trends obtained are very close to those obtained for Ra, which would justify the previously described trends. Thus, it can be seen that a combination of a pressure of 420 MPa, an abrasive flow of 385 g/min, and a traverse speed of 100 mm/min generate the minimum values of *Ra* and *Rz*.

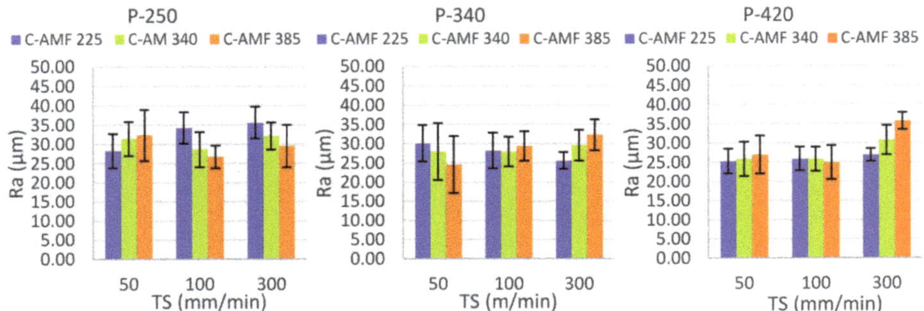

Figure 6. Maximum profile height (*Rz*) values for CFRTP in CFRTP/Steel configuration.

In addition, the surface quality of the steel in terms of Ra is shown in Figure 7 and in terms of Rz in Figure 8. It can be seen that lower values are obtained in the metal alloy compared to the composite material. This is due to the composition of both materials. The anisotropy of the composite material causes each layer to behave differently when interacting with the water jet [30].

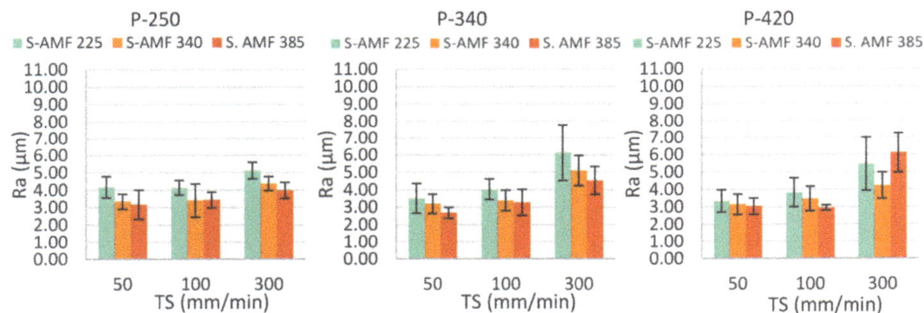

Figure 7. Steel surface quality results in terms of *Ra* as a function of the cut-off parameters set for the CFRTP/Steel configuration.

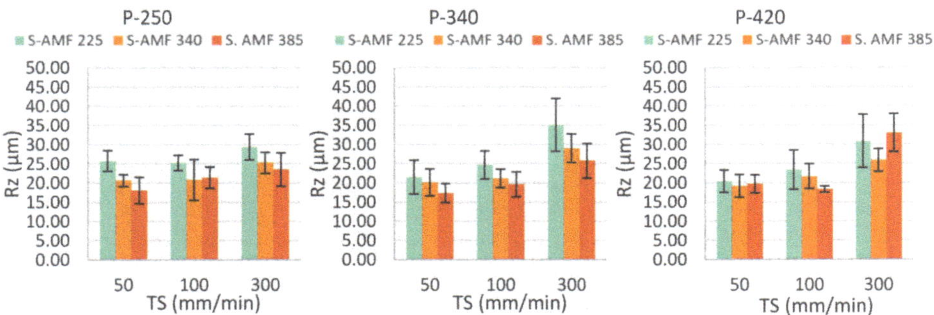

Figure 8. *Rz* values for steel in CFRTP/Steel configuration.

Furthermore, the different machinability of the reinforcement and matrix in combination with the dispersion of the water jet produces an effect known as hydrodistortion [21] (Figure 9). In addition, the main removal mechanism in the composite material was by microbending and fracture and in the matrix was by erosion. This results in a transversal removal of the matrix, leaving the reinforcement unprotected and generating a worsening of the surface quality (Figure 10).

Due to the monolithic composition of the steel, a more homogeneous machining has been obtained. This produces a smoother surface compared to composite material. However, as the second material is machined in the CFRTP/Steel configuration, a reverse situation is generated. A combination of a *TS* of 300 mm/min and an *AMF* of 340 g/min shows this effect. The divergence of the water jet between the inlet and outlet region is very high due to the destabilization of the water jet at this travel speed. This is enhanced by the difference in machinability between the two materials. In other words, the water jet is not capable of machining both materials consistently at the same time.

In addition, the reduced amount of abrasive particles minimizes the actual machinability of the water jet, resulting in a rougher surface [22].

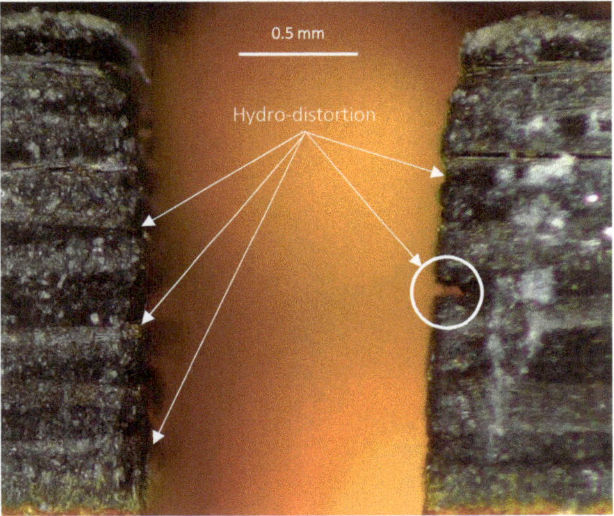

Figure 9. Hydrodistortion defect at a pressure of 250 MPa, an abrasive mass flow of 225 g/min, and a travel speed of 50 mm/min.

Figure 10. Loss of thermoplastic matrix leaving the reinforcement unprotected at 1000× (*P* of 250 MPa, *TS* of 300 mm/min, and *AMF* of 385 g/min).

Due to the low machinability of steel, the increased abrasive mass flow has a positive effect on the surface quality. An increase in particles enhances the erosive capacity of the water jet allowing it to penetrate the steel more easily due to greater stabilization in the cut [38]. This can be seen in most tests where an increase in this parameter generates a reduction in both *Ra* and *Rz*.

The trend of each cutting parameter in the surface quality generated in terms of Ra for each material is shown in Figure 11. In terms of hydraulic pressure, steel is the most important material as it is more difficult to machine and is the second material in the structure. An increase in this parameter improves the penetration capacity of the water jet, facilitating shear impacts on the surface and obtaining a better surface quality. An increase in the amount of abrasive particles reduces the resistance of the material when machined.

Nevertheless, depending on the level of pressure and traverse speed, it can become a negative aspect. On the contrary, the traverse speed seems to be the most critical parameter for surface quality. The increase from 50 to 300 mm/min produces a 40% increase in both materials due to the destabilization of the water jet and the inability to machine both materials at the same time.

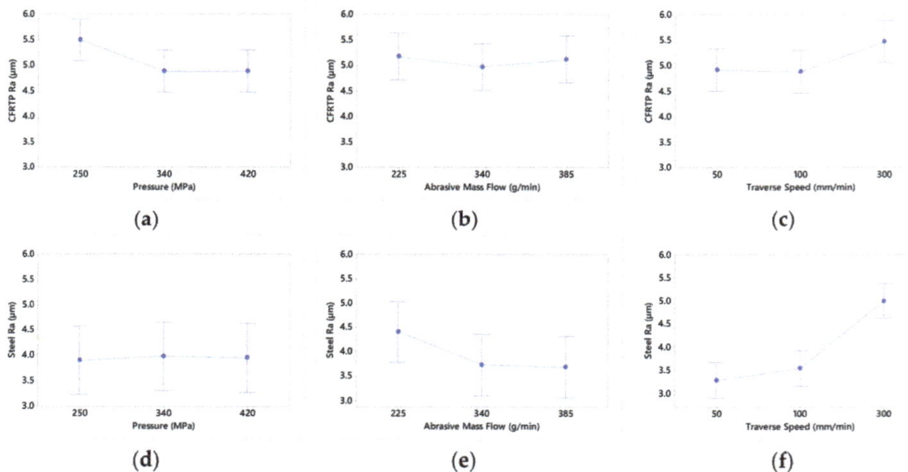

Figure 11. Cutting parameter trends in surface quality (*Ra*) in CFRTP/Steel configuration. (**a**) CFRTP Ra in function of *P*, (**b**) CFRTP Ra in function of *AMF*, (**c**) CFRTP Ra in function of *TS*, (**d**) Steel Ra in function of *P*, (**e**) Steel Ra in function of *AMF*, and (**f**) Steel Ra in function of *TS*.

3.2. Steel/CFRTP

The results obtained in the reverse stacking order for the composite material are shown in Figures 12 and 13.

Compared to the CFRTP/Steel configuration, the results obtained are slightly higher. This may be due to the positioning within the hybrid structure. When machining the metal alloy, a large part of the kinetic energy of the water jet is absorbed by this material, reducing the ability to penetrate the composite material [22,26]. This is especially outstanding when the pressure is minimal (250 MPa) and the amount of abrasive particles is insufficient. This produces an increase in the hydrodistortion effect between reinforcement and matrix resulting in a very rough surface where the reinforcement is not properly machined, generating high deviations (Figure 14).

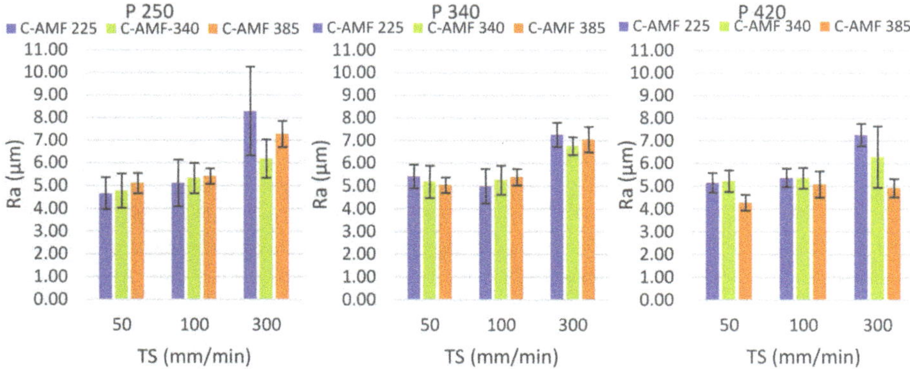

Figure 12. CFRTP surface quality results in terms of *Ra* as a function of the cutting parameters set for the inverse Steel/CFRTP configuration.

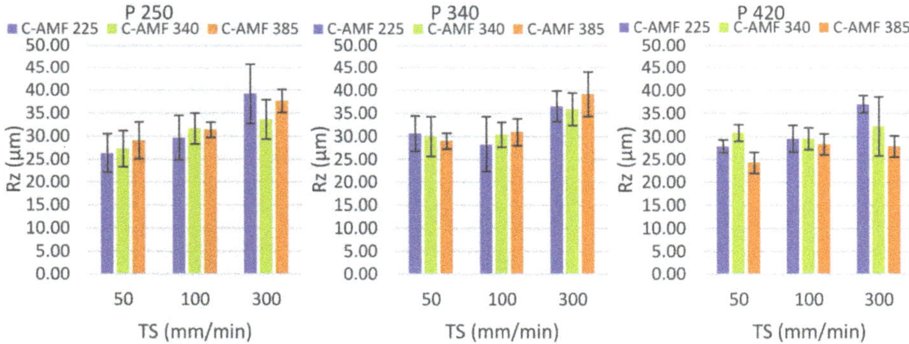

Figure 13. *Rz* values for CFRTP in Steel/CFRTP configuration.

Figure 14. Reinforcements that are not machined, resulting in a worse surface quality.

The formation of a turbulent jet in the interlayer and the consequent loss of power of the water jet results in an unstable flow that generates a very rough area. This, in combination with the low cohesion between the reinforcement and the thermoplastic matrix, generates a separation between both leaving the reinforcement unprotected and increasing the final roughness [39]. This can be seen in Figure 15.

Again, similar trends are observed in both *Ra* and *Rz*. However, in contrast to the CFRTP/Steel configuration, both hydraulic pressure and abrasive mass flow do not seem to have such a significant effect on surface quality. Only when the speed is maximum, an increase in these parameters generates a noticeable difference. In the CFRTP/Steel configuration, the influence of these parameters is more noticeable because the water jet dispersion is lower. In this sense, the increase in pressure minimizes the hydrodistortion defect in the composite material, minimizing the loss of kinetic energy prior to steel machining.

Thus, the machining capacity of the water jet is more constant and a variation in these parameters is more relevant. On the contrary, when the first material to be machined is steel, this energy loss is greater because it presents a greater difficulty to be machined, minimizing the effect of the cutting parameters in the composite material.

In contrast, the travel speed seems to have a more prominent effect in this configuration due to the dispersion of the water jet. When the jet starts machining at a very high speed and the first material (Steel) has a worse machinability, an excessive delay is generated between the machining of this and the second material (CFRTP) and an increase in the hydraulic pressure enhances the penetration capacity of the water jet improving the surface integrity in spite of obtaining very high *Ra* values [13].

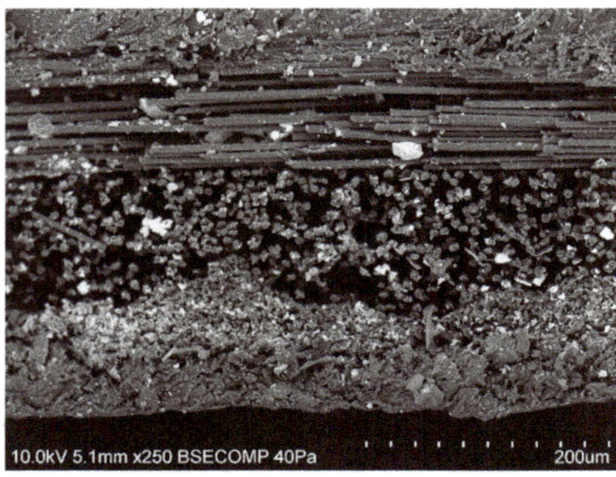

Figure 15. Total loss of thermoplastic matrix leaving the reinforcement unprotected in 0° and 90° stacking orientation at 250× (*P* of 250 MPa, *TS* of 300 mm/min, and *AMF* of 385 g/min).

On the other hand, the results obtained in steel machining in terms of *Ra* (Figure 16) and *Rz* (Figure 17) are very close to those obtained in the reverse configuration. This would indicate that steel is the most decisive material in the machining of this structure. In terms of surface quality, the positioning of the steel does not affect the results obtained, but it does directly affect the final quality generated in the composite material.

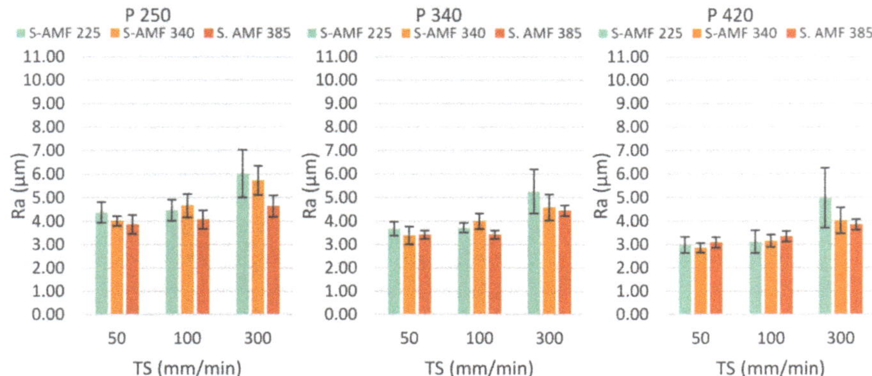

Figure 16. Steel surface quality results in terms of *Ra* as a function of the cutting parameters set for the inverse Steel/CFRTP configuration.

The three cutting parameters generate an improvement in the quality obtained. When the traverse speed is between 50 and 100 mm/min, it seems that the most dominant parameter is the hydraulic pressure by reducing the Ra values from 4 to 3 μm and minimizing the deviations obtained, which would indicate that the surface is very homogeneous. This is corroborated by the *Rz* results, which would indicate that the surface is smooth with constant surface variations. On the other hand, as with the other results, the increase at a speed of 300 mm/min significantly worsens the surface quality due to the destabilization of the water jet, which leads to an increase in the lag defect in the RCR region.

With regard to the abrasive flow, its effect is more noticeable when the speed is higher than 300 mm/min due to the loss of kinetic energy. An increase in this parameter improves the tearing of the steel by shear forces. In addition, flows of 385 g/min offer a very reduced deviation in both Ra and *Rz*, which would indicate minimal variations in the surface obtained. Thus, for this level of abrasive mass flow and a pressure of 420 MPa, very close surface quality values are obtained for speeds of 100 and 300 mm/min, allowing an increase in productivity in the machining of hybrid structures.

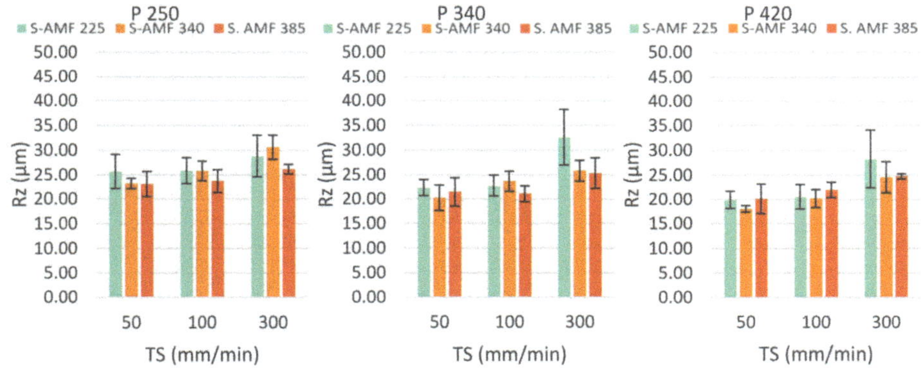

Figure 17. *Rz* values for steel in the Steel/CFRTP configuration.

The trends for each cutting parameter are shown in Figure 18. A great influence is observed on the composite material by increasing the hydraulic pressure and traverse speed. Being the second material to be machined in this stack directly affects the results obtained. Steel has a higher resistance to be machined and makes it difficult to stabilize the water jet prior to the machining of the composite material. This, in combination with the turbulence that can be generated in the interlayer, affects the surface quality by varying these parameters.

On the contrary, different trends have been observed in the surface quality of the steel compared to the CFRTP/Steel configuration. Thus, the pressure does not produce a significant variation in the results and the abrasive flow seems to slightly reduce Ra values that improves the surface quality. However, the trend of the traverse speed is constant and similar in both materials and stacking configurations with an increase in the results due to the delay in the water jet and not using the same amount of particles per unit area.

The machining of materials of different machinability reflects that the surface quality can be very disparate and that the machined part does not fulfill its function. The ratio between the Ra values for the CFRTP/Steel configuration is shown in Figure 19 and for the Steel/CFRTP configuration in Figure 20.

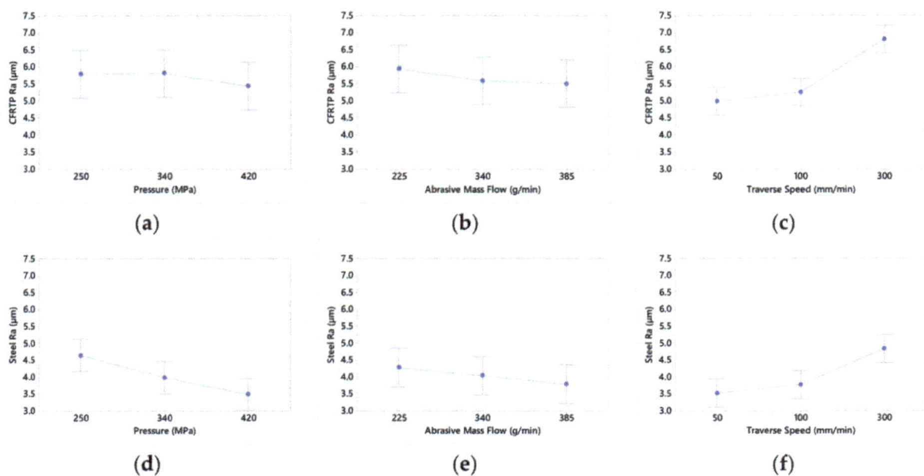

Figure 18. Cutting parameter trends in surface quality (*Ra*) in the Steel/CFRTP configuration. (**a**) CFRTP Ra in function of *P*, (**b**) CFRTP Ra in function of *AMF*, (**c**) CFRTP Ra in function of *TS*, (**d**) Steel Ra in function of *P*, (**e**) Steel Ra in function of *AMF*, and (**f**) Steel Ra in function of *TS*.

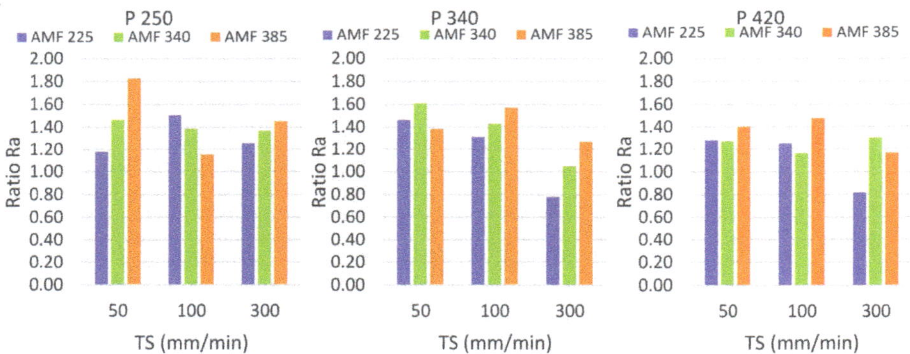

Figure 19. Ratio values *Ra* CFRTP and *Ra* Steel for CFRTP/Steel configuration.

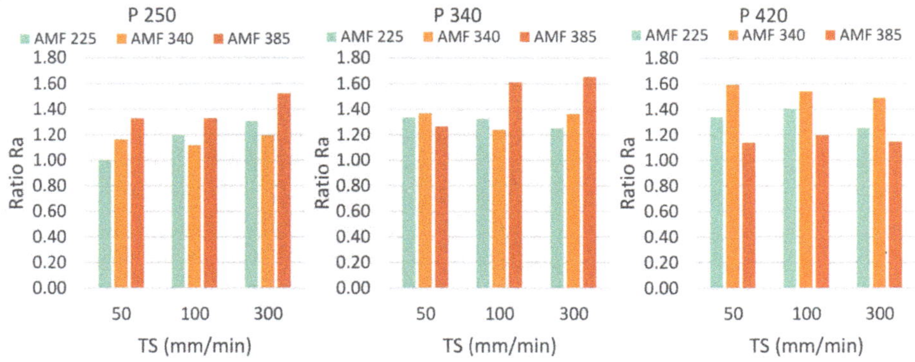

Figure 20. Ratio values *Ra* CFRTP and *Ra* Steel for Steel/CFRTP configuration.

In general, the results obtained in other studies are corroborated, where the quality obtained in the composite material is superior to steel with ratios greater than 1. In turn, in both configurations, an increase in the abrasive flow parameter produces a greater dispersion between both materials.

This can be seen especially in the CFRTP/Steel configuration, due to the fact that a greater number of abrasive particles increase the detachment of the thermoplastic matrix, causing the reinforcement to be free and worsening the surface quality. On the contrary, this increase improves the penetration capacity of the water jet allowing a stable cut in the steel and reducing the *Ra* values compared to the composite material.

It should be noted that, although the traverse speed is a parameter that worsens the surface quality considerably, its tendency is very close in both materials, which generates close ratios.

In terms of cutting parameters, ratios close to 1 are obtained for a pressure of 420 MPa because the loss of kinetic energy of the water jet is not significant, especially in the Steel/CFRTP configuration. Thus, the stacking configuration that offers the closest values of surface quality in terms of Ra is Steel/CFRTP.

It should be noted that, in both stacking configurations, gaps between materials have not been observed (Figure 21).

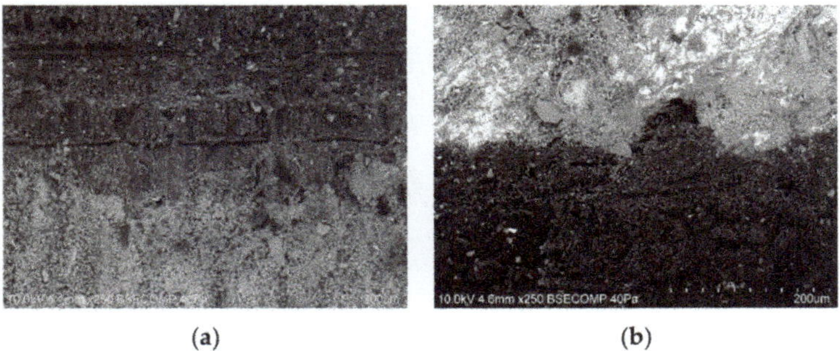

(a) (b)

Figure 21. Final quality of the bond between materials after machining at 250× (*P* of 250 MPa, *TS* of 300 mm/min, and *AMF* of 385 g/min): (**a**) CFRTP/Steel and (**b**) Steel/CFRTP.

In both cases, abrasive particles have remained adhered to the two materials and can affect the final surface quality. In the CFRTP/Steel configuration, remains of the thermoplastic matrix can be seen that have been pulled and adhered to the surface of the steel. On the contrary, in the inverse configuration, a cleaner surface can be seen in the interlayer.

3.3. Statistical Analysis and Contour Diagrams

The percentage contribution of each cutting parameter in the surface quality for each material and stacking configuration obtained by ANOVA analysis is shown in Figure 22.

It is confirmed that the traverse speed is the most determining parameter according to the results obtained. This is particularly evident in the Steel/CFRTP configuration. An increase in this parameter generates a greater destabilization in the water jet generating a rougher surface. In addition, due to the loss of kinetic energy, the RCR region increases, leading to the formation of grooves [25,40]. On the other hand, another key factor is the hydraulic pressure. An increase in this parameter improves the penetration and machining capacity of the water jet by facilitating the removal mechanism [41].

A balance between traverse speed and hydraulic pressure has been observed in the CFRTP/Steel configuration. A correct selection of these cutting parameters reduces the hydrodistortion defect in the layers of the composite material and minimizes the detachment of the thermoplastic matrix. This results in a less rough surface and improved surface quality.

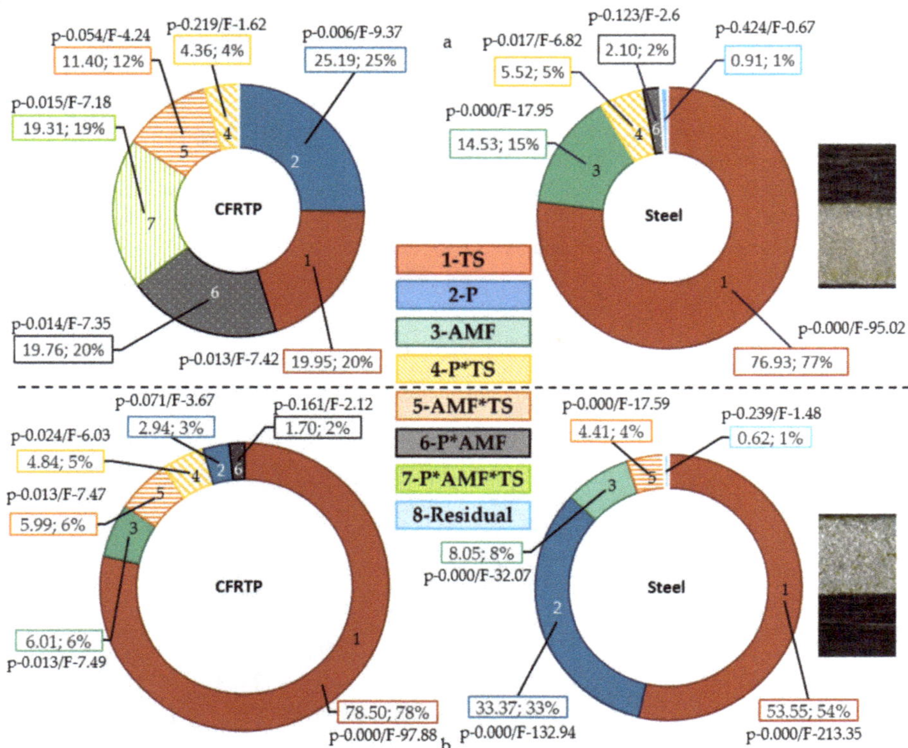

Figure 22. Percentage contribution of cutting parameters on surface quality for: (**a**) CFRTP/Steel and (**b**) Steel/CFRTP.

In parallel, with the experimental results obtained, a series of predictive second-order polynomial models have been generated that relate surface quality in terms of Ra with cutting parameters for applications in the industrial sector.

The models obtained for the CFRTP/Steel configuration are shown in (1) and (2) with values of R2 of 67.48% and 86.98%, respectively, and the models for the Steel/CFRTP configuration in (3) and (4) with adjustments of 85.73% and 95.25%, respectively. It should be noted that, due to the anisotropy and the reduced thickness of the composite material, it generates a randomness in the surface quality that reduces the adjustment obtained.

$$Ra\ (CFRTP) = 5.37 + 0.00211{\cdot}P + 0.0050{\cdot}AMF + 0.0350{\cdot}TS - 0.000028{\cdot}P{\cdot}AMF - 0.000119{\cdot}P{\cdot}TS - 0.000123{\cdot}AMF{\cdot}TS \quad (1)$$

$$Ra\ (Steel) = 6.18 - 0.00568{\cdot}P - 0.00615{\cdot}AMF + 0.0116{\cdot}TS + 0.000005{\cdot}P{\cdot}AMF - 0.000011{\cdot}P{\cdot}TS - 0.000045{\cdot}AAMF{\cdot}TS \quad (2)$$

$$Ra\ (CFRTP) = -0.20 + 0.0123{\cdot}P + 0.0128{\cdot}AMF + 0.0299{\cdot}TS - 0.000032{\cdot}P{\cdot}AMF - 0.000037{\cdot}P{\cdot}TS - 0.000042{\cdot}AMF{\cdot}TS \quad (3)$$

$$Ra\ (Steel) = 7.91 - 0.01441{\cdot}P - 0.00765{\cdot}AMF + 0.000025{\cdot}P{\cdot}AMF + 0.000024{\cdot}P{\cdot}TS + 0.000003{\cdot}AMF{\cdot}TS \quad (4)$$

And the corresponding contour diagrams are shown in Figures 23 and 24, which relates the surface quality (*Ra*) to the cutting parameters for both stacking configurations. Values close to 4.5 μm in the composite material in the CFRTP/Steel configuration and close to 5 μm in the Steel/CFRTP structure are obtained by combining a pressure of 420 MPa, an AMF of 385 g/min, and a *TS* of 50 mm/min in the CFRTP/Steel configuration.

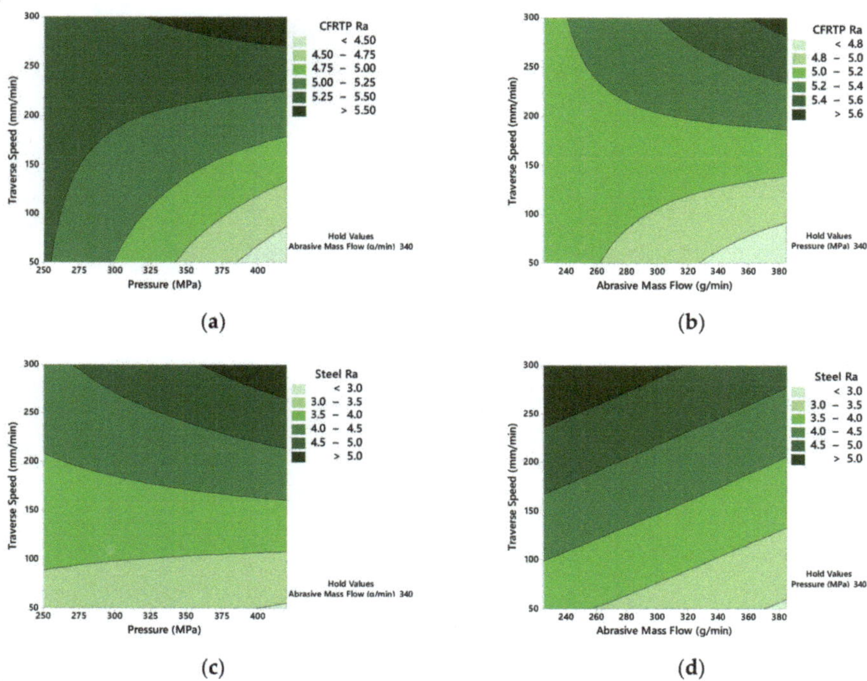

Figure 23. Contour diagrams for the CFRTP/Steel configuration: (**a**) Composite *TS* vs. *P*, (**b**) Composite *AMF* vs. *TS*, (**c**) Steel *TS* vs. *P*, and (**d**) Steel *AMF* vs. *TS*.

In this way, there is a direct relationship between the roughness generated and the ratio between the power of the water jet and the penetration depth ($\dot{E}/h$). Very high values of this parameter indicate surfaces with low roughness. Pahuja et al. [30] explain that the composite/metal configuration shows a high initial roughness and suffers a very fast decrease in the composite material and slower in the titanium due to the loss of kinetic energy of the water jet.

Furthermore, regardless of the material, the second material to be machined suffers an increase in *Ra* values compared to the reverse configuration. Thus, it is corroborated that no matter the composite material (thermoplastic or thermoset) and the metal alloy used, the roughness in a stacked configuration is mainly governed by the characteristics of the jet. This is enhanced when the pressure is minimal due to the reduction in machining capacity.

Conversely, lower values are obtained for the metal alloy with very similar results in both stacking configurations close to 3.5 μm. These values are achieved by combining a *P* between 320 and 420 MPa, an *AMF* of 385 g/min, and a *TS* of 50 mm/min.

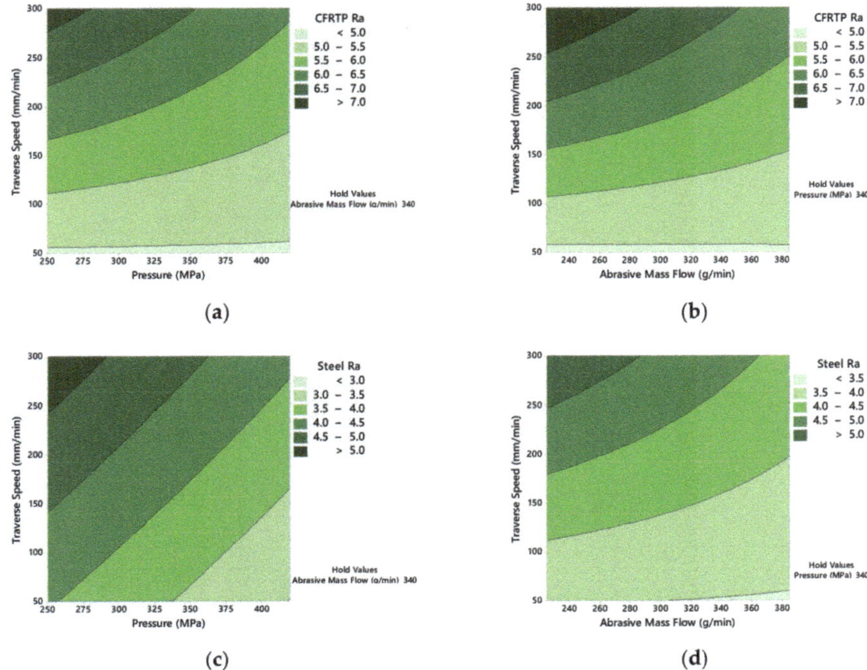

Figure 24. Contour diagrams for the Steel/CFRTP configuration: (**a**) Composite *TS* vs. *P*, (**b**) Composite *AMF* vs. *TS*, (**c**) Steel *TS* vs. *P*, and (**d**) Steel *AMF* vs. *TS*.

4. Conclusions

Surface quality in machining processes is a key parameter in terms of functional performance. Abrasive water jet machining of hybrid structures of dissimilar materials generates a highly variable surface quality that depends directly on the correct selection of cutting parameters and stacking order.

Typical defectology in abrasive water jet machining of thermoset composite materials has been identified in thermoplastic composites. Small delamination and matrix loss have been detected leaving the reinforcement unprotected.

Stacking order is a key factor. Lower *Ra* and *Rz* values are obtained in the CFRTP/Steel configuration due to better conservation of the kinetic energy of the water jet. This allows for a better cutting capacity of the water jet, especially in the composite material by minimizing matrix loss and reducing fiber pull-out defectology. In contrast, in the Steel/CFRTP configuration, due to the difference in machinability, the steel absorbs much of the energy of the water jet reducing the ability to penetrate into the composite material and resulting in a rougher and more random surface.

With regard to cutting parameters, the traverse speed is the most critical factor. In both materials and stacking configurations, an increase in this parameter generates a notable growth in the *Ra* and *Rz* values due to the divergence of the water jet and the offset that is generated between the first material and the second during machining. Thus, smoother surfaces are obtained with a traverse speed close to 50 mm/min.

The lowest values of surface quality have been obtained by combining a traverse speed of 50 mm/min, a hydraulic pressure of 420 MPa, and an abrasive mass flow of 385 g/min, maximizing the machining capacity of the water jet.

Finally, a series of predictive mathematical models have been obtained with good fits that relate the surface quality in terms of *Ra* in both materials and stacking configurations to the cutting parameters and which may be of interest and application in current industry.

Author Contributions: F.B. and A.S. developed machining tests. F.B. developed data treatment. F.B. and A.S. analyzed the influence of the parameters involved. F.B. and J.S. wrote the manuscript, figures, and tables. M.B., B.S., and J.S. contributed to the experimental design and critical comments on the final manuscript. All authors have read and agreed to the published version of the manuscript.

Funding: This work has been developed under support of a predoctoral industrial fellow financed by NANOTURES SL, mechanical engineering and industrial design department, and Vice-rectorate of Transference and Technological Innovation of the University of Cadiz.

Acknowledgments: The authors would like to thank the Laboratory of Corrosion and Protection TEP-231 (Labcyp) of the University of Cadiz for the support with scanning electron microscopy.

Conflicts of Interest: The authors declare no conflict of interest.

Glossary of Terms

CFRTP	Carbon fiber-reinforced thermoplastics
IDR	Initial damage region
SCR	Smooth cutting region
RCR	Rough cutting region
ANOVA	Analysis of variance
P	Hydraulic pressure
AMF	Abrasive mass flow
TS	Traverse speed
AWJM	Abrasive water jet machining
TPU	Thermoplastic polyurethane

References

1. Goto, K.; Imai, K.; Arai, M.; Ishikawa, T. Shear and tensile joint strengths of carbon fiber-reinforced thermoplastics using ultrasonic welding. *Compos. Part A Appl. Sci. Manuf.* **2019**, *116*, 126–137. [CrossRef]
2. Ishikawa, T.; Amaoka, K.; Masubuchi, Y.; Yamamoto, T.; Yamanaka, A.; Arai, M.; Takahashi, J. Overview of automotive structural composites technology developments in Japan. *Compos. Sci. Technol.* **2018**, *155*, 221–246. [CrossRef]
3. Christmann, M.; Medina, L.; Mitschang, P. Effect of inhomogeneous temperature distribution on the impregnation process of the continuous compression molding technology. *J. Thermoplast. Compos. Mater.* **2017**, *30*, 1285–1302. [CrossRef]
4. Biron, M. Outline of the actual situation of plastics compared to conventional materials. In *Thermoplastics and Thermoplastic Composites*; William Andrew: Norwich, NY, USA, 2018; Volume i, pp. 1–32, ISBN 9780081025017.
5. Bu, H.; Li, Y.; Yang, H.; Wang, L.; Zhan, X. Investigation of laser joining process of CFRTP and aluminum alloy. *Mater. Manuf. Process.* **2020**, 1–8. [CrossRef]
6. Haider, D.R.; Krahl, M.; Liebsch, A.; Kupfer, R.; Haider, D.R.; Krahl, M.; Koshukow, W.; Wolf, M.; Liebsch, A.; Kupfer, R.; et al. Adhesion Studies of Thermoplastic Fibre-Plastic Composite Hybrid Components Part 2: Thermoplastic-Metal-Composites. In Proceedings of the Hybrid Materials and Structures 2018, Bremen, Germany, 18–19 April 2018.
7. Sheng, L.; Jiao, J.; Du, B.; Wang, F.; Wang, Q. Influence of Processing Parameters on Laser Direct Joining of CFRTP and Stainless Steel. *Adv. Mater. Sci. Eng.* **2018**, *2018*, 1–15. [CrossRef]
8. Sheng, L.Y.; Wang, F.Y.; Wang, Q.; Jiao, J.K. Shear Strength Optimization of Laser-Joined Polyphenylene Sulfide-Based CFRTP and Stainless Steel. *Strength Mater.* **2018**, *50*, 824–831. [CrossRef]
9. Jiao, J.; Xu, Z.; Wang, Q.; Sheng, L.; Zhang, W. CFRTP and stainless steel laser joining: Thermal defects analysis and joining parameters optimization. *Opt. Laser Technol.* **2018**, *103*, 170–176. [CrossRef]
10. Mamalis, D.; Obande, W.; Koutsos, V.; Blackford, J.R.; Ó Brádaigh, C.M.; Ray, D. Novel thermoplastic fibre-metal laminates manufactured by vacuum resin infusion: The effect of surface treatments on interfacial bonding. *Mater. Des.* **2019**, *162*, 331–344. [CrossRef]
11. Tanaka, K.; Yamashiro, T.; Katayama, T. Internal Damage Evaluation of Cfrtp Cut By a Circular Saw. *Mater. Contact Charact.* **2017**, *116*, 345–351.
12. Masek, P.; Zeman, P.; Kolar, P. Development of a Cutting Tool for Composites With Thermoplastic Matrix. *Sci. J.* **2013**, 423–427. [CrossRef]

13. Kakinuma, Y.; Ishida, T.; Koike, R.; Klemme, H.; Denkena, B.; Aoyama, T. Ultrafast Feed Drilling of Carbon Fiber-Reinforced Thermoplastics. *Procedia CIRP* **2015**, *35*, 91–95. [CrossRef]
14. Zitoune, R.; Krishnaraj, V.; Collombet, F. Study of drilling of composite material and aluminium stack. *Compos. Struct.* **2010**, *92*, 1246–1255. [CrossRef]
15. Bañon, F.; Sambruno, A.; Batista, M.; Simonet, B.; Salguero, J. Study of the surface quality of carbon fiber–reinforced thermoplastic matrix composite (CFRTP) machined by abrasive water jet (AWJM). *Int. J. Adv. Manuf. Technol.* **2020**, *107*, 3299–3313. [CrossRef]
16. Ramulu, M.; Pahuja, R.; Hashish, M.; Isvilonanda, V. Abrasive Waterjet Machining Effects on Kerf Quality in Thin Fiber Metal Laminate. In Proceedings of the WJTA-IMCA Conference and Expo, New Orleans, LA, USA, 2–4 November.
17. El-Hofy, M.; Helmy, M.O.; Escobar-Palafox, G.; Kerrigan, K.; Scaife, R.; El-Hofy, H. Abrasive Water Jet Machining of Multidirectional CFRP Laminates. *Procedia CIRP* **2018**, *68*, 535–540. [CrossRef]
18. Perec, A. Environmental aspects of abrasive water jet cutting. *Rocz. Ochr. Sr.* **2018**, *20*, 258–274.
19. Pahuja, R.; Ramulu, M. Abrasive water jet machining of Titanium (Ti6Al4V)–CFRP stacks – A semi-analytical modeling approach in the prediction of kerf geometry. *J. Manuf. Process.* **2019**, *39*, 327–337. [CrossRef]
20. Pahuja, R.; Ramulu, M.; Hashish, M. Abrasive Water jet machining (AWJ) of hybrid Titanium/Graphite composite laminate: Preliminary results. In Proceedings of the 22nd International Conference on Water Jetting, Haarlem, The Netherlands, 3–5 September 2014; pp. 83–95.
21. Pahuja, R.; Ramulu, M.; Hashish, M. Abrasive waterjet profile cutting of thick Titanium/Graphite fiber metal laminate. In *ASME International Mechanical Engineering Congress and Exposition;* American Society of Mechanical Engineers: New York, NY, USA, 2016; pp. 1–11.
22. Li, M.; Huang, M.; Chen, Y.; Gong, P.; Yang, X. Effects of processing parameters on kerf characteristics and surface integrity following abrasive waterjet slotting of Ti6Al4V / CFRP stacks. *J. Manuf. Process.* **2019**, *42*, 82–95. [CrossRef]
23. Alberdi, A.; Artaza, T.; Suárez, A.; Rivero, A.; Girot, F. An experimental study on abrasive waterjet cutting of CFRP/Ti6Al4V stacks for drilling operations. *Int. J. Adv. Manuf. Technol.* **2016**, *86*, 691–704. [CrossRef]
24. Sambruno, A.; Bañon, F.; Salguero, J.; Simonet, B.; Batista, M. Kerf Taper Defect Minimization Based on Abrasive Waterjet Machining of Low Thickness Thermoplastic Carbon Fiber Composites C/TPU. *Materials* **2019**, *12*, 4192. [CrossRef]
25. Dumbhare, P.A.; Dubey, S.; Deshpande, Y.V.; Andhare, A.B.; Barve, P.S. Modelling and multi-objective optimization of surface roughness and kerf taper angle in abrasive water jet machining of steel. *J. Braz. Soc. Mech. Sci. Eng.* **2018**, *40*, 259. [CrossRef]
26. Ruiz-Garcia, R.; Mayuet Ares, P.; Vazquez-Martinez, J.; Salguero Gómez, J. Influence of Abrasive Waterjet Parameters on the Cutting and Drilling of CFRP/UNS A97075 and UNS A97075/CFRP Stacks. *Materials* **2018**, *12*, 107. [CrossRef] [PubMed]
27. Bañon, F.; Sambruno, A.; Ruiz-Garcia, R.; Salguero, J.; Mayuet, P.F. Study of the influence of cutting parameters on surface quality in AWJM machining of thermoplastic matrix composites. *Procedia Manuf.* **2019**, *41*, 233–240. [CrossRef]
28. Mayuet Ares, P.F.; Rodríguez-Parada, L.; Gómez-Parra, A.; Batista, M. Characterization and Defect Analysis of Machined Regions in Al-SiC Metal Matrix Composites Using an Abrasive Water Jet Machining Process. *Appl. Sci.* **2020**, *10*, 1512. [CrossRef]
29. Zhang, S.; Ji, L.; Wu, Y.; Chen, M.; Zhou, W. Exploring a new method to obtain the 3D abrasive water jet profile. *Int. J. Adv. Manuf. Technol.* **2020**, 4797–4809. [CrossRef]
30. Pahuja, R.; Ramulu, M.; Hashish, M. Surface quality and kerf width prediction in abrasive water jet machining of metal-composite stacks. *Compos. Part B Eng.* **2019**, *175*, 107134. [CrossRef]
31. Chithirai, M.; Selvan, P.; Mohana, N.; Raju, S. Assessment of Process Parameters in Abrasive Waterjet Cutting of Stainless Steel. *Int. J. Adv. Eng. Technol.* **2011**, *1*, 34.
32. Bañon, F.; Sambruno, A.; Batista, M.; Simonet, B.; Salguero, J. Surface Quality and Free Energy Evaluation of s275 Steel by Shot Blasting, Abrasive Water Jet Texturing and Laser Surface Texturing. *Metals* **2020**, *10*, 290.
33. Mathia, T.G.; Pawlus, P.; Wieczorowski, M. Recent trends in surface metrology. *Wear* **2011**, *271*, 494–508. [CrossRef]

34. Masek, P.; Kolar, P.; Zeman, P. Optimization of trimming operations for machining carbon fibre reinforced thermoplastic composite. In Proceedings of the International Conference on Advanced Manufacturing Engineering and Technologies (NEWTECH 2013), Stockholm, Sweden, 27–30 October 2013. Available online: https://www.researchgate.net/publication/272576514_Optimization_of_trimming_operations_for_machining_carbon_fibre_reinforced_thermoplastic_composite (accessed on 23 June 2020).
35. Mayuet Ares, P.F.; Girot Mata, F.; Batista Ponce, M.; Salguero Gñmez, J. Defect Analysis and Detection of Cutting Regions in CFRP Machining Using AWJM. *Materials* **2019**, *12*, 4055. [CrossRef]
36. Pahuja, R.; Ramulu, M. Surface quality monitoring in abrasive water jet machining of Ti6Al4V–CFRP stacks through wavelet packet analysis of acoustic emission signals. *Int. J. Adv. Manuf. Technol.* 2019. Available online: https://www.researchgate.net/publication/334756766_Surface_quality_monitoring_in_abrasive_water_jet_machining_of_Ti6Al4V-CFRP_stacks_through_wavelet_packet_analysis_of_acoustic_emission_signals (accessed on 23 June 2020).
37. Kumar, S.; Laxminarayana, P. Optimization of Process Parameters on Kerf Width & Taper Angle on En-8 Carbon Steel by Abrasive Water Jet Machining. Available online: https://www.researchgate.net/publication/332171967_Optimization_of_Process_Parameters_on_Kerf_Width_Taper_Angle_on_En-8_Carbon_Steel_by_Abrasive_Water_Jet_Machining (accessed on 23 June 2020).
38. Chithirai, M.; Selvan, P.; Mohana, N.; Raju, S. A Machinability Study of Kevlar-Phenolic Composites Using Abrasive Waterjet Cutting Process. *CLEAR IJRET* **2012**, *1*, 46–57. Available online: https://www.semanticscholar.org/paper/A-Machinability-Study-of-Kevlar-Phenolic-Composites-Selvan-Raju/18f1f090fd279616eb814562f177db664b5461ca (accessed on 23 June 2020).
39. Köhler, T.; Röding, T.; Gries, T.; Seide, G. An Overview of Impregnation Methods for Carbon Fibre Reinforced Thermoplastics. *Key Eng. Mater.* **2017**, *742*, 473–481. [CrossRef]
40. Rao, M.S. Parametric Optimization of Abrasive Waterjet Machining for Mild Steel: Taguchi Approach. *Int. J. Curr. Eng. Technol.* **2014**, *2*, 28–30. [CrossRef]
41. Murugabaaji, V.; Kannan, A.; Nagarajan, N. Experimental Investigation on Abrasive Waterjet Machining of Stainless Steel 304. 2015. Available online: https://www.researchgate.net/publication/330577552_Experimental_Investigation_on_Abrasive_Waterjet_Machining_of_Stainless_Steel_304 (accessed on 23 June 2020).

Article

Effects of Machining Parameters on the Quality in Machining of Aluminium Alloys Thin Plates

Irene Del Sol [1,*], Asuncion Rivero [2] and Antonio J. Gamez [1]

[1] Department of Mechanical Engineering and Industrial Design, School of Engineering, University of Cadiz, Av. Universidad de Cádiz 10, Puerto Real (Cádiz) E-11519, Spain

[2] Fundación Tecnalia Research & Innovation, Scientific and Technological Park of Guipuzkoa, Paseo Mikeletegui, Donostia-San Sebastián (Guipuzcoa) 7 E-20009, Spain

* Correspondence: irene.delsol@uca.es; Tel.: +34-956-483-513

Received: 23 July 2019; Accepted: 22 August 2019; Published: 24 August 2019

Abstract: Nowadays, the industry looks for sustainable processes to ensure a more environmentally friendly production. For that reason, more and more aeronautical companies are replacing chemical milling in the manufacture of skin panels and thin plates components. This is a challenging operation that requires meeting tight dimensional tolerances and differs from a rigid body machining due to the low stiffness of the part. In order to fill the gap of literature research on this field, this work proposes an experimental study of the effect of the depth of cut, the feed rate and the cutting speed on the quality characteristics of the machined parts and on the cutting forces produced during the process. Whereas surface roughness values meet the specifications for all the machining conditions, an appropriate cutting parameters selection is likely to lead to a reduction of the final thickness deviation by up to 40% and the average cutting forces by up to a 20%, which consequently eases the clamping system and reduces machine consumption. Finally, an experimental model to control the process quality based on monitoring the machine power consumption is proposed.

Keywords: thin plates; thin-wall; machining; aluminium; cutting forces; roughness

---

1. Introduction

Aluminium fuselage skin panel machining is considered a challenging operation due to its dimensional and surface requirements. These parts are lightened by machining superficial pockets in order to increase the fuel efficiency of aircrafts by reducing their structure weight. These pockets have historically been machined using chemical milling operations, although green manufacturing approaches have been focused on the study of mechanical machining for this purpose [1]. In fact, different projects and research studies have invested hundreds of thousands of euros to remove chemical milling, designing specific clamping systems to ensure surface quality and dimensional requirements while maintaining clamping flexibility. These systems are focused on twin-machining heads [2,3], magnetic slaves [4] or flexible vacuum beds [5] that control the deflection of the part avoiding overcut during the operation.

Additionally, the conventional machining of low stiffness parts presents dynamic and static problems [6,7]. On the one hand, dynamic stability of machining strongly depends on system stiffness, its natural frequency response, and the selected cutting parameters. Vibrations—chatter and forced vibrations—can directly affect the final roughness of parts, increasing their value and forcing manufacturers to make reprocessing steps, therefore increasing the operational cost [8,9]. In order to avoid them, chatter influence is studied using stability lobe diagrams (SLD), a representation tool that commonly relates the stability areas of machining with the feed rate, the spindle speed, the depth of cut, or the tool position [10–14], and forced vibrations can be studied through dynamic models. In this case, the applied force is studied to reduce the dynamic deflection of the part. On the other hand, quasi-static deflection can take place when the elastoplastic behaviour of the workpiece, combined with a failure on the clamping, is not enough to counter the machining force effect, reducing the real depth of cut [15–17]. This fact was proved experimentally by Yan et al. [18], who optimized the depth of cut depending on the cutting force, reducing the part deflection and increasing the process efficiency. Similarly, Sonawane et al. [19] used a statistical approach to model workpiece deflection, addressing it to the machining parameters and cutting tool orientation. In contrast, Oliveira et al. [20] established that the most influencing factor on the real depth of cut was the milling strategy (up or down milling), while other studies [21,22] have focused on the analysis of the toolpath effect on the final quality of slim parts.

However, few studies have been focused on the analysis of the parameters effect into the quality characteristics of thin plates. The literature review has shown that most of the studies have focused on thin-wall machining rather than thin-plates or thin floors [7]. Few of them have focused on the analysis of the parameters effect into the final quality characteristics. For this reason, this paper focuses on the study of thin plates in order to simulate the machining of large skin panels to evaluate the effect of the machining parameters on the final thickness, surface roughness and cutting forces of the part.

2. Materials and Methods

2.1. Machining Tests

Two type of machining tests were performed. The first type was used to analyse the thin plates' behaviour through the machining of 50 × 50 mm^2 pockets on samples of 80 × 80 mm^2 and 2 mm thickness. Parts were screwed to a faced mill plate that was housed on a dynamometer. The samples were dry machined while keeping the axial distance constant, and the depth of cut (a_p), the feed rate per tooth (f_z) and the cutting speed (V_c) were variable. The values of each machining conditions are compiled in the levels listed in Table 1, in which the spindle speed (S) is also shown. The depth of cut was selected based on the geometrical requirements of industrial parts. Feed rate and cutting speed were selected based on the literature [14,23,24] and aerospace recommendations. The chosen strategy was down milling, following a toolpath from the centre of the workpiece up to the outside of it.

Table 1. Machining parameters and levels for skin sample tests.

Parameter	Level 1	Level 2	Level 3	Level 4
a_p (mm)	0.2	0.4	0.8	1.0
f_z (mm/tooth)	0.08	0.1	-	-
S (rpm)	4,000	12,000	18,000	-
V_c (m/min)	126	378	566	-

The second test aimed to obtain the cutting force values following a mechanistic approach. In this approach, the specific force coefficients of the combination pair tool-material had to be experimentally defined by performing different slots on a rigid aluminium workpiece. Additionally, in order to avoid chatter and ensure the rigid behaviour of the samples, SLD were calculated using an impact hammer test. The maximum depth of cut for stable machining was calculated using the procedure described by Altintas and Budak [25]. Following this procedure, the frequency responses of the tool and the workpiece at four different steps of the cutting operation were obtained. This combination provided the SLD diagram at the four stages in order to analyse possible changes during the machining operation.

Each machining test was performed in a 5-axis NC centre ZV 25U600 EXTREME (Ibarmia Innovatek S.L.U., Azkoitia, Spain). The material used on the sample parts was aluminium 2024-T3, and the tool was a torus end mill KENDU 4400.60 (Kendu, Segura, Spain) with a 10 mm diameter, 30° helix angle, 18° rake angle, 16° clearance angle for the secondary edge and 9° angle for the primary edge.

Forces and accelerations on the workpiece were monitored using a dynamometer Kistler 9257B (Kistler Group, Winterthur, Switzerland) and an accelerometer Kistler 8728A500 (Kistler Group, Winterthur, Switzerland), connected to National Instruments acquisition boards NI 9215(National Instruments, Austin, TX, USA) and NI 9234 (National Instruments, Austin, TX, USA), respectively. The power consumed by the spindle speed and the whole machine was also recorded using a Fanuc Servoguide system (Fanuc Corporation, Oshino-mura, Japan). The test configuration and monitoring system are shown in Figure 1.

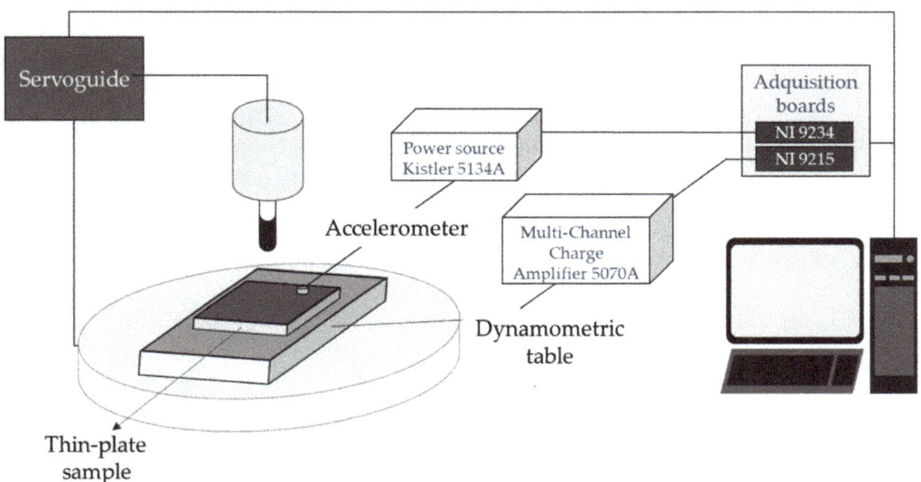

Figure 1. Scheme of the monitoring system.

2.2. Force Mechanistic Model

The expected forces could be calculated through a mechanistic approach [26,27] Tangential (*t*), axial (*a*) and radial (*r*) forces could be considered as a function of the cutting coefficient (K_{qc}) and the friction coefficient $\left(K_{qe}\right)$.

$$\partial F_q(\varphi, z) = K_{qe}\partial S + K_{qc} f_z \sin\varphi(\varphi_i, z)\,\partial z,\ q = \{t, r, a\} \tag{1}$$

where ∂S is the differential chip edge length and φ is the applied rotation angle which depends on the instant depth of cut (z), the number of teeth engaged (*j*), the total number of teeth (*N*), and the helix angle (β).

$$\varphi(\varphi_i, z) = \varphi_i - (j-1)\frac{2\pi}{N} - \beta \tag{2}$$

The cutting forces were converted to Cartesian coordinates using Equation (3), with κ being the angle referred to the torus part of the mill.

$$\frac{\partial F_{x,y,z}}{\partial z} = \begin{bmatrix} -\cos\varphi & -\sin\varphi\sin\kappa & -\sin\varphi\cos\kappa \\ \sin\varphi & -\cos\varphi\sin\kappa & -\cos\varphi\cos\kappa \\ 0 & \cos\kappa & -\sin\kappa \end{bmatrix} \begin{bmatrix} \partial F_t \\ \partial F_r \\ \partial F_a \end{bmatrix} \tag{3}$$

The cutting and friction coefficients were obtained by solving the equation using the force values obtained in the slot test performed in a rigid part. The coefficients were considered constant for all the f_z, but the effect of the V_c was taken into account. The test conditions are shown in Table 2. The results were used to predict the SLD.

Table 2. Machining parameters and levels for cutting coefficient calculation tests.

Parameter	Level 1	Level 2	Level 3
a_p (mm)	0.2	0.6	1.0
f_z (mm/tooth)	0.06	0.08	0.1
S (rpm)	4,000	12,000	18,000
V_c (m/min)	126	378	566

2.3. *Quality Evaluation*

The quality of the parts was established through the final thickness distribution and roughness. Typical tolerance values in the industry were really tight, about ±0.1 mm for final thickness and under 1.6 µm for the roughness average (*Ra*).

Final thicknesses were measured using a single coordinate measurement machine with an electronic comparator set. Nine points of the sample were evaluated, as shown in Figure 2a. *Ra* was measured using a Mahr Perthometer Concept PGK120 roughness measure station (Mahr technology, Göttingen, Germany) on five different areas of the part (Figure 2b). The areas for the roughness measures were selected in order to study the whole machining process and to cover both machine x and y axes. Roughness measurements were taken following the standard ISO 4288:1996 [28].

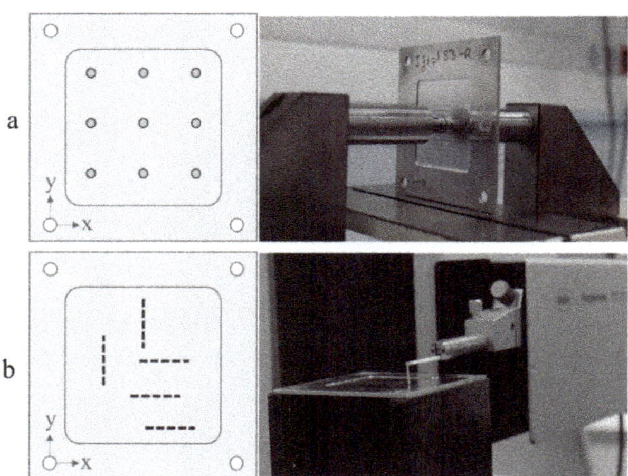

Figure 2. Measure procedure. (**a**) points selected for the thickness analysis and (**b**) areas studied for the roughness analysis.

3. Results and Discussion

3.1. *Final Thickness Error*

The final thickness error can be defined as the difference between the experimental thickness and the expected one; this parameter measures the real part dimension. In addition to the static and dynamic phenomena that occur during low rigidity parts machining [7], aspects such as the machine positioning error and the thermal expansion of the spindle have influence over this parameter. Other possible thermal errors can be discarded due to the short length of the machining test, which allowed us to underestimate the effect on the accuracy of the machine due to temperature changes in its surroundings.

The analysis of the results showed that the feed rate had a negligible effect on the average final thickness error, while the cutting speed seems to have had a significant effect on this parameter. The higher the cutting speed, the greater the geometric error was, and the piece became thinner (Figure 3).

Nevertheless, the thermal expansion of the spindle increased with the revolutions and therefore with the cutting speed [29]. This fact explains the part thinning, rejecting the direct effect of the cutting speed. Though the thermal expansion of the spindle was identified as an influencing parameter for the increase of the final average thickness error of the part, this error was easy to compensate when studying the elongation curves of the spindle and considering them in the CAM design [30] or implementing error compensation rules in the machine control [5].

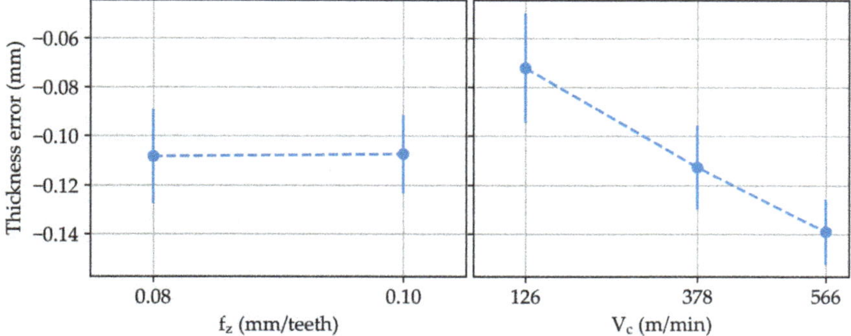

Figure 3. Average final thickness error depending on the feed per teeth (f_z) and the cutting speed (V_c).

Once the average error—not directly related to the cutting parameters—can be compensated, the target is to get a homogeneous thickness distribution. Higher values of feeds per teeth combined with higher values of spindle speed led to a reduction up to 40% of the standard deviation of the thickness errors measured in a test sample. This decrease was due to a better behaviour of the process in terms of dynamics. If the part was machined at higher rotation speed where cutting forces excited higher frequency vibration modes and created lower vibration amplitudes (Figure 4), therefore leading to a more homogeneous thickness distribution

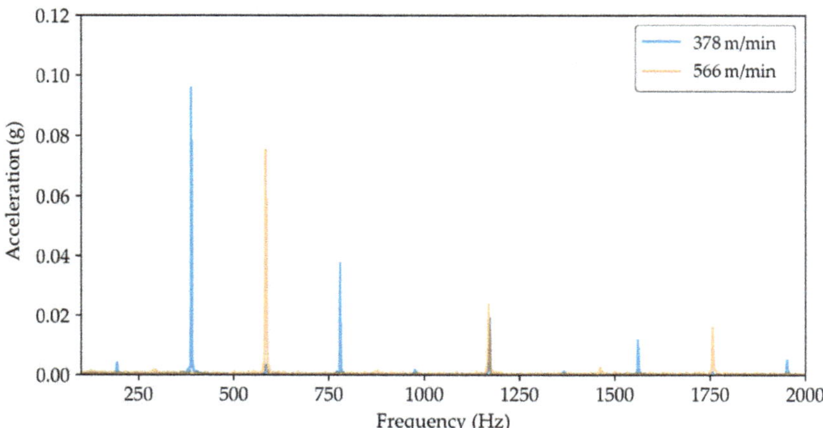

Figure 4. Fast Fourier Transform of the part acceleration signal for test at $fz = 0.1$ and $ap = 1.0$ mm, under two different cutting speeds.

3.2. Roughness

Roughness results were not significantly affected by any of the studied parameters in average deviation, with all of them being inside the most restrictive tolerance values (1.6 μm) required in chemical milling process (Figure 5a).

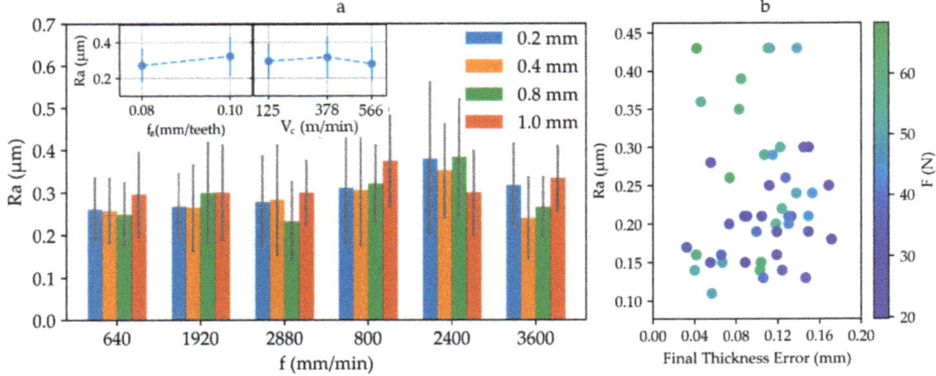

Figure 5. (**a**) Average of the roughness values obtained for different feed rates ($f = N{\cdot}f_z{\cdot}S$) as a function of the depth of cut. The inlet shows the partial effect of the feed rate per teeth and the cutting speed. (**b**) Average roughness values of each performed test against the depth of cut error considering the force module.

Measured forces and depths of cut errors did not have any impact on this quality characteristic (Figure 5b). However, lower cutting force values were related to roughness values under 0.3 µm. In fact, lower forces caused less tool deflection and vibrations of smaller amplitudes, leading to more stable machining processes that allowed us to produce more homogenous surfaces [31]. This revealed that the surface quality can be kept constant for any depth of cut under similar machining conditions. For this reason, the selection of parameters that decrease the forces should be considered.

3.3. Forces and Power Models

According to the SLD (Figure 6), thin plates almost behaved like a rigid part. This fact ensures that the tests were performed under stable machining conditions, proving that any variation occurring on the final thickness of the part was not produced by chatter issues.

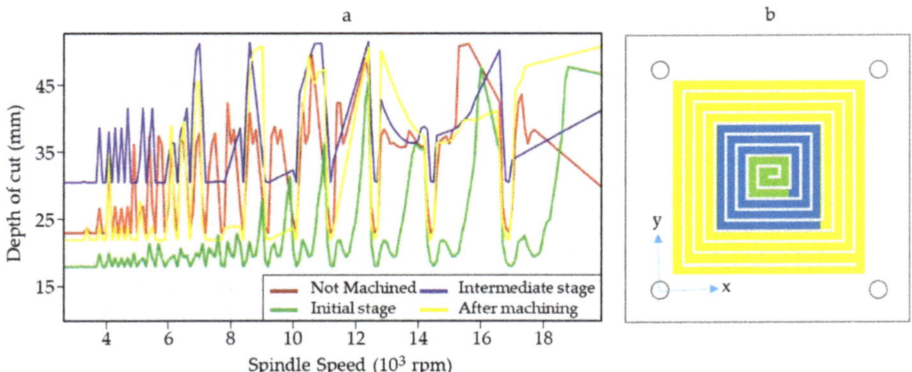

Figure 6. (**a**) Stability Lobe Diagrams (SLD) variation depending on the machining stage. Red line, before the machining; green line, in the first instants of the machining; blue line, intermediate stage; and yellow line, after the whole machining operation was performed. (**b**) Scheme of the areas of material removed in each stage with corresponding colours.

However, the forces obtained on the tests performed on a rigid body did not completely agree with those measured on the flexible parts (Figure 7). The depth of cut was linearly related to Fx and Fy, but for medium and high cutting speed values, the Fz initiated a constant trend at cutting depths of 0.8 and 0.1 mm, respectively. Even though the machining operation was not in a high speed machining regimen for aluminium alloys, there was a decrease of the force values expected for the higher depths of cut. This fact had been previously observed by López de Lacalle et al. [32], where, due to the reduction of stiffness, the cutting forces decreased, obtaining behaviours closer to high speed machining at lower cutting speed.

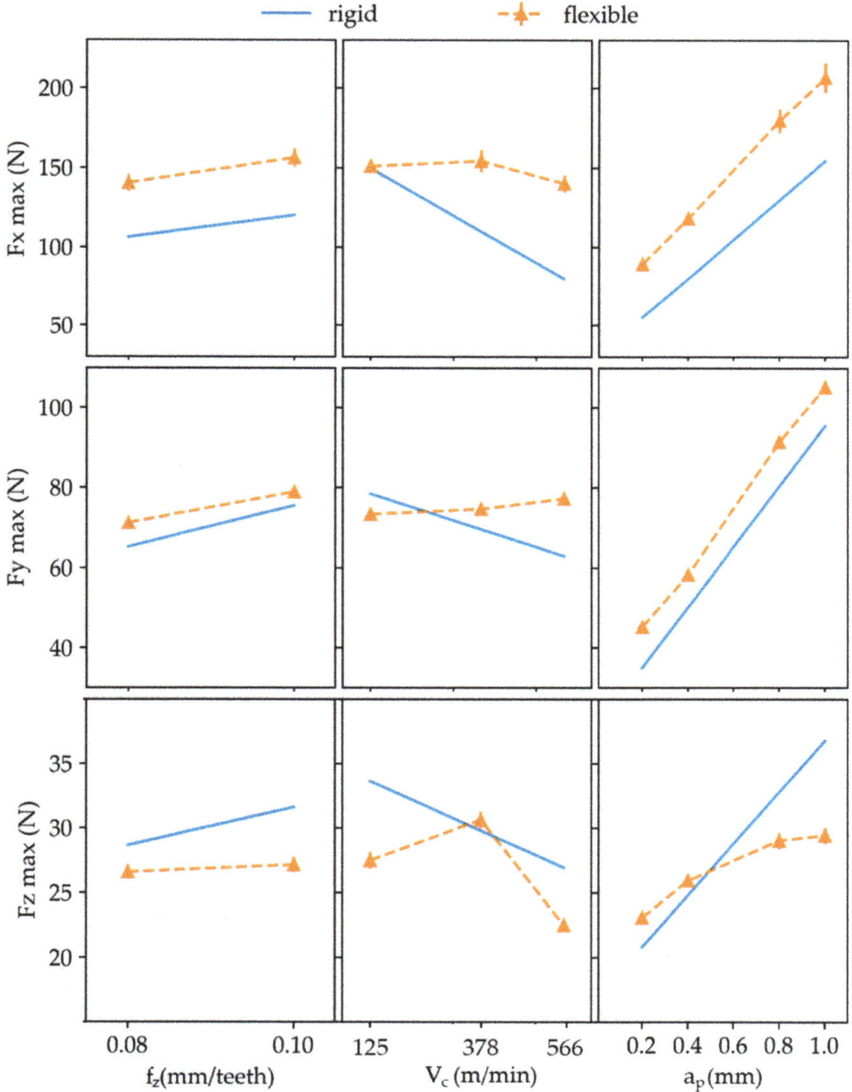

Figure 7. Average forces for the x, y and z axes under rigid (solid blue line) and flexible (dashed orange line) consideration.

Aiming to provide a suitable model to relate the machining parameters to the process performance, the forces have been also studied by following a statistical approach. In this research, the force module and the material removal rate (*MRR*), which can be calculated as is shown in Equation (4), were correlated.

$$MRR = f_z \cdot S \cdot a_p \cdot N \cdot a_r \quad (4)$$

A potential regression was chosen for the model in order to ensure an R^2 close to 0.95. The model equation is shown in Equation (5), where e is the Euler number, N is the number of teeth and a_r is the radial depth of cut.

$$F = MRR^{0.63} e^{-0.44-7.78 \cdot S \cdot 10^{-5}} \quad (5)$$

This model relates the force module (F) to the machining parameters through the MRR, and, as such, the concept of process productivity is introduced. This approach makes it easier to select higher efficient parameters. Both experimental and predicted data are shown in Figure 8. The reduction of forces at high spindle speeds for the same *MRR* is remarkable. This fact can be explained by a working regime close to high speed machining, as was referred to in previous paragraphs. The increase of the process temperature reduced the effort needed to cut the material [33]. Additionally, the heat generated in the process was quickly evacuated, which could have reduced induced residual stresses [34] and tool wear [33].

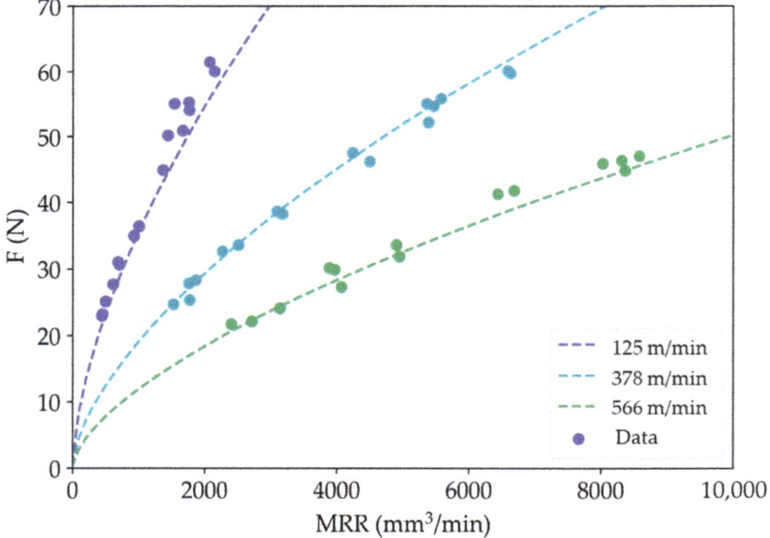

Figure 8. Average force module predicted (dashed lines) and experimental data against the material removal rate (*MRR*) as a function of the cutting speed (V_c).

The force predicted by the proposed model, as a function of the cutting parameters, allowed the machining operation to be monitored using the electrical power registered by the ServoGuide System. Machining cutting power involved all the cutting parameters together, thus giving a closer idea of the overall interactions in the cutting process [35]. This monitoring option could be used as an input for adaptive control systems, in which the instant depth of cut can be controlled and modified online, as an alternative to others online depth of cut control based on ultrasound measurements [36]. This can reduce the number of overcuts and defective parts produced. In this case, the empirical model followed Equation (4), where e is the Euler number and D the tool diameter.

$$Power = \left(\frac{F \cdot S \cdot D \cdot \pi}{60}\right)^{3.5} a_p{}^{-2.2} e^{-16.7} \quad (6)$$

This equation was empirically obtained following an ANOVA approach, using the data represented on Figure 9. Combined relations between the different variables were neglected during the analysis. The R^2 for the final statistical model was 0.953.

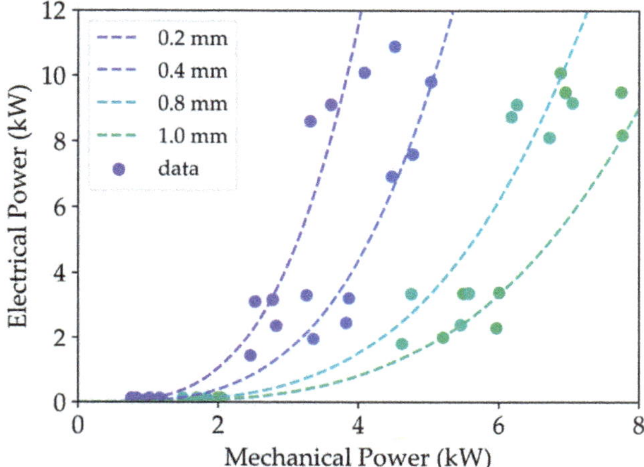

Figure 9. Model correlation between the mechanical power, calculated based on the recorded forces, and the electrical power obtained on the ServoGuide system as a function of the a_p.

4. Conclusions

The machining of aluminium skin panels is used as a sustainable alternative for chemical milling process in the aerospace industry; therefore, it requires tight quality tolerances. This work presents a study of how cutting parameters influence the final thickness, surface roughness and cutting forces of thin plate aluminium parts in order to pursue two main objectives: To ensure the final quality of the part and to find an easy way to monitor the process.

The influence of the cutting speed on the final thickness error map of the machined thin plates has been proven. Higher values of cutting speeds tended to reduce the standard deviation of thickness error values measured in the test samples. The higher the cutting speed, the lower the cutting force module and the higher its excitation frequency, leading to an increase of process stability and a reduction of the results variability. Consequently, an improvement by up to a 40% of the implicit process tolerances has been achieved using a cutting speed of 566 m/min. This fact suggests that these parts could present even more homogenous results in terms of final thickness if higher cutting speeds were used. Roughness values are always under the more restrictive requirements for chemical milling. Lower values of cutting forces under stable machining conditions could ensure roughness values under 0.3 µm.

Furthermore, forces are affected by the low rigidity of the part that obtain lower average values for the z axis than those expected, based on rigid body experiments. A statistical analysis of the tests also revealed high cutting speed parameters as the more efficient ones based on *MRR*, providing a force model that includes all cutting parameter effects.

Finally, this model was used to relate mechanical power and electrical power consumption, allowing us to control online the depth of cut. This model is proposed as an alternative method to implement adaptive control techniques in other to avoid overcuts on aeronautical panels, reducing defective parts at the final stages of the process chain when their value is very high.

Author Contributions: Conceptualization, I.D.S., A.R. and A.J.G.; methodology, I.D.S. and A.R.; formal analysis, I.D.S. and A.J.G.; investigation, I.D.S. and A.R.; resources, A.R.; writing—original draft preparation, I.D.S.; writing–review and editing, A.R. and A.J.G.

Funding: This research was funded by University of Cadiz, grant number University training plan UCA/REC01VI/2016.

Acknowledgments: The authors acknowledge the support given by the Fraunhofer Joint Laboratory of Excellence on Advanced Production Technology (Fh–J_LEAPT Naples).

Conflicts of Interest: The authors declare no conflict of interest.

References

1. Bi, Q.; Huang, N.; Zhang, S.; Shuai, C.; Wang, Y. Adaptive machining for curved contour on deformed large skin based on on-machine measurement and isometric mapping. *Int. J. Mach. Tools Manuf.* **2019**, *136*, 34–44. [CrossRef]
2. Mtorres Surface Milling Machining. Available online: http://www.mtorres.es/en/aeronautics/products/metallic/torres-surface-milling (accessed on 2 June 2016).
3. Dufieux Dufieux Industrie–Modular. Available online: http://www.dufieux-industrie.com/en/mirror-milling-system-mms (accessed on 2 June 2016).
4. Mahmud, A.; Mayer, J.R.R.; Baron, L. Magnetic attraction forces between permanent magnet group arrays in a mobile magnetic clamp for pocket machining. *CIRP J. Manuf. Sci. Technol.* **2015**, *11*, 82–88. [CrossRef]
5. Rubio, A.; Rivero, A.; Del Sol, I.; Ukar, E.; Lamikiz, A.; Rubio-Mateos, A.; Rivero–Rastrero, A.; Del Sol-Illana, I.; Ukar-Arrien, E.; Lamikiz-Mentxaka, A. Capacitation of flexibles fixtures for its use in high quality machining processes: An application case of the Industry 4.0 paradigm. *DYNA* **2018**, *93*, 608–612. [CrossRef]
6. Campa, F.J.; López de Lacalle, L.N.; Celaya, A. Chatter avoidance in the milling of thin floors with bull-nose end mills: Model and stability diagrams. *Int. J. Mach. Tools Manuf.* **2011**, *51*, 43–53. [CrossRef]
7. Del Sol, I.; Rivero, A.; Norberto, L.; Gamez, A.J. Thin–Wall Machining of Light Alloys: A Review of Models and Industrial Approaches. *Materials* **2019**, *12*, 2012. [CrossRef] [PubMed]
8. Elbestawi, M.A.; Sagherian, R. Dynamic modeling for the prediction of surface errors in the milling of thin-walled sections. *J. Mater. Process. Technol.* **1991**, *25*, 215–228. [CrossRef]
9. Tian, W.; Ren, J.; Wang, D.; Zhang, B. Optimization of non-uniform allowance process of thin-walled parts based on eigenvalue sensitivity. *Int. J. Adv. Manuf. Technol.* **2018**, *96*, 2101–2116. [CrossRef]
10. Olvera, D.; Urbikain, G.; Elías-Zuñiga, A.; López de Lacalle, L. Improving Stability Prediction in Peripheral Milling of Al7075T6. *Appl. Sci.* **2018**, *8*, 1316. [CrossRef]
11. Urbikain, G.; Olvera, D.; de Lacalle, L.N.L. Stability contour maps with barrel cutters considering the tool orientation. *Int. J. Adv. Manuf. Technol.* **2017**, *89*, 2491–2501. [CrossRef]
12. Liu, X.; Gao, H.; Yue, C.; Li, R.; Jiang, N.; Yang, L. Investigation of the milling stability based on modified variable cutting force coefficients. *Int. J. Adv. Manuf. Technol.* **2018**, *96*, 2991–3002. [CrossRef]
13. Qu, S.; Zhao, J.; Wang, T. Three–dimensional stability prediction and chatter analysis in milling of thin-walled plate. *Int. J. Adv. Manuf. Technol.* **2016**, *86*, 2291–2300. [CrossRef]
14. Campa, F.J.; López de Lacalle, L.N.; Lamikiz, A.; Sánchez, J.A. Selection of cutting conditions for a stable milling of flexible parts with bull–nose end mills. *J. Mater. Process. Technol.* **2007**, *191*, 279–282. [CrossRef]
15. Ding, H.; Ke, Y. Study on Machining Deformation of Aircraft Monolithic Component by FEM and Experiment. *Chinese J. Aeronaut.* **2006**, *19*, 247–254. [CrossRef]
16. Tuysuz, O.; Altintas, Y. Frequency Domain Updating of Thin-Walled Workpiece Dynamics Using Reduced Order Substructuring Method in Machining. *J. Manuf. Sci. Eng.* **2017**, *139*, 071013. [CrossRef]
17. Lin, X.; Wu, D.; Yang, B.; Wu, G.; Shan, X.; Xiao, Q.; Hu, L.; Yu, J. Research on the mechanism of milling surface waviness formation in thin-walled blades. *Int. J. Adv. Manuf. Technol.* **2017**, *93*, 2459–2470. [CrossRef]
18. Yan, Q.; Luo, M.; Tang, K. Multi–axis variable depth-of-cut machining of thin-walled workpieces based on the workpiece deflection constraint. *CAD Comput. Aided Des.* **2018**, *100*, 14–29. [CrossRef]
19. Sonawane, H.A.; Joshi, S.S. Modeling of machined surface quality in high-speed ball-end milling of Inconel-718 thin cantilevers. *Int. J. Adv. Manuf. Technol.* **2015**, *78*, 1751–1768. [CrossRef]

20. De Oliveira, E.L.; de Souza, A.F.; Diniz, A.E. Evaluating the influences of the cutting parameters on the surface roughness and form errors in 4–axis milling of thin–walled free–form parts of AISI H13 steel. *J. Braz. Soc. Mech. Sci. Eng.* **2018**, *40*, 334. [CrossRef]
21. Del Sol, I.; Rivero, A.; Salguero, J.; Fernández–Vidal, S.R.; Marcos, M. Tool–path effect on the geometric deviations in the machining of UNS A92024 aeronautic skins. *Procedia Manuf.* **2017**, *13*, 639–646. [CrossRef]
22. Seguy, S.; Campa, F.J.; López de Lacalle, L.N.; Arnaud, L.; Dessein, G.; Aramendi, G. Toolpath dependent stability lobes for the milling of thin-walled parts. *Int. J. Mach. Mach. Mater.* **2008**, *4*, 377–392. [CrossRef]
23. Mahmud, A.; Mayer, J.R.R.; Baron, L. Determining the minimum clamping force by cutting force simulation in aerospace fuselage pocket machining. *Int. J. Adv. Manuf. Technol.* **2015**, *80*, 1751–1758. [CrossRef]
24. Li, B.; Jiang, X.; Yang, J.; Liang, S.Y. Effects of depth of cut on the redistribution of residual stress and distortion during the milling of thin–walled part. *J. Mater. Process. Technol.* **2015**, *216*, 223–233. [CrossRef]
25. Altintaş, Y.; Budak, E. Analytical Prediction of Stability Lobes in Milling. *CIRP Ann. Manuf. Technol.* **1995**, *44*, 357–362. [CrossRef]
26. Budak, E. Analytical models for high performance milling. Part I: Cutting forces, structural deformations and tolerance integrity. *Int. J. Mach. Tools Manuf.* **2006**, *46*, 1478–1488. [CrossRef]
27. Gradišek, J.; Kalveram, M.; Weinert, K. Mechanistic identification of specific force coefficients for a general end mill. *Int. J. Mach. Tools Manuf.* **2004**, *44*, 401–414. [CrossRef]
28. *ISO, EN. 4288—Geometrical Product Specifications (GPS)—Surface Texture: Profile Method–Rules and Procedures for the Assessment of Surface Texture*; International Organization for Standardization: Geneva, Switzerland, 1996.
29. Chen, J.S.; Hsu, W.Y. Characterizations and models for the thermal growth of a motorized high speed spindle. *Int. J. Mach. Tools Manuf.* **2003**, *43*, 1163–1170. [CrossRef]
30. Ratchev, S.; Liu, S.; Huang, W.; Becker, A.A. Milling error prediction and compensation in machining of low–rigidity parts. *Int. J. Mach. Tools Manuf.* **2004**, *44*, 1629–1641. [CrossRef]
31. Bolar, G.; Das, A.; Joshi, S.N. Measurement and analysis of cutting force and product surface quality during end-milling of thin-wall components. *Meas. J. Int. Meas. Confed.* **2018**, *121*, 190–204. [CrossRef]
32. López De Lacalle, L.N.; Lamikiz, A.; Sánchez, J.A.; Bustos, I.F. de Recording of real cutting forces along the milling of complex parts. *Mechatronics* **2006**, *16*, 21–32. [CrossRef]
33. Brinksmeier, E.; Preuss, W.; Riemer, O.; Rentsch, R. Cutting forces, tool wear and surface finish in high speed diamond machining. *Precis. Eng.* **2017**, *49*, 293–304. [CrossRef]
34. Masoudi, S.; Amini, S.; Saeidi, E.; Eslami-Chalander, H. Effect of machining-induced residual stress on the distortion of thin-walled parts. *Int. J. Adv. Manuf. Technol.* **2014**, *76*, 597–608. [CrossRef]
35. Ma, Y.; Feng, P.; Zhang, J.; Wu, Z.; Yu, D. Energy criteria for machining-induced residual stresses in face milling and their relation with cutting power. *Int. J. Adv. Manuf. Technol.* **2015**, *81*, 1023–1032. [CrossRef]
36. Rubio, A.; Calleja, L.; Orive, J.; Mújica, Á.; Rivero, A. Flexible Machining System for an Efficient Skin Machining. *SAE Int.* **2016**, *1–12*. [CrossRef]

Article

Influence of Elastomer Layers in the Quality of Aluminum Parts on Finishing Operations

Antonio Rubio-Mateos [1,*], Asuncion Rivero [1], Eneko Ukar [2] and Aitzol Lamikiz [2]

1 TECNALIA, Basque Research and Technology Alliance, Paseo Mikeletegi 7, E-20009 Donostia-San Sebastian, Spain; asun.rivero@tecnalia.com

2 Department of Mechanical Engineering, University of the Basque Country, C/ Alameda de Urquijo s/n, E-48013 Bilbao, Spain; eneko.ukar@ehu.es (E.U.); aitzol.lamikiz@ehu.es (A.L.)

* Correspondence: antonio.rubio@tecnalia.com; Tel.: +34-667-119-712

Received: 31 January 2020; Accepted: 20 February 2020; Published: 22 February 2020

Abstract: In finishing processes, the quality of aluminum parts is mostly influenced by static and dynamic phenomena. Different solutions have been studied toward a stable milling process attainment. However, the improvements obtained with the tuning of process parameters are limited by the system stiffness and external dampers devices interfere with the machining process. To deal with this challenge, this work analyzes the suitability of elastomer layers as passive damping elements directly located under the part to be machined. Thus, exploiting the sealing properties of nitrile butadiene rubber (NBR), a suitable flexible vacuum fixture is developed, enabling a proper implementation in the manufacturing process. Two different compounds are characterized under axial compression and under finishing operations. The compression tests present the effect of the feed rate and the strain accumulative effect in the fixture compressive behavior. Despite the higher strain variability of the softer rubber, different milling process parameters, such as the tool feed rate, can lead to a similar compressive behavior of the fixture regardless the elastomer hardness. On the other hand, the characterization of these flexible fixtures is completed over AA2024 floor milling of rigid parts and compared with the use of a rigid part clamping. These results show that, as the cutting speed and the feed rate increases, due to the strain evolution of the rubber, the part quality obtained tend to equalize between the flexible and the rigid clamping of the workpiece. Due to the versatility of the NBR for clamping different part geometries without new fixture redesigns, this leads to a competitive advantage of these flexible solutions against the classic rigid vacuum fixtures. Finally, a model to predict the grooving forces with a bull-nose end mill regardless of the stiffness of the part support is proposed and validated for the working range.

Keywords: vibrations; part quality; flexible vacuum fixture; AA2024 floor milling

1. Introduction

Monolithic aluminum components are widely employed in the aeronautical sector due to their good strength-to-weight ratio [1]. The final quality of these parts is normally obtained or improved in the finishing operation and it is influenced by static and dynamic phenomena [2].

On the one hand, from the static point of view, cutting forces and part clamping produce elastic deformation that can lead to deteriorating the final dimension and the surface of the workpiece [3]. On the other hand, vibrations increase the roughness of the parts. These dynamic instabilities become frequent in the milling operation and are produced by the lack of dynamic stiffness in one or more components of the system [4]. The most characteristic vibrations appeared in the milling operation are the self-excited vibrations or chatter [5,6]. However, even in the absence of chatter, it almost always exists forced vibrations derived from the periodic excitation of the intermittent cutting engagement of the milling cutter on the workpiece [7].

Thus, in order to improve the part quality different approach has been studied. First, the tuning of the cutting parameters can lead to static and dynamic improvements. Thus, a process forces reduction leads to a decrease of the elastic deformation of the workpiece that it is reflected in the part accuracy. For instance, Perez et al. [8] obtained a machining forces reduction and an improvement of the compressive residual stresses with the increase of the cutting speed. On the other hand, in terms of vibrations, different surveys have been developed for the improvement of the system dynamic stability and the obtained surface quality. Different stability models have been developed for the milling operations of compliant systems [9,10] and the effect of the cutting parameters on the process damping have been analyzed [11]. However, these solutions are limited for the inherent stiffness of the system.

The key element for the increase of the system stiffness is the workpiece clamping. Thus, different part-fixture systems have been employed to guarantee a suitable part positioning and fixing [12]. Due·to the lack of dynamic stability of some of these solutions, different damping features have been implemented in the system in order to improve the machining process. Thus, different active features based on the use of eddy currents [13,14], pivot mechanism [15,16] and piezoelectric dampers [17] have been studied. These solutions are cost efficient and their implementation is limited to certain applications.

The use of passive damping elements is increasing for milling operations as they are more cost efficient compared to the active developments. The passive damping systems are based in the implementation of different elements or fluids with outstanding damping properties to stabilize dynamically the system. Thus, by employing electrorheological [18] or magnetorheological [19] fluids, the vibration amplitude of the cutting processes varies and the part quality is improved. Moreover, in order to increase the narrow vibration band of these passive dampers, Yang et al. [20] developed a tunable passive devices. However, the industrial implementation of these passive solutions are challenging as they interfere with the clamping of the workpiece and the machining process.

In the present study, the use of an elastomer layer employed as a passive damping element is proposed and characterized. Elastomers, particularly rubber materials, are ideal materials for vibrations isolation as they are low in cost with high internal friction [21]. Moreover, the industrial implementation of these compounds for machining applications is feasible as they are employed as passive control of vibration [22–24]. In fact, the damping properties of these elastomers have been analyzed for low frequency [25] and high frequency applications [26], including under certain machining operations. For instance, Kolloru et al. [27] employed neoprene layers combined with torsion springs to reduce up to eight times the vibration in the milling process of circular thin-wall components. On the other hand, Liu et al. [28] implemented a viscoelastic material in the toolholder to increase by 99% its damping ratio. Nevertheless, there is no study of the direct application of rubber materials as the clamping element of workpieces in milling operations.

In this case, in order to combine a fixture and a passive damping system the use of a nitrile butadiene rubber (NBR) layer is proposed. This sort of elastomers is one of the most employed seal component in the oil and gas industry [29], and the proposed development benefitted from these outstanding sealing properties [30,31] to transform a flexible layer into a suitable vacuum table. Hence, these solutions enable milling in aggressive environments, with capacity to clamp different geometries. Moreover, as the passive damping element act as a fixture, its industrial implementation is feasible as the interference with the rest of the machining system is reduced.

In order to characterize the behavior of these elastic polymers under the machining processes loads, compression tests and milling tests have been performed. Thus, these flexible solutions have been characterized in terms of chatter and forced vibrations performed by the milling tool.

Regardless the milling strategy, the most aggressive machining zone is the entrance of each pocketing where the tool machines with an axial pitch equal to its diameter. Thus, the analysis is focused in the grooving application with depth of cuts defined by finishing operations. The suitability survey is performed in terms of part quality. First, the machined depth is measured to quantify the groove thickness error. Then, the floor roughness is measured and analyzed. Finally, the dynamic behavior of each system is characterized, and a universal force model is developed for the grooving operation in finishing applications.

2. Materials and Methods

Two different NBR layers were selected for the analysis as these passive elements are defined by different vibration bands. The mechanical properties of both vulcanized rubber materials are shown on Table 1. Besides the hardness and the density, the compounds ingredients are given, where the carbon black is a form of elemental carbon that is used to increase the resistance of rubber and also to improve the tensile strength [32].

Table 1. Materials mechanical properties.

Properties	Rubber A	Rubber B
Hardness (Sh·A)	65	90
Density (g/cm^3)	1.45	1.43
Polymer (wt.%)	37.3	54.6
Carbon black (wt.%)	3.5	14.4
Other inorganic charges (wt.%)	59.2	31.0

The mechanical behavior of rubber depends on the amplitude, feed rate and frequency of the applied load, combined with the temperature of the material [33]. In the case of milling operations, the amplitude and feed rate of the applied forces are completely defined by the machining conditions.

Similarly, the load frequencies suffered by the part are generated by the milling tool rotation and by the workpiece fundamental modes. Finally, the temperature of the material is influenced by the heat generated on the cutting zone and the room temperature.

Based on the load application strategies employed on this survey, some simplifications were considered. For instance, the decrease in stiffness during the first few cyclic loads, the so-called Mullins effect [34], was neglected. Therefore, different loads prior to each test were performed over each elastomer layer. The characteristics of these loads were defined in terms of the test to perform. Thus, for compression tests, a compression load was performed prior to each test. Likewise, prior to each milling test, a previous groove was performed to reduce the Mullins effect on the rubber and to level the upper side of each slot.

Finally, due to the reduced compression loads during the machining operation and the wide part area in contact with the elastomer layer, the expected strain amplitudes are minimal. Therefore, it is not considered a rubber heat up due to material damping derived from large harmonic loads [33]. Hence, due to the part thickness located between the cutting zone and the rubber layer, the temperature of the rubber was considered as the room value.

Tests were performed in a standard 5-axis numeric control (NC) center. The selected geometry for the elastomer layers was a 300 × 300 mm^2. The mean value of the thickness for both cases was 14.2 mm with a tolerance of ±5%. In order to guarantee a uniform contact and clamping conditions between the part and the elastic material, a slot grid was machined in each rubber layer (Figure 1a). Thus, the vacuum clamping force was distributed along the contact area by means of the channels. Then, the air was removed through a unique orifice and the part could be safely clamped during the machining operation, as shown in Figure 1b.

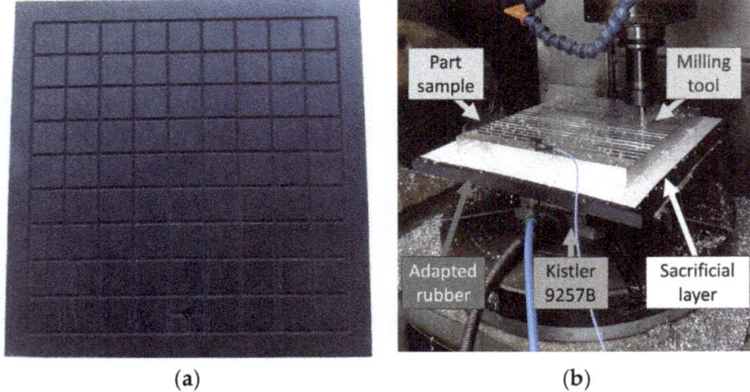

Figure 1. Adapted rubber: (**a**) vacuum channels distribution. (**b**) Rubber layer implementation as a vacuum fixture on milling tests.

Both, the compression of the rubber and the part profile before and after each milling tests were monitored with a GT1000 type linear variable differential transformer (LVDT) gauging transducer (RDP Group, Wolverhampton, UK). The forces were registered with Kistler 9257B measurement equipment (Kistler Ibérica S.L., Granollers, Spain). In each test, following another similar set-up [35], the part and the elastic element were attached to the force sensor with a synthetic rubber adhesive. This double-sided filmic tape TESA 64620 (Tesa Tape S.A., Argentona, Spain) guaranteed a homogeneous clamping due to the compressive nature of the axial loads in compression and milling tests.

2.1. Compression Tests

For most of engineering rubbers, material damping is caused by two different mechanism, resulting in amplitude and rate dependent behavior [36]. Thus, the objective of these compression tests is characterizing the effect of the feed rate on the strain and comparing the influence of the strain cycles on each rubber. In general, compression tests on rubber materials are performed with circular samples [37]. However, in order to include the effect of the slots in the material deformation, the tests are implemented in the same elastic layer employed as vacuum fixture, see Figure 2.

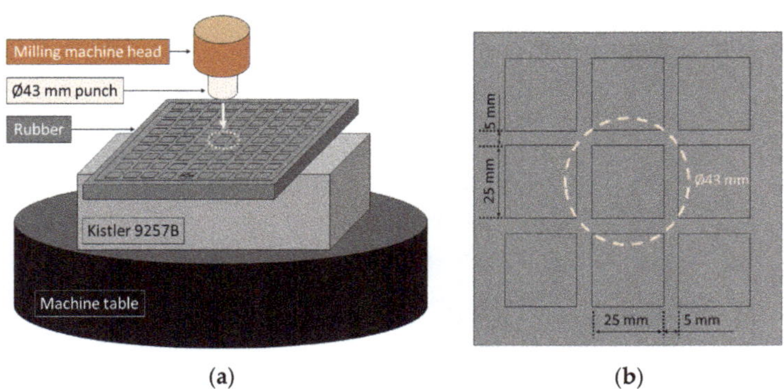

Figure 2. Compression test procedure: (**a**) set-up scheme. (**b**) Load application zone.

The axial loads were applied by the machine head by means of a cylindrical punch and its position was monitored with a LVDT. The feed rates conditions were selected based on the most extreme cases tested in the milling tests. Three repetitions were performed for each condition and, in order to evaluate the strain accumulative effect, a 15 min relaxation period was guaranteed between the successive tests.

2.2. Milling Tests

For the milling survey, the same adapted rubber layers were employed. The part samples to be machined were 20 mm thick AA2024-T3 rigid blocks. These parts were 240 × 240 mm^2 wide and were located in the center of the elastic support. Hence, any rubber edge effect could be neglected, and its local thickness tolerance was diminished from ±5% to ±3%. In order to reduce the vacuum leaks a 290 × 290 × 0.7 mm^3 sacrificial porous layer was included between the elastic element and the specimen to be machined. Hence, the vacuum leaks depended on the part area and it was not influenced by the part contour. Thus, different part geometries could be clamped without any fixture redesign.

The air from the channels was removed through the hole with a standard Venturi guaranteeing a proper vacuum union between the rubber and the aluminum part sample for all the working range, see Figure 3. Then, the rubber was held to the dynamometric table with the double-sided filmic tape. In the tests with no rubber, the part sample was stuck directly to the Kistler by the same token.

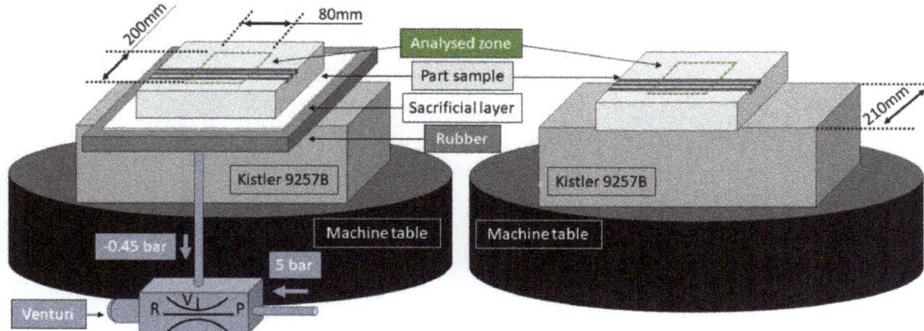

Figure 3. Milling test set-up with and without a rubber layer.

Groove milling was the selected machining operation. These slots were dry machined side to side, in two steps. First, a previous 0.2 mm groove was performed in order to guarantee the same initial profile between tests. Then, the test with each condition was milled. The separation between each groove was 10 mm.

The selected tool was a two flutes bull-nose end-mill Kendu 4400 (Kendu, Segura, Spain), with a diameter of 10 mm and a 2.5 mm edge radius. Table 2 shows the tests conditions. Therefore, the effect of the depth of cut (a_p), feed per tooth (f_z), spindle speed (n) or cutting speed (v_c) and the tool feed rate (f) on both elastic polymers could be studied and compared with the use of a rigid clamping. Two different depth of cuts were selected based on the values employed in finishing operations in the aeronautic field [38]. On the other hand, three different tool rotation values were studied in order to reduce vibrations generated out of the tool-part system while two feed per tooth values were selected for maintaining a suitable milling process of aluminum parts [9,38]. Hence, based on the feed per tooth and spindle speed configurations, six different milling conditions were analyzed for each depth of cut.

Table 2. Machining tests conditions.

Parameters	Level 1	Level 2	Level 3
Clamping material	Rubber A	Rubber B	No rubber
a_p(mm)	0.2	0.8	-
f_z (mm/tooth)	0.06	0.1	-
n (rpm)	2000	4000	6000
v_c (m/min)	63	126	189
f (mm/min)	240; 400	480; 800	720; 1200

For each milling condition, three repetitions were performed in different random positions relative to the center of the part. The analyzed zone was restricted to each groove middle area. Then, the real machined thickness was evaluated by measuring the part profile beforehand and afterward each milling operation. This measure was performed with the previously presented LVDT attached to the machine tool head with an adaptor. Besides, the roughness of the floor of each slot was measured in four different zones equally separated by 20 mm. Thus, the R_a value was evaluated with a Mitutoyo Surftest SV-2000 (Mitutoyo, Kawasaki, Japan) roughness measure station. As a reference, the typical tolerance values in the aeronautical industry were about ±0.1 mm for final thickness and under 1.6 µm for the R_a [39].

2.3. Force Mechanistic Model in the Tool Axis Direction with a Bull-Nose Mill

In order to evaluate the stability for each set-up, stability lobe diagrams (SLD) were calculated in the specimen center with an impact hammer test. A uniaxial PCB accelerometer model 352C22 (PCB Piezotronics, Inc, Depew, NY, USA) with a measuring range from 1 kHz to 10 kHz and a sensitivity of 1.0 mV/(m/s^2) was employed to register the tool and part vibration. The maximum acceptable a_p in the stable regime is calculated with the model described by Altintas and Budak [10]. The cutting forces (tangential (t), radial (r) and axial (a)) over the cutting edge i could be considered as a function of the friction coefficients (K_{te}, K_{re} and K_{ae}) and the shearing cutting coefficients (K_{tc}, K_{rc} and K_{ac})

$$\left\{\begin{array}{c}\partial F_t\\ \partial F_r\\ \partial F_a\end{array}\right\}=\left\{\begin{array}{c}K_{te}\\ K_{re}\\ K_{ae}\end{array}\right\}\times\partial S+\left\{\begin{array}{c}K_{tc}\\ K_{rc}\\ K_{ac}\end{array}\right\}\times f_z\times sin\phi_i\times\partial z \quad (1)$$

In this equation, ∂S is the length of the differential chip edge, ϕ_i is the angular position of the cutting edge i measured from axis Y, perpendicular to the tool feed direction, and ∂z is the depth of cut.

Compared with other tool geometries, bull-nose end mills have a variable radius and helix angle along the tool axis. Likewise, the lead angle increases its value from 0° to 90° in the toroidal part, and then kept constant and equal to 90° all along the flank [40].

This geometrical variation combined with cutting speed and the depth of cut leads to variable cutting coefficients. This nonlinearity could be solved using a linear model to calculate the SLD [9]. However, Altintas [41], simplified a circular insert geometry taking an average edge angle of 45°.

In this case, the model was oriented to the floor finishing application. For these cases machined depths were usually focused in a range between 0.2 mm and 1.2 mm, mainly in low stiffness parts. This means that the edge angle was located between 11° and 29°. Thus, for this survey, the average edge angle was defined as 20°.

The friction and shearing cutting coefficients were obtained by solving the equation by using the force values obtained in the milling tests. These coefficients were considered constant for all the milling conditions. The model results were employed to predict the SLD for each flexible fixture and compared with the use of a rigid clamping underneath the part sample. Moreover, with the axial forces obtained in these tests, a model was proposed regardless of the hardness of the support.

3. Results

3.1. Rubbers Compressive Behaviour

The differences in the composition of each tested rubber led to a completely different stress-strain behavior. In the Figure 4 it can be observed the strain variation of each rubber based on the stress and feed rate evolution. This evolution is presented with a fifth-degree interpolation in order to visualize the more linear behavior of rubber B compared with rubber A.

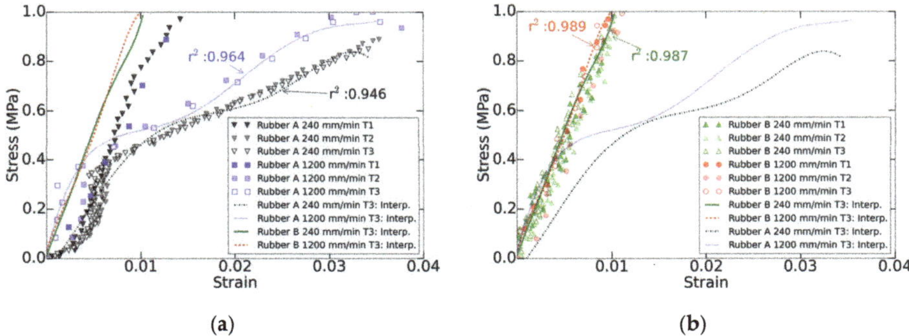

Figure 4. Accumulative stress and feed rate increase effects on both rubber materials: (**a**) rubber A and (**b**) rubber B.

Both elastomers increased their elastic modulus as the feed rate rose. The rate effect was usually attributed to the resistance in reorganizing the polymeric chains during the loading period. Since this reorganization cannot occur instantaneously, the loss of energy is rate dependent [33]. Moreover, the polymeric chains in the rubber A lacked time in the relaxing period to return to the original state and, thus, the elastic modulus decreased for the second and third trial. This effect was not noticeable in the rubber B. However, it can be observed that, as the feed rates increased, the behavior of both elastomers tended to match for stresses under 0.4 MPa, which was within the work range for the milling tests. Hence, regardless of the rubber composition, these flexible fixtures presented stress-strain behavior influenced by the load amplitude and feed rate transmitted by the tool and by the previous deformation implemented on the rubber.

3.2. Thickness Error

The thickness error is defined as the difference between the experimental and theoretical thickness. In addition to the static and dynamic phenomena that occurred when applying loads over the elastic layers, other effects such as the machine precision, repeatability and the thermal expansion of the spindle had an influence over the real machined thickness. For instance, the repeatability for the rubber A was within ±9 µm, for the rubber B was within ±19 µm and for the use of no rubber was within ±8 µm.

In order to analyze this parameter, an analysis of variance (ANOVA) was employed. Thus, the influence of the main machining parameters in the thickness error was evaluated. Therefore, first, the normal distribution of the data was checked by the Anderson-Darling (AD) test, and the variance homogeneity with the Bartlett's test. In both cases, the confidence interval of 95% ($\alpha = 0.05$) was employed. As it can be observed in the Table 3, for all the tests their *p*-values were over α and, thus, were suitable for an ANOVA.

Table 3. Analysis of the suitability of the thickness error data.

Analysis	Parameter	Rubber A	Rubber B	No Rubber
Normal distribution	AD	0.276	0.322	0.447
	p-value	0.628	0.510	0.257
Homogeneity of variance	Bartlett's	4.480	5.160	2.310
	p-value	0.723	0.640	0.941

On the other hand, a variance analysis was performed to determine the main parameters affecting the machined depth inaccuracy. In this case, the null hypothesis was that the factors or their combination have no influence over the thickness error. As it is detailed in the Table 4, from this survey it was obtained that, with a 95% confidence, the null hypothesis was proved to be true. The only exception was the effect of spindle speed in the case of the rubber A as its p-value was under α, see Table 4 values in bold.

Table 4. Analysis of variance of the thickness error data.

Factor	Parameter	Rubber A	Rubber B	No Rubber
n	F-value	7.040	3.200	1.990
	p-value	0.017	0.093	0.178
f_z	F-value	0.680	0.430	0.350
	p-value	0.420	0.520	0.562
a_p	F-value	1.540	1.940	0.050
	p-value	0.233	0.182	0.830
$n \times f_z$	F-value	0.000	0.080	0.090
	p-value	0.990	0.787	0.770
$n \times a_p$	F-value	0.300	0.090	0.140
	p-value	0.593	0.774	0.716
$f_z \times a_p$	F-value	0.540	0.280	0.060
	p-value	0.475	0.606	0.804
$n \times f_z \times a_p$	F-value	1.200	0.010	0.350
	p-value	0.289	0.936	0.563

This result was coherent with the compression tests, as the rubber A was the most sensible to strain changes. Furthermore, as it can be observed in the Figure 5, there was a global decrease in the thickness mean error as the v_c increased. In this case, the positives values meant that the system was compressed, and the depth of cut was lower than programmed. This effect, as expected, was more noticeable with the use of rubber as a support. In the other hand, if the thickness error had a negative value, it meant that the tool machined more depth than expected. This last effect was mainly caused by the thermal expansion of the spindle, as it increased combined with the revolutions [42]. This error can be compensated previous to the machining [43] or even with in-process tool position adjustments [44,45].

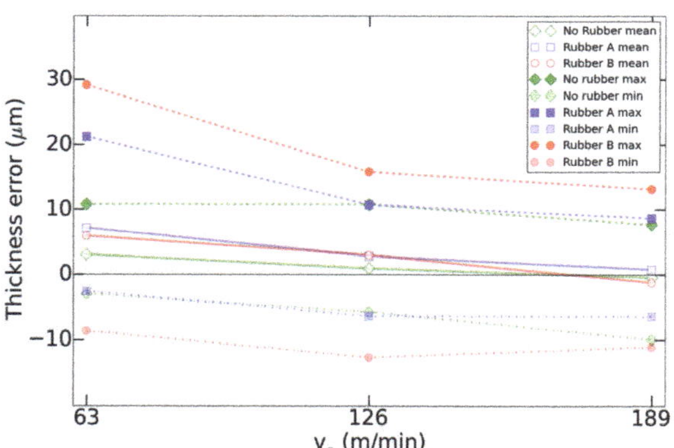

Figure 5. Thickness error evolution based on the v_c variation.

Another noticeable effect was that, as the cutting speed increased, the thickness errors tended to equalize. This effect matched with the fact that, due to the cutting conditions tested, as the cutting speed rose, the feed rates increased accordingly. Then, as it is observed in the compression analysis, the elastic modulus of the rubbers rose as the feed rates increased, leading to a more rigid-like support. This is aligned with the industrial implementation of this system in milling operations of aluminum, as the productivity of these applications tend to the employment of these or even higher cutting speeds.

Finally, there was a thickness error component that was caused by the system vibration and that produced the difference in the variability of each system. This vibration was analyzed in terms of roughness in the next section.

3.3. *Roughness*

In order to analyze the effect of the machining parameters on the floor R_a on the slots, another ANOVA was performed. Likewise in the thickness error analysis, the data suitability was analyzed with an AD and a Bartlett's test. As shown in the Table 5, it was demonstrated that data met, with 95% of confidence, the requirements for a valid ANOVA.

Table 5. Analysis of the suitability of the roughness data.

Analysis	Parameter	Rubber A	Rubber B	No Rubber
Normal distribution	AD	0.251	0.497	0.305
	p-value	0.694	0.181	0.528
Homogeneity of variance	Bartlett's	3.400	12.750	5.510
	p-value	0.846	0.078	0.599

With this data, a variance analysis was replicated based on the effect of the machining parameter on the roughness obtained on the groove floor. As it is can be observed in the Table 6 in bold, compared with the thickness error, more parameters and their combination affected the part vibration. This influence was more noticeable in the rubber A, as it happened with the thickness error, due to a higher sensibility to strain variations.

Table 6. Analysis of variance of the thickness error data.

Factor	Parameter	Rubber A	Rubber B	No Rubber
n	F-value	54.100	2.880	17.360
	p-value	0.000	0.128	0.003
f_z	F-value	42.240	13.910	4.850
	p-value	0.000	0.006	0.059
a_p	F-value	0.820	2.690	0.790
	p-value	0.391	0.140	0.399
$n \times f_z$	F-value	16.740	2.280	2.990
	p-value	0.003	0.170	0.122
$n \times a_p$	F-value	1.070	0.110	0.590
	p-value	0.331	0.746	0.464
$f_z \times a_p$	F-value	8.320	0.460	0.040
	p-value	0.020	0.519	0.844
$n \times f_z \times a_p$	F-value	0.450	0.000	0.100
	p-value	0.521	0.994	0.755

Despite the dependence on the machining parameters of the roughness, there was no direct influence of a single parameter into the behavior of the three systems at a time. However, it was clear that the most influential parameters were the spindle speed and the feed rate. Thus, in the Figure 6, it can be observed the evolution of the roughness with the increase of the cutting speed and the feed per tooth. In this case the repeatability for the rubber A was within ±0.13 μm, for the rubber B was within ±0.17 μm and for the use of no rubber was within ±0.11 μm.

Results show that roughness obtained with rubber A tended to match the one obtained with the part robustly clamped to the machine as the cutting speed increased. This effect, as explained in the case of the thickness error, was caused by the increase of the elastic modulus. However, this increase in the cutting speed has to be balanced with the feed rate in order to maintain the feed per tooth.

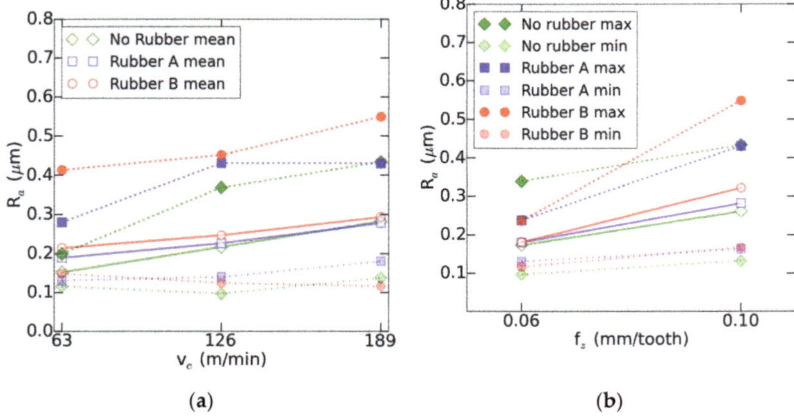

(**a**) (**b**)

Figure 6. Roughness evolution: (**a**) based on the cutting speed and (**b**) based on the feed per tooth.

On the other hand, rubber B tests suffered higher roughness and wider variability. As the stiffness of rubber B was above the rubber A's, the instability must be caused by the vacuum union between the part and the rubber. Thus, due to the higher hardness of the rubber B, the contact with the part did not perform proper clamping conditions as the rubber A.

This analysis indicates that cutting loads applied by the machining tool did not affect exclusively the rubber compression but the clamping suitability as well. Despite rubber A having more variable compression behavior, its lower hardness improved the fixture clamping capacity and the obtained part quality.

3.4. Force Model

The SLD performed over the three systems, as shown in the Figure 7, presented the identical behavior of them. The reason was the combination of a hammer shot at a high feed rate and a wide supporting area of the rigid part. Then, as observed in the compression tests, these fixtures based on elastomers, at high feed rates behaved as a rigid system in terms of chatter vibrations. Thus, these results proved that there was no chatter on the performed milling tests as the maximum depths of the cut were below these curves.

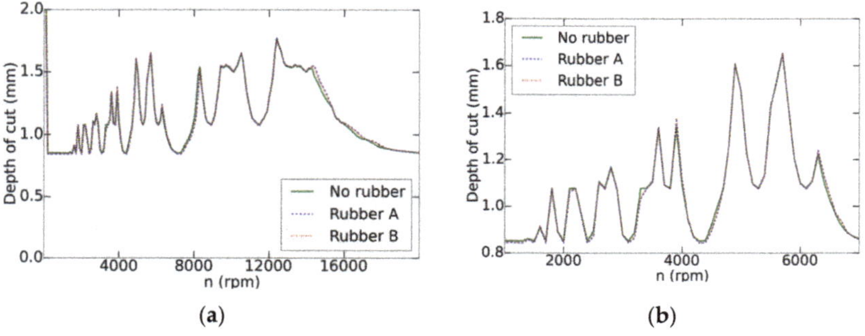

(**a**) (**b**)

Figure 7. Stability Lobe Diagrams (SLD) variation: (**a**) complete and (**b**) zoomed on the studied zone.

The analysis of the force harmonics, see Figure 8, confirmed that the vibration was mainly influenced by the tool cutting loads. The case of rubber A and no rubber had similar behavior, with lower amplitudes and with the cutting per tooth as the main driver of the vibration. However, the forced vibrations at different harmonics were higher in rubber B. Once again, this evidence confirmed that the union between rubber B and the part was not completely suitable.

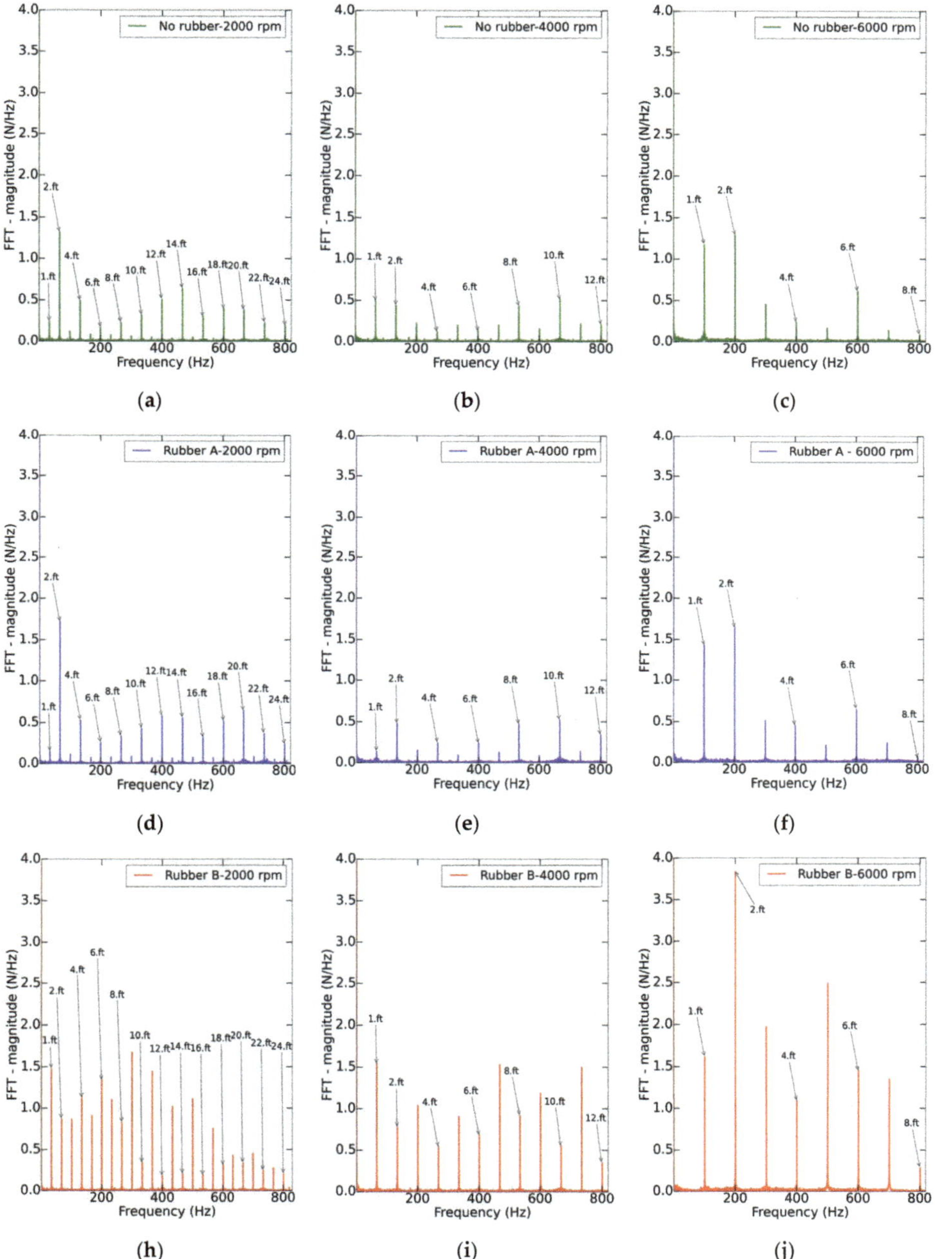

Figure 8. Fast Fourier Transform of the F_z signal for the test at f_z = 0.1 mm/tooth and a_p = 0.8 mm, under different spindle speeds: (**a**) No rubber - 2000 rpm, (**b**) No rubber - 4000 rpm, (**c**) No rubber - 6000 rpm, (**d**) Rubber A - 2000 rpm, (**e**) Rubber A - 4000 rpm, (**f**) Rubber A - 6000 rpm, (**h**) Rubber B - 2000 rpm, (**i**) Rubber B - 4000 rpm and (**j**) Rubber B - 6000 rpm.

Then, the main cutting forces in the axial direction were studied following an empirical approach in order to provide a suitable model able to relate them with the machining parameters regardless the part support. The repeatability for all the tests was within ±2 N. The process parameters were grouped around the material removal rate (MRR), where N is the number of teeth, and a_r is the radial depth of cut:

$$MRR = n \times f_z \times a_p \times N \times a_r \quad (2)$$

The model is based on a potential regression, see Equation (3). The R^2 of this model is 0.984. This equation emphasized, once again, the strong influence of the cutting speed on the machining process

$$F_z = 67.22 \times n^{-0.58} \times MRR^{0.49} \quad (3)$$

Figure 9 presents how the model fits with the analyzed data. As it can be observed, the main forces could be modeled regardless of the part support. López de Lacalle et al. [46] noticed that the cutting forces decreased due to the reduction of stiffness. However, by using a rubber underneath a high stiffness part sample, the system flexibility can be considered not compromised as the cutting forces are maintained.

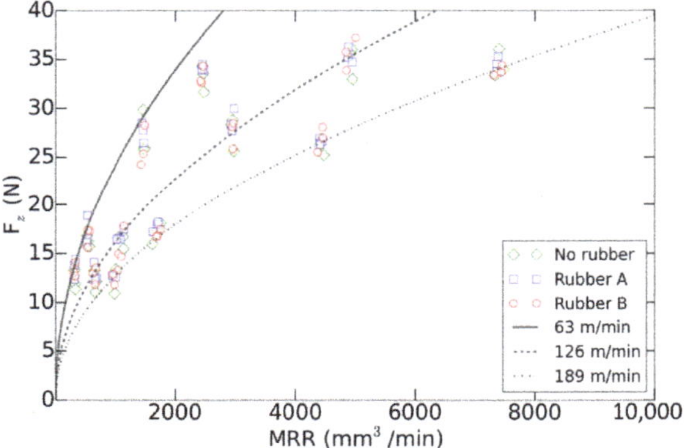

Figure 9. Axial mean loads predicted (lines) and experimental data against the material removal rate (MRR) as a function of the cutting speed (v_c).

Finally, this model demonstrated that the process axial forces in finishing of the aluminum parts did not depend on the material hardness or the accumulative strain state of the rubber. This facilitates the implementation of these flexible fixtures in the industry and provides a calculation tool for the improvement of the milling process productivity.

4. Conclusions

In this paper the effect of clamping high stiffness aluminum part samples over elastomer layers was analyzed. The machining application was groove milling, simulating finishing conditions in the aeronautic field. First, by a compression test the influence of stress amplitude, feed rate and cycles were examined. Thus, the rise of the elastic modulus as the strain rates increased and the strong dependence of the stress cycles were proved, especially for the soft rubber.

Then, the effect of cutting speed, tool feed and depth of the cut were analyzed in terms of the machined thickness error, roughness and axial forces. The results show that, as the cutting speed increases combined with the feed rate, the rubbers tended to behave like a rigid support, guaranteeing the thickness and roughness tolerances required in certain aeronautic applications. Moreover, as on these applications high speed machining operations were performed with higher cutting speeds and feed rates, the results of this solutions could improve compared to the actual rigid clamping solutions.

In terms of hardness, the softer rubber tended to provide more stable machining conditions due to a better clamping capacity.

Finally, an axial force model was developed and validated regardless the support stiffness and accumulative strain. This could lead to facilitate the implementation and improve the productivity of this solution into certain industrial applications, including the milling of aeronautical aluminum parts.

Author Contributions: Conceptualization, A.R.-M., A.R. and A.L.; methodology, A.R.-M., E.U. and A.L.; software, A.R.-M.; validation A.R.-M. and E.U.; formal analysis, A.R.-M.; investigation A.R.-M.; resources, A.R. and A.L.; data curation, A.R.-M.; writing—original draft preparation, A.R.-M.; writing—review and editing, A.R.-M. and E.U.; visualization, A.R.-M.; supervision, A.R., E.U. and A.L.; project administration, A.R. and A.L; funding acquisition, A.R. All authors have read and agreed to the published version of the manuscript.

Funding: This research was funded by Basque Government (Eusko Jaurlaritza) under the ELKARTEK Program, SMAR3NAK project, grant number KK-2019/00051.

Conflicts of Interest: The authors declare no conflict of interest. The funders had no role in the design of the study; in the collection, analyses, or interpretation of data; in the writing of the manuscript, or in the decision to publish the results.

References

1. Herranz, S.; Campa, F.J.; De Lacalle, L.L.; Rivero, A.; Lamikiz, A.; Ukar, E.; Sánchez, J.A.; Bravo, U. The milling of airframe components with low rigidity: A general approach to avoid static and dynamic problems. *Proc. Inst. Mech. Eng. Part B J. Eng. Manuf.* **2005**, *219*, 789–801. [CrossRef]
2. Del Sol, I.; Rivero, A.; López de Lacalle, L.N.; Gamez, A.J. Thin-Wall Machining of Light Alloys: A Review of Models and Industrial Approaches. *Materials* **2019**, *12*, 2012. [CrossRef]
3. Chen, W.; Xue, J.; Tang, D.; Chen, H.; Qu, S. Deformation prediction and error compensation in multilayer milling processes for thin-walled parts. *Int. J. Mach. Tools Manuf.* **2009**, *49*, 859–864. [CrossRef]
4. Yue, C.; Gao, H.; Liu, X.; Liang, S.Y.; Wang, L. A review of chatter vibration research in milling. *Chin. J. Aeronaut.* **2019**, *32*, 215–242. [CrossRef]
5. Eynian, M. Vibration frequencies in stable and unstable milling. *Int. J. Mach. Tools Manuf.* **2015**, *90*, 44–49. [CrossRef]
6. Gurdal, O.; Ozturk, E.; Sims, N.D. Analysis of Process Damping in Milling. *Procedia CIRP* **2016**, *55*, 152–157. [CrossRef]
7. Huang, C.-Y.; Wang, J.-J.J. A pole/zero cancellation approach to reducing forced vibration in end milling. *Int. J. Mach. Tools Manuf.* **2010**, *50*, 601–610. [CrossRef]
8. Perez, I.; Madariaga, A.; Cuesta, M.; Garay, A.; Arrazola, P.J.; Ruiz, J.J.; Rubio, F.J.; Sanchez, R. Effect of cutting speed on the surface integrity of face milled 7050-T7451 aluminium workpieces. *Procedia CIRP* **2018**, *71*, 460–465. [CrossRef]
9. Campa, F.J.; Lopez de Lacalle, L.N.; Celaya, A. Chatter avoidance in the milling of thin floors with bull-nose end mills: Model and stability diagrams. *Int. J. Mach. Tools Manuf.* **2011**, *51*, 43–53. [CrossRef]
10. Altintaş, Y.; Budak, E. Analytical Prediction of Stability Lobes in Milling. *CIRP Ann.* **1995**, *44*, 357–362. [CrossRef]
11. Huang, C.-Y.; Wang, J.-J.J. Effects of cutting conditions on dynamic cutting factor and process damping in milling. *Int. J. Mach. Tools Manuf.* **2011**, *51*, 320–330. [CrossRef]
12. Gameros, A.; Lowth, S.; Axinte, D.; Nagy-Sochacki, A.; Craig, O.; Siller, H.R. State-of-the-art in fixture systems for the manufacture and assembly of rigid components: A review. *Int. J. Mach. Tools Manuf.* **2017**, *123*, 1–21. [CrossRef]
13. Butt, M.A.; Yang, Y.; Pei, X.; Liu, Q. Five-axis milling vibration attenuation of freeform thin-walled part by eddy current damping. *Precis. Eng.* **2018**, *51*, 682–690. [CrossRef]
14. Yang, Y.; Xu, D.; Liu, Q. Milling vibration attenuation by eddy current damping. *Int. J. Adv. Manuf. Technol.* **2015**, *81*, 445–454. [CrossRef]
15. Fei, J.; Lin, B.; Yan, S.; Ding, M.; Xiao, J.; Zhang, J.; Zhang, X.; Ji, C.; Sui, T. Chatter mitigation using moving damper. *J. Sound Vib.* **2017**, *410*, 49–63. [CrossRef]
16. Matsubara, A.; Taniyama, Y.; Wang, J.; Kono, D. Design of a support system with a pivot mechanism for suppressing vibrations in thin-wall milling. *CIRP Ann.* **2017**, *66*, 381–384. [CrossRef]

17. Zhang, Y.; Sims, N.D. Milling workpiece chatter avoidance using piezoelectric active damping: A feasibility study. *Smart Mater. Struct.* **2005**, *14*, N65. [CrossRef]
18. Wang, M.; Fei, R. Chatter suppression based on nonlinear vibration characteristic of electrorheological fluids. *Int. J. Mach. Tools Manuf.* **1999**, *39*, 1925–1934. [CrossRef]
19. Ma, J.; Zhang, D.; Wu, B.; Luo, M.; Chen, B. Vibration suppression of thin-walled workpiece machining considering external damping properties based on magnetorheological fluids flexible fixture. *Chin. J. Aeronaut.* **2016**, *29*, 1074–1083. [CrossRef]
20. Yang, Y.; Xie, R.; Liu, Q. Design of a passive damper with tunable stiffness and its application in thin-walled part milling. *Int. J. Adv. Manuf. Technol.* **2017**, *89*, 2713–2720. [CrossRef]
21. Shoyama, T.; Fujimoto, K. Direct measurement of high-frequency viscoelastic properties of pre-deformed rubber. *Polym. Test.* **2018**, *67*, 399–408. [CrossRef]
22. Chung, D.D.L. Review: Materials for vibration damping. *J. Mater. Sci.* **2001**, *36*, 5733–5737. [CrossRef]
23. Ge, C.; Rice, B. Impact damping ratio of a nonlinear viscoelastic foam. *Polym. Test.* **2018**, *72*, 187–195. [CrossRef]
24. Albooyeh, A.R. The effect of addition of Multiwall Carbon Nanotubes on the vibration properties of Short Glass Fiber reinforced polypropylene and polypropylene foam composites. *Polym. Test.* **2019**, *74*, 86–98. [CrossRef]
25. Zhao, X.; Yang, J.; Zhao, D.; Lu, Y.; Wang, W.; Zhang, L.; Nishi, T. Natural rubber/nitrile butadiene rubber/hindered phenol composites with high-damping properties. *Int. J. Smart Nano Mater.* **2015**, *6*, 239–250. [CrossRef]
26. Shit, S.C.; Shah, P. A Review on Silicone Rubber. *Natl. Acad. Sci. Lett.* **2013**, *36*, 355–365. [CrossRef]
27. Kolluru, K.; Axinte, D. Novel ancillary device for minimising machining vibrations in thin wall assemblies. *Int. J. Mach. Tools Manuf.* **2014**, *85*, 79–86. [CrossRef]
28. Liu, Y.; Liu, Z.; Song, Q.; Wang, B. Analysis and implementation of chatter frequency dependent constrained layer damping tool holder for stability improvement in turning process. *J. Mater. Process. Technol.* **2019**, *266*, 687–695. [CrossRef]
29. Patel, H.; Salehi, S.; Ahmed, R.; Teodoriu, C. Review of elastomer seal assemblies in oil & gas wells: Performance evaluation, failure mechanisms, and gaps in industry standards. *J. Pet. Sci. Eng.* **2019**, *179*, 1046–1062. [CrossRef]
30. Mitra, S.; Ghanbari-Siahkali, A.; Almdal, K. A novel method for monitoring chemical degradation of crosslinked rubber by stress relaxation under tension. *Polym. Degrad. Stab.* **2006**, *91*, 2520–2526. [CrossRef]
31. Da Rocha, E.B.D.; Linhares, F.N.; Gabriel, C.F.S.; De Sousa, A.M.F.; Furtado, C.R.G. Stress relaxation of nitrile rubber composites filled with a hybrid metakaolin/carbon black filler under tensile and compressive forces. *Appl. Clay Sci.* **2018**, *151*, 181–188. [CrossRef]
32. Mallipudi, P.K.; Ramanaiah, N. Effect of Carbon Black on the Performance of Nitrile Rubber For Analyzing Free Layered Surface Damping Treatment. *Mater. Today Proc.* **2019**, *18*, 3371–3379. [CrossRef]
33. Olsson, A.K. *Finite Element Procedures in Modelling the Dynamic Properties of Rubber*; Department of Construction Sciences, Structural Mechanics, Lund University: Lund, Sweden, 2007.
34. Mullins, L. Softening of Rubber by Deformation. *Rubber Chem. Technol.* **1969**, *42*, 339–362. [CrossRef]
35. Balasubramanian, P.; Ferrari, G.; Amabili, M. Identification of the viscoelastic response and nonlinear damping of a rubber plate in nonlinear vibration regime. *Mech. Syst. Signal Process.* **2018**, *111*, 376–398. [CrossRef]
36. Austrell, P.-E.; Olsson, A.K. Modelling procedures and properties of rubber in rolling contact. *Polym. Test.* **2013**, *32*, 306–312. [CrossRef]
37. ASTM D 395. Standard test method for rubber. In *Philadelphia: Annual Book of ASTM Standards*; American Society for Testing and Materials: West Conshohocken, PA, USA, 1955.
38. Del Sol, I.; Rivero, A.; Salguero, J.; Fernández-Vidal, S.R.; Marcos, M. Tool-path effect on the geometric deviations in the machining of UNS A92024 aeronautic skins. *Procedia Manuf.* **2017**, *13*, 639–646. [CrossRef]
39. Del Sol, I.; Rivero, A.; Gamez, A.J. Effects of Machining Parameters on the Quality in Machining of Aluminium Alloys Thin Plates. *Metals* **2019**, *9*, 927. [CrossRef]
40. Engin, S.; Altintas, Y. Mechanics and dynamics of general milling cutters.: Part I: Helical end mills. *Int. J. Mach. Tools Manuf.* **2001**, *41*, 2195–2212. [CrossRef]

41. Altintas, Y. Analytical Prediction of Three Dimensional Chatter Stability in Milling. *JSME Int. J. Ser. C Mech. Syst. Mach. Elem. Manuf.* **2001**, *44*, 717–723. [CrossRef]
42. Chen, J.-S.; Hsu, W.-Y. Characterizations and models for the thermal growth of a motorized high speed spindle. *Int. J. Mach. Tools Manuf.* **2003**, *43*, 1163–1170. [CrossRef]
43. Ratchev, S.; Liu, S.; Huang, W.; Becker, A.A. Milling error prediction and compensation in machining of low-rigidity parts. *Int. J. Mach. Tools Manuf.* **2004**, *44*, 1629–1641. [CrossRef]
44. Rubio-Mateos, A.; Rivero, A.; del Sol, I.; Ukar, E.; Lamikiz, A. Capacitation of flexibles fixtures for its use in high quality machining processes: An application case of the industry 4.0. paradigm. *DYNA* **2018**, *93*, 608–612. [CrossRef]
45. Bi, Q.; Huang, N.; Zhang, S.; Shuai, C.; Wang, Y. Adaptive machining for curved contour on deformed large skin based on on-machine measurement and isometric mapping. *Int. J. Mach. Tools Manuf.* **2019**, *136*, 34–44. [CrossRef]
46. De Lacalle, L.N.L.; Lamikiz, A.; Sánchez, J.A.; de Bustos, I.F. Recording of real cutting forces along the milling of complex parts. *Mechatronics* **2006**, *16*, 21–32. [CrossRef]

Article

Feasibility Study of Hole Repair and Maintenance Operations by Dry Drilling of Magnesium Alloy UNS M11917 for Aeronautical Components

Fernando Berzosa [1], Beatriz de Agustina [1], Eva María Rubio [1,*] and J. Paulo Davim [2]

1. Department of Manufacturing Engineering, Industrial Engineering School, Universidad Nacional de Educación a Distancia (UNED), St/Juan del Rosal 12, E28040 Madrid, Spain
2. Department of Mechanical Engineering, University of Aveiro, 3810-193 Aveiro, Portugal

* Correspondence: erubio@ind.uned.es; Tel.: +34-913-988-226

Received: 26 May 2019; Accepted: 28 June 2019; Published: 30 June 2019

Abstract: Magnesium alloys are increasingly used due to the reduction of weight and pollutants that can be obtained, especially in the aeronautical, aerospace, and automotive sectors. In maintenance and repair tasks, it is common to carry out re-drilling processes, which must comply with the established quality requirements and be performed following the required safety and environmental standards. Currently, there is still a lack of knowledge of the machining of these alloys, especially with regards to drilling operations. The present article studies the influence of different cutting parameters on the surface quality obtained by drilling during repair and/or maintaining operations. For this propose, an experimental design was established that allows for the optimization of resources, using the average roughness (*Ra*) as the response variable, and it was analyzed through the analysis of variance (ANOVA). The results were within the margins of variation of the factors considered: the combination of factor levels that keep the *Ra* within the established margin, those that allow for the minimization of roughness, and those that allow for the reduction of machining time. In this sense, these operations were carried out in the most efficient way.

Keywords: magnesium alloy; UNS M11917; AZ91D; hole repair; surface roughness; dry drilling; re-drilling

---

1. Introduction

Recently, the need to reduce energy consumption as well as environmental pollution has been highlighted, especially in the transport sector, which includes industries such as the aeronautical, aerospace, and automotive industries. This need has led to a constant search for the reduction of the weight of components by using lighter materials, which allow for mass reduction and, therefore, lower consumption of fuel and polluting emissions. In this context, there has been increasing interest in extending the use of materials such as magnesium, which has excellent specific mechanical properties and whose full potential has not yet been reached, due in part to the insufficient knowledge about magnesium compared to other materials such steel and aluminum [1–9].

The main advantages of magnesium alloys are their low density, high availability, high recyclability, and good properties for foundry and machining, such as high specific strength and good weldability under a controlled atmosphere. Nevertheless, there are also a few disadvantages such as low creep resistance above 100 °C, low resistance to corrosion, hardness, and they are difficult to form at room temperature [10,11]. Magnesium's high chemical reactivity is another drawback that is closely related to problems during machining [12].

Magnesium alloys are mainly formed by casting, of which about 70% is processed by casting in permanent molds, producing near net shape parts. After that, machining operations are necessary in most cases [13,14]. Magnesium is considered to have excellent machinability. This is due to its low specific cutting strength, low tool wear, excellent surface quality, short and brittle chips due to its hexagonal crystal structure, and high thermal conductivity, which maintains low temperature increases even using dry machining, allowing for high cutting speeds and feed rates [15,16]. As a result, all common machining operations such as turning, milling, drilling, threading, reaming, or grinding can be performed with these alloys without major problems. The fundamental difference between magnesium and other structural materials is the ability to use higher feed rates and depths of cuts for magnesium, to give low roughness and closed tolerances [17].

The published literature shows that research on magnesium alloy machining focuses on cutting speed, feed rates, depth of cut, precision, and quality of the machining surfaces, also on the formation of adhesions, mainly build-up edge (BUE) and build-up layer (BUL) [18]. In turning and milling processes, researchers pay attention mainly to cutting forces, surface roughness, tool materials, tool wear, lubricant-cooling systems, temperature, chip morphology, and hardness. The cutting conditions for turning used in previous experimental works were as follows: cutting speed from 75 to 2400 m/min; feed rate from 0.05 to 0.65 mm/rev; and depth of cut from 0.2 to 5 mm. For milling, the parameter values were the same order of magnitude [1,10,19–22].

In the aeronautical industry, the drilling process is key due to the high number of joints by riveting, threaded joints, and mechanical fasteners made in the whole vehicle. In fact, the operation that consumes the most time during the assembly of a plane is the pre-assembly operation in the fuselage. An important cause of problems in the structural integrity of the fuselage is the growth of cracks in the drilled holes. For this reason, effective hole drilling is fundamentally important. In the case of commercial aircraft, the number of drilled holes can reach up to 3 million. Twist drills are used for most metals, using High Speed Steels (HSS) for aluminum and magnesium alloys [5].

In most studies on the drilling of magnesium alloys, the cutting speeds were around 50 m/min and the feed rates ranged between 0.1 and 0.7 mm/rev. In these studies, the influence of machining conditions on variables such as surface quality, force, torque, and tool wear, among others, was studied. The majority of studies used average roughness (*Ra*) to evaluate the surface quality of the obtained surfaces [3,23–27].

Weinert et al. [28] carried out a study on magnesium drilling using wide cutting parameters, reaching cutting speeds up to 1100 m/min and extending feed rates to 1.2 mm/rev. In that study they found that the surface quality, quantified by the maximum height of the profile, *Rz*, remained approximately constant by varying the cutting speed between 100 and 1100 m/min, while keeping the feed rate constant at 0.2 mm/rev; increasing the feed increased the roughness. In addition, the mechanical load on the tool did not vary significantly in the range of cutting speeds from 100 to 700 m/min, being determined by feed rate.

Other studies focused on machining parameters that are not high performance, using average roughness (*Ra*) as a variable to quantify the surface quality [3,23,26]. There are potential risks in the machining of magnesium alloys; on the one hand, there is the danger of ignition when the chips reach temperatures of 450 °C, and on the other hand, with the use of water-based lubrication there is danger of a reaction between water and magnesium, which forms a hydrogen atmosphere that is flammable [20,29]. Considering these reasons, it was decided to carry out the present study using dry machining.

As discussed above, there are still gaps in our knowledge about magnesium alloys. There are not many scientific works that discuss the problems during solid drilling of these alloys, and we found only one work specifically about re-drilling or core drilling operations in magnesium alloys: Rubio et al. [30] studied this process, but for hybrid Mg-Ti-Mg components. This type of machining is used in the repair process of damaged holes, which is common in the aeronautical sector where the holes are machined to a larger diameter to insert oversized rivets. These repairs must be carried out with great care to avoid damage to the machined parts [30]. These types of operations can be framed as low performance operations since productivity is a secondary objective.

In machining plants, drilling operations have traditionally been carried out in two steps: first drilling and then enlarging the diameter of the holes. These operations are executed with a solid base of knowledge of the materials and operations. However, in maintenance and/or repair operations, this is not always the case, especially considering that occasionally a smaller increase in the diameter of the hole is sought in order to not weaken the pieces. In this aspect, there is a certain lack of understanding and, therefore, such drilling operations can be improved to increase the safety and quality of the holes.

The aim of this work is to analyze the feasibility of carrying out repair and maintenance operations on pre-drilled parts used in the aeronautical industry. To do this, a pre-drilled test piece was used to simulate the repair of housings in covers that are joined by elements such as rivets. This joint type is widely used in aeronautics and can be the origin of fatigue cracks, which can lead to catastrophic failures in the pieces if they are not repaired in time.

This paper presents the analysis of the surface roughness, in terms of *Ra*, obtained by drilling holes to a slightly larger diameter in magnesium alloys UNS M11917 (AZ91D) at low cutting parameters. In this way, the behavior of these alloys in maintenance and/or repair operations was studied. The use of twist drills in these operations has certain advantages compared to reamers, which have less availability in the machining sections, generally have a higher price, and have a smaller variety in terms of the machined final diameters. The final aim is to establish if it is feasible to carry out such operations under environmentally sustainable conditions, maintaining the surface roughness requirements within a range of values established for the aeronautical industry, that is, from 0.8 to 1.6 μm [31,32].

To achieve this goal, the experimental design was established taking into consideration the three most important factors, feed rate, cutting speed, and type of tool, at two and three levels according to the recommendations of the manufacturers and prior knowledge of drilling operations. In addition, the small variations of the diameters to be drilled and the depth of the holes where the roughness measurements were to be taken were considered factors in the experimental design. Blocks were considered for quantification, and if obtained surface roughness was constant along the machined surface, a replicate was performed in order to quantify the error. The statistical method used to study the results obtained was the analysis of variance (ANOVA).

The tests were carried out in two stages by machining a pre-drill and then re-drilling, maintaining a constant depth of 0.125 mm. The final diameters were drilled to 7 and 7.5 mm, in the first and second stages, respectively. The last stage served firstly to corroborate the data obtained initially and secondly to check if small differences in the diameter affected to the variables studied.

2. Materials and Methods

The UNS M11917 magnesium alloy is produced by the die casting method and was supplied as an ingot with a length of approximately 500 mm and a section of 118 × 60 mm. A rectangular parallel-piped block was milled using low machining parameters so as not to raise the temperature of the piece, until reaching the measurements of 110 × 62 × 50 mm, maintaining surface roughness below 2 μm. This alloy has a chemical composition of mass of 90% Mg, 8.30–9.70% Al, 0.35–1% Zn, ≥0.13% Mn, ≤0.1% Si, ≤0.03% Cu, ≤0.005% Fe, and ≤0.002% Ni and presents a microstructure formed by an α-phase matrix and an intermetallic β-phase whose composition is $Mg_{17}Al_{12}$ and is located at the boundaries of the grains [33]. The main mechanical properties of this alloy are shown in Table 1.

Table 1. Mechanical properties of die casting magnesium alloy UNS M11917.

Brinell Hardness	63 HB
Tensile strength	230 MPa
Yield strength	173 MPa
Elongation	3%
Modulus of elasticity	44.8 GPa
Charpy impact	2.7 J

The block of magnesium was positioned in the hydraulic jaw of the machining center, aligning it so that the upper face was parallel to the plane of the machine table. To do this, a 3D tester (Haimer GmbH, Igenhausen, Hollenbach, Germany) was used. To avoid bias, pre-drills and drills were carried out without moving the specimen of the clamping jaw. The clamping of drills was done by a collet ER25 suitable for the drill diameter.

Two types of tool were used for the performance of the tests. They were both manufactured by Phantom (Van Ommen B.V., Beekbergen, Gelderland, Netherlands) in HSS, and are called 11.130 (type A), and 11.160 (type B), Figure 1.

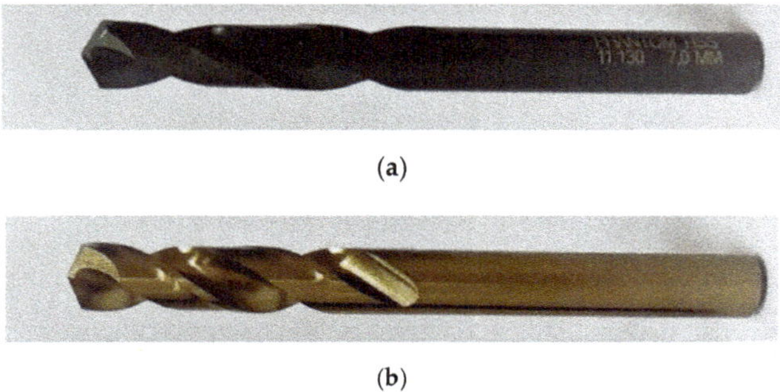

(**a**)

(**b**)

Figure 1. Twist drills for the tests; (**a**) Tool type A; (**b**) Tool type B.

These drills are suitable for drilling depths up to three times the diameter. They are manufactured according to DIN 1897 [34], with a straight shank and two flutes of 34 mm. They have a helix shape normal type N according to DIN 1836 [35], and both drill points are sharpened using split form C point in accordance with DIN 1412 [36].

Drilling tests were carried out using a computer numerical control (CNC) controlled vertical machining center Lagun L650 (Lagun Machinery S.L.L., Legutio, Álava, Spain) under dry conditions. The cutting parameters were selected based on solid drilling operations, taking into consideration the values given by the manufacturer for the group of non-ferrous materials and those used in the previous published studies. Keeping in mind that the present work was focused on repair and/or maintenance, we selected the following test values: cutting speed (S): 60 and 120 m/min; and feed rate (f): 0.2, 0.4, and 0.8 mm/rev; the cutting depth was kept constant at 0.125 mm. All blind holes were drilled to a depth of 20 mm from the top face of the specimen.

In the first stage, drilling tests were performed to enlarge holes from a diameter of 6.75 mm to a diameter of 7 mm, for all the combination of cutting conditions pointed out above, whereupon the machined surface roughness was measured. In a second stage, drilling tests were carried out under the same combination of cutting conditions, in this case from a diameter of 7.25 mm to a diameter 7.50 mm, in order to also evaluate the influence of the diameter of the drill on the surface roughness.

Between each drilling operation, the upper surface of the specimen and its surroundings were cleaned using a brush and pressurized air. Before carrying out the drilling, the periphery of the block was covered with paper, with the purpose of collecting samples of the fragile chips produced in the machining, as shown in the Figure 2. Once the process was finished, the hole and the used drill were marked with a number, and a photographic record of the obtained chips was taken using a Nikon Coolpix P510 digital camera (Nikon, Tokyo, Japan).

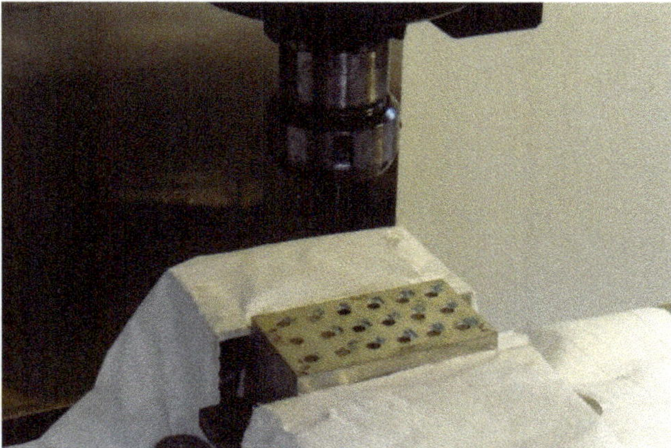

Figure 2. Detail of the method for collecting chips.

The arithmetical mean deviation of the assessed profile, *Ra*, was used as a response variable to quantify the surface roughness of the machined surface, which according to the standard ISO 4287:1997 [37] is defined as the "arithmetical mean of the absolute ordinate values $Z(x)$ within a sampling length". The range, that a priori would be expected to be the value of the measured *Ra*, should be between 0.1 and 2 μm according to ISO 4288:1996 [38], which also established the sampling length (*lr*) at 0.8 mm and the evaluation length (*ln*) at 4 mm. Subsequently, after the measurements of the roughness were made, these assumptions were confirmed.

In each of the drilled holes, the roughness was measured on eight different lines. That is, measurements of the *Ra* were taken along four lines equi-angularly separated by 90° in two cylindrical zones at different distances from the upper surface of the specimen. The first one, named the top plane (TP), was at a distance of 5.5 mm from top face and the second one, named the bottom plane (BP), was at a distance of 15 mm from top face, as can been seen in Figure 3. The measurement length along each one of the eight lines was of 4 mm.

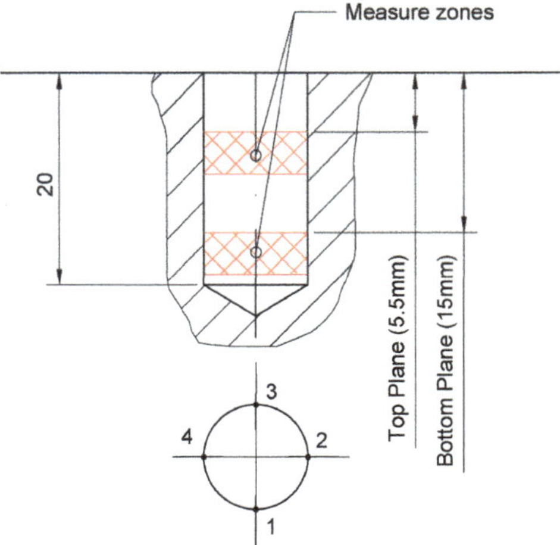

Figure 3. Cylindrical sections inside the holes where the measurements of roughness were taken for the evaluation of average roughness (*Ra*) (striped in red) and, in each one of them, the specific zones to take the measurements are defined by the four points separated to 90° and marked as 1, 2, 3, and 4.

Ra was measured using a contact roughness surface tester Zeiss Handysurf E-35A (Carl Zeiss AG, Oberkochen, Baden-Wurtemberg, Germany). This model has the possibility of exporting the measured data to a computer or displaying it directly on a display panel; it includes several types of parameters of roughness, among them the *Ra*. To carry out the measurements, the roughness meter and the test block were placed on a surface plate. The probe of the roughness tester is portable, so to perform the eight measurements in each hole, the probe was placed in a tool coupled to a height gauge, allowing for vertical positioning at the desired height, then rotation of the specimen four times to measure the equi-angularly located points, and then positioning of the probe tip at the two established depths. Figure 4 shows the surface tester, probe tip, and positioning system.

Figure 4. Zeiss Handysurf E-35A roughness tester, positioning system, and probe tip.

Based on all of this, the experimental design was determined, and its objective was to determine the influence of the factors considered in the response variable, that is, the surface roughness studied by the *Ra*. The experimental design selected was a full factorial design with three factors at two levels, one factor at three levels, and a block at two levels, which is the measurement depth from the upper face of the block, including the performance of a replica, which supposes a total of 96 experimental runs. The considered factors and their levels are included in Table 2.

Table 2. Factors and their levels.

Factor	Levels Values
Cutting speed, *S*, (m/min)	60, 120
Feed rate, *f*, (mm/rev)	0.2, 0.4, 0.8
Type of tool, *T*	A, B
Diameter, *D*, (mm)	7, 7.5
Measurement depth, *MD*, (mm)	Top plane (TP), bottom plane (BP)

Before carrying out the statistical analysis, the assumption of normality was checked; an Anderson–Darling and Kolmogorov-Smirnov tests data were carried out, which was not overcome, hence a Johnson transformation was carried out. Once the normal data was obtained, the next step was the statistical analysis in order to study the influential factors and interactions on the surface quality measured by *Ra*. This analysis was performed by the ANOVA.

3. Results and Analysis

3.1. Results

The experiment consisted of carrying out 96 experiments and, for each of them, measuring the roughness on four separate zones at 90° in two cylindrical sections located inside the holes as can be seen in the striped red zones in the Figure 3. To study the measured roughness, it was evaluated considering the average of the four values of *Ra* measured in each one of the two cylindrical sections located at the different distances from the top face of the specimen in each experimental run. Before carrying out the statistical analysis, the assumption of normality of the average *Ra* was checked. To do this, normality was assessed using the Anderson–Darling and Kolmogorov–Smirnov tests. The results obtained in both tests are shown in Table 3, and indicate the non-normality of the data and therefore the need to carry out its transformation prior to the statistical study.

Table 3. The *p*-values of tests for normality of *Ra*.

Factor	Anderson–Darling	Kolmogorov–Smirnov
Original Data	<0.005	<0.010
Transformed data	0.736	>0.150

Johnson transformation was used to convert the original non-normal data into a standard normal distribution. The best adjustment of the transformation was obtained using the equality function collected in Equation (1), where the transformed roughness values are designated by Ra^t, being the *Ra* of the average roughness initial values. The probability plot for the population before and after the transformation can be seen in Figure 5.

$$Ra^t = -1.05518 + 1.01828 \times \sinh^{-1}\left(\frac{Ra - 0.562553}{0.128499}\right) \quad (1)$$

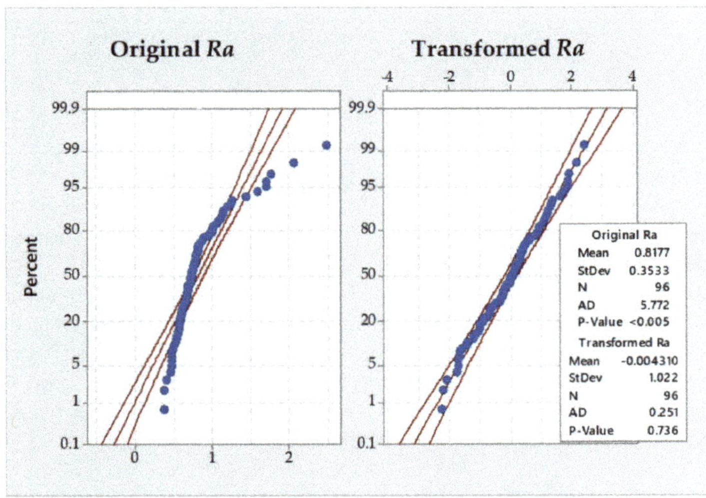

Figure 5. Probability plots of original non-normal data and Johnson transformed data of the *Ra*.

Once the measured roughness data were normalized, it was possible to analyze them using the ANOVA statistical method. For this, the arithmetic mean of the four measured values of the *Ra* in each elementary run was made. The results obtained in the experimental design are those included in Table 4.

Table 4. Replicas of original and transformed data of *Ra* in μm, at different measurement depths.

S [m/min]	f [mm/rev]	T	D [mm]	Ra				Ra^t			
				TP		BP		TP		BP	
60	0.2	A	7.0	1.76	0.85	1.71	0.77	1.88	0.25	1.92	0.53
60	0.2	A	7.5	0.57	0.78	0.47	0.82	−1.70	0.42	−0.99	0.29
60	0.2	B	7.0	0.61	0.67	0.58	0.77	−0.89	0.22	−0.68	−0.26
60	0.2	B	7.5	0.38	0.80	0.74,	0.56	0.12	−1.03	−2.21	0.34
60	0.4	A	7.0	1.26	1.12	0.79	1.06	0.31	1.04	1.39	1.17
60	0.4	A	7.5	0.56	0.71	0.51	0.67	−1.40	−0.26	−1.01	−0.02
60	0.4	B	7.0	0.63	0.82	0.58	0.96	−0.91	0.84	-0.50	0.42
60	0.4	B	7.5	0.67	0.41	0.68	0.37	−0.20	−2.23	−0.26	−2.08
60	0.8	A	7.0	0.67	0.90	0.81	0.73	0.40	0.08	−0.23	0.66
60	0.8	A	7.5	0.66	1.15	0.69	1.24	−0.14	1.35	−0.34	1.21
60	0.8	B	7.0	0.56	0.71	0.74	0.74	0.11	0.12	−1.01	−0.02
60	0.8	B	7.5	0.61	0.49	0.54	0.48	−1.21	−1.65	−0.66	−1.58
120	0.2	A	7.0	0.87	1.07	0.61	0.78	−0.63	0.28	0.58	1.07
120	0.2	A	7.5	1.01	2.06	0.79	2.49	0.32	2.41	0.94	2.15
120	0.2	B	7.0	0.47	0.69	0.68	0.58	−0.19	−0.89	−1.72	−0.12
120	0.2	B	7.5	0.83	0.61	0.59	0.69	−0.78	−0.13	0.46	−0.68
120	0.4	A	7.0	0.66	1.70	0.64	1.20	−0.45	1.29	−0.31	1.88
120	0.4	A	7.5	1.00	0.68	0.88	0.75	0.62	0.13	0.92	−0.19
120	0.4	B	7.0	0.55	0.73	1.01	0.79	0.95	0.31	−1.11	0.08
120	0.4	B	7.5	0.95	0.54	0.51	0.73	−1.44	0.05	0.81	−1.21
120	0.8	A	7.0	0.65	0.46	0.71	0.60	−0.01	−0.70	−0.35	−1.75
120	0.8	A	7.5	1.14	1.44	1.11	1.58	1.15	1.76	1.20	1.61
120	0.8	B	7.0	0.47	0.74	0.62	0.80	−0.59	0.36	−1.68	0.10
120	0.8	B	7.5	0.71	0.81	0.77	0.99	0.23	0.90	−0.02	0.40

*S = cutting speed; f = feed rate; T = type of tool; D = diameter; Ra^t = transformed Ra.

3.2. Analysis and Discussion

From the data obtained, it is clear that the obtained surface quality achieved improved results, independent of the parameters tested. All the roughness values were between 0.38 and 2.49 μm. Only five of the 96 *Ra* values were above the 1.6 μm that is set as the upper limit in the aeronautical sector. In consideration of the 16%-rule established in the ISO 4288:1996 standard [38], the surface is considered acceptable because only 5.2% exceeded the upper limit.

For the statistical analysis of the transformed *Ra* values, the Minitab 17 computer program was used. The model was reduced to include only the significant factors, for which a stepwise procedure was followed that eliminates or adds terms to the model using a significance level $\alpha = 0.05$, starting with an empty model and then adding or removing a term for each step. Table 5 shows the influential factors and interactions, in other words, cutting speed, feed rate, type of tool, diameter, sum of squares, degrees of freedom (DF), mean square, F value, and *p*-value.

Table 5. Analysis ANOVA of the transformed *Ra*. DF = degrees of freedom.

Effect	DF	Sum of Squares	Mean Square	F-Value	p-Value
S*	1	2.832	2.8319	4.63	0.034
T	1	18.347	18.3470	29.98	0.000
S*D	1	14.621	14.6205	23.89	0.000
f*D	2	8.259	4.1297	6.75	0.002
Error	90	55.081	0.6120	-	-
Total	95	99.140	-	-	-

The analysis showed that among the main factors, only the cutting speed, *S*, and the type of tool, *T*, were statistically significant for *Ra*. Regarding the interactions of the factors, there were only two significant second-order interactions: the cutting speed with the diameter, *S*D*, and the feed rate with the diameter, *f*D*. The contribution of each effect to the variability is shown in Figure 6. The most important effect was the type of tool with a percentage of 45.25%, the second most important was the interaction of cutting speed with the diameter of the drill with 36.06%, the third was the interaction between feed rate and diameter with 10.18% and the last was the cutting speed with 6.98%.

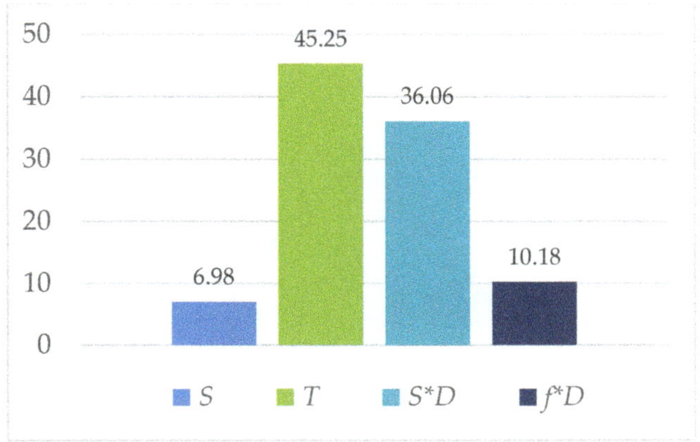

Figure 6. Percentage of contribution to the variability of the ANOVA model for each effect.

In Figure 7, it can be seen that the use of a twist drill type B enhances the surface quality of the machined surfaces with respect to those gained from the use of type A. From the untransformed values obtained in the tests, tool A obtained a *Ra* of 0.97 μm, while type B was 0.67 μm, which represents a considerable improvement. An explanation of this behavior could be the appearance of radial stresses that would produce deformation of the drill and therefore affect the roughness, this would be caused by the small depth of cut of only 0.125 μm in conjunction with the use of a non-high rigidity drill and relatively high feed rates and cutting speed values. Regarding the cutting speed, it is the factor that has the least influence, producing a higher *Ra* with increasing speed. This result is concordant with those obtained by Weinert et al. for the case of solid drilling [13].

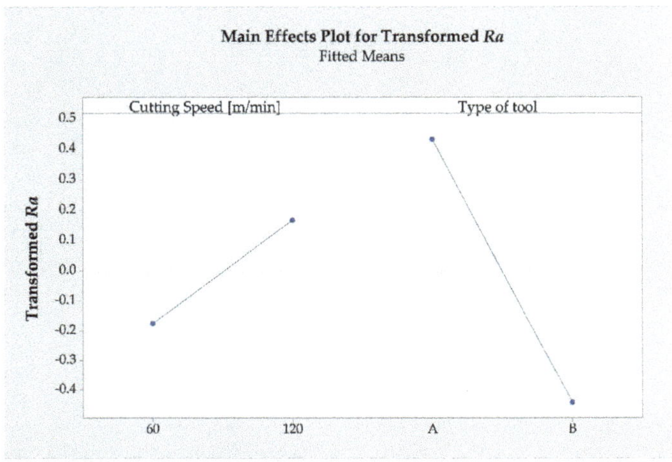

Figure 7. Effects on the transformed *Ra* of the significant factors.

The second effect in terms of importance is the interaction between the cutting speed and the diameter of the drill, $S*D$, although the diameter itself is not a significant effect. Figure 8 shows that increasing the cutting speed in drills with a diameter of 7 mm causes a lower superficial roughness in contrast to drills with a 7.5 mm diameter.

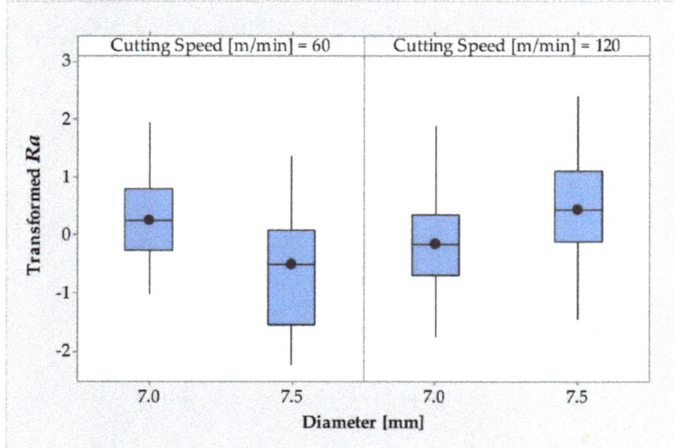

Figure 8. Box and whiskers plot graph for interactions between cutting speed and diameter, $S*D$.

The influence of the feed rate on the *Ra* was the opposite and this is explained by the third most important effect: the interaction between feed rate and the diameter of the drill, $f*D$, as can be seen in the Figure 9. This can be seen most clearly for feed rates of 0.4 and 0.8 mm/rev. Use of 7 mm tools at higher feed rates caused lower *Ra* values; in contrast, for 7.5 mm tools, the lower *Ra* values were obtained with lower feed rates.

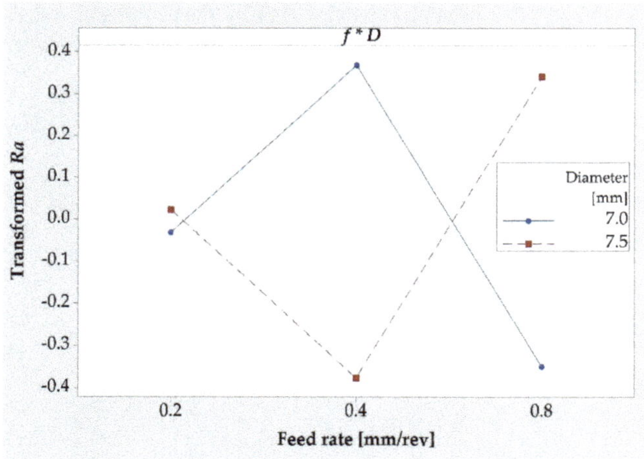

Figure 9. Interaction graph between feed rate and diameter, $f*D$.

The results regarding these two interactions of the cutting speed and the feed rate with the diameter of the drill on magnesium alloys are remarkable because the difference between the diameters tested was so small. Further studies should be carried out to clarify this point, considering greater values in the drill diameters studied.

A model of surface roughness was developed to predict the variability in the transformed data, Ra^t, through Equation (2). This equation uses the significant factors identified by the ANOVA, in other words, cutting speed, type of tool, feed rate, and diameter of the drill, where s, t, f, and d represent their effects, μ is the term to adjust the mean, and ε is the error. The estimation parameters of the equation are included in Table 6.

$$Ra^t_{ijkl} = \mu + s_i + t_j + sd_{ik} + fd_{lk} + \varepsilon_{ijkl} \quad (2)$$

Table 6. Estimation parameters of the predictive model.

Parameter	Designation	Estimation	Parameter	Designation	Estimation
60 (m/min)	s_1	−0.1718	0.2 (mm/rev)*7 mm	fd_{11}	−0.027
120 (m/min)	s_2	0.1718	0.2 (mm/rev)*7.5 mm	fd_{12}	0.027
A	t_1	0.4372	0.4 (mm/rev)*7 mm	fd_{21}	0.372
B	t_2	−0.4372	0.4 (mm/rev)*7.5 mm	fd_{22}	−0.372
60 (m/min)*7 mm	sd_{11}	0.3903	0.8 (mm/rev)*7 mm	fd_{31}	−0.345
60 (m/min)*7.5 mm	sd_{12}	−0.3903	0.8 (mm/rev)*7.5 mm	fd_{32}	0.345
120 (m/min)*7 mm	sd_{21}	−0.3903	Intercept term	μ	−0.0043
120 (m/min)*7.5 mm	sd_{22}	0.3903	-	-	-

The residuals of the model were obtained by the difference between measured and predicted values and were used for checking the model hypotheses. As can be seen in Figure 10, the residuals satisfy the normality and homoscedasticity hypothesis; also, no patterns were found in the model.

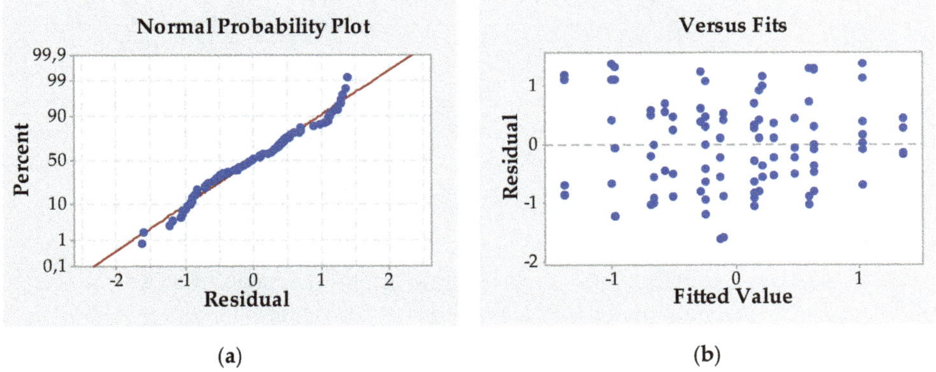

(**a**) (**b**)

Figure 10. (**a**) Probability plot and (**b**) residuals versus predicted values for the model.

Once the validity of the model was verified, the inverse of the Johnson transformation used in Equation (1), of the fitted values according to Equation (2), was carried out to predict the surface quality by *Ra* in re-drilling operations of UNS M11917 magnesium alloys. For this, Equation (3) was used.

$$Ra_{ijkl} = 0.562553 + 0.128499 \times \sinh\left(\frac{\mu + s_i + t_j + sd_{ik} + fd_{lk} + \varepsilon_{ijkl} + 1.05518}{1.01828}\right) \quad (3)$$

From Equation (3) and considering both the significant factors identified in the ANOVA, as well as their levels, it was possible to calculate the predicted values of *Ra* for the different combinations. These values are shown in Table 7 along with the *Ra* obtained from the values measured in the tests, and the absolute error between the predicted and measured *Ra*.

Table 7. Predicted *Ra*, measured *Ra*, and absolute error for levels and effects combinations.

*S** (m/min)	*T*	*f* (mm/rev)	*D* (mm)	*Ra* Predicted (μm)	*Ra* Measured (μm)	Abs. Error (μm)
60	B	0.4	7.5	0.52	0.67	0.15
120	B	0.8	7	0.57	0.48	0.09
60	B	0.2	7.5	0.57	0.38	0.19
120	B	0.2	7	0.61	0.47	0.14
60	B	0.8	7.5	0.61	0.61	0.00
60	B	0.8	7	0.63	0.57	0.06
60	A	0.4	7.5	0.64	0.57	0.07
120	B	0.4	7	0.67	0.56	0.11
120	B	0.4	7.5	0.67	0.96	0.28
60	B	0.2	7	0.68	0.61	0.07
120	A	0.8	7	0.70	0.66	0.04
60	A	0.2	7.5	0.70	0.57	0.13
120	B	0.2	7.5	0.75	0.84	0.08
60	B	0.4	7	0.75	0.64	0.12
120	A	0.2	7	0.76	0.87	0.11
60	A	0.8	7.5	0.77	0.66	0.11
60	A	0.8	7	0.79	0.68	0.11
120	B	0.8	7.5	0.83	0.72	0.12
120	A	0.4	7	0.87	0.67	0.21
120	A	0.4	7.5	0.88	1.00	0.12
60	A	0.2	7	0.88	1.76	0.88
120	A	0.2	7.5	1.05	1.01	0.04
60	A	0.4	7	1.05	1.27	0.22
120	A	0.8	7.5	1.23	1.15	0.09

The surface roughness values obtained by the model predicted values between 0.52 and 1.23 μm. The absolute error between the measured and predicted values was less than 0.28 μm in all cases except one, in which the error reached 0.88 μm. The minimum *Ra* value was obtained with the combination of a cutting speed of 60 m/min, type of tool B, feed rate of 0.4 mm/rev, and by using a drill with a 7.5 mm diameter.

4. Technological Discussion

For the framework in which the present study was set up, that of enlarging holes with a low depth of cut by drilling in magnesium alloy UNS M11917, it is important to highlight some technological aspects with implications to the practical application of these operations, mainly in the aerospace sector.

Considering all the tests carried out and according the 16%-rule established in the ISO 4288:1996 standard [38], it can be affirmed that within the margins of the levels and factors tested, the surface quality would be within the quality requirements established in this sector (0.8 μm < *Ra* < 1.6 μm) [31,32]. Of all the 96 measured *Ra* values, only five of them were above the upper limit, and those five were drilled using the type A drill. From this it follows, in addition to be the main significant factor, the importance of the type of drill in the performance of these operations.

It is of great importance to carry out these maintenance and/or repair operations in the shortest time possible, meeting the quality standards required for the parts, so it is important to optimize the rate of material removal (RMM). According to Astakhov [39], this rate is directly proportional to the product of the feed rate by the cutting speed, *f**S*. Therefore, a way of increasing productivity and improving process time is to select the highest feed rate and cutting speed values, that is, 0.8 mm/rev and 120 m/min, respectively. Using these operating parameters, the average *Ra* values of 0.6 μm for the type B tool and 0.9 μm for type A were obtained.

The type of tool is the most important factor in terms of its influence on the *Ra*. The Anderson–Darling test shows clearly that for the case of type B drills, the population follows a normal distribution. However, for the case of the type A drill, with a value of $p < 0.005$, the

Anderson–Darling test confirms that it follows a non-normal distribution with a strong asymmetry and positive kurtosis, as seen in Figure 11 with its displacement to the right. An explanation of this phenomenon could be in the appearance of greater radial efforts that, due to the low rigidity of the drills, give rise to deformations that affect the surface quality. Subsequent studies must to be carried out to confirm this hypothesis.

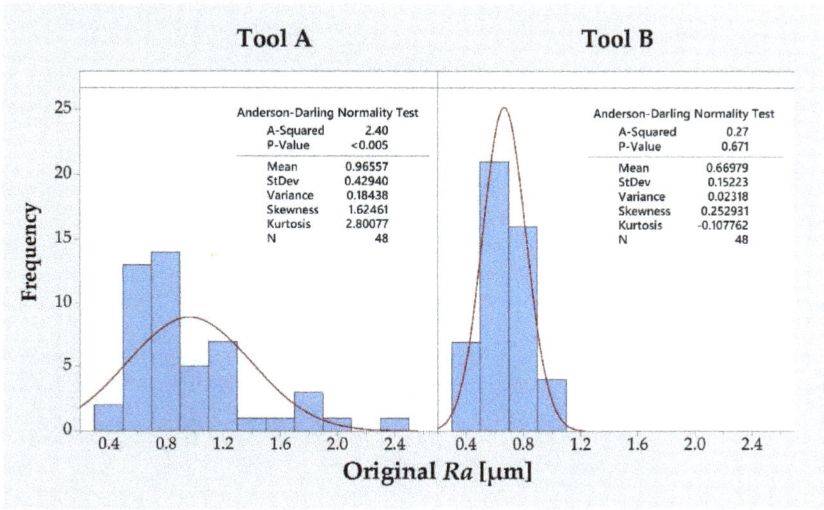

Figure 11. Histogram of data with an overlaid normal curve of the original *Ra* for tool types A and B.

5. Conclusions

This experimental study on the small-scale re-drilling operations in magnesium alloy UNS M11917 within the maintenance and/or repair processes of pieces in the aerospace sector, confirms that it is possible to perform such operations in a way that satisfies the requirements of the surface quality and safety and, at the same time, under environmentally friendly conditions, that is, using dry machining or without the use of lubricant coolants. The most important factor to consider is the type of tool used, obtaining the best results with type B drills, which have a point angle of 135°, compared to type A drills, which have a point angle of 118°. However, further studies have to corroborate if, besides the point angle, other variables of the drill, such as the type of coating, affect the surface quality. Using that type of drill and choosing the highest values in the cutting parameters, that is, a feed rate of 0.8 mm/rev and a cutting speed of 120 m/min, a surface roughness was obtained of approximately half of the maximum limit considered within this sector, in other words, 0.8 μm.

The depth from the upper surface of the specimens did not present a statistical influence on the roughness, so it can be considered constant throughout the depth of the drilling holes, which in this study was 20 mm. Contrary to what was initially assumed, feed rate did not have an influence on surface roughness. This result is consistent with some other work on solid drilling operations in similar magnesium alloys, however, there are other works that show contrary data. Feed rate does have a second-order influence in its interaction with the diameter of the drill, but its influence value is small in relation to the other significant factors.

Author Contributions: Conceptualization, E.M.R., B.d.A., F.B., and J.P.D.; Methodology, E.M.R., B.d.A., F.B., and J.P.D.; Software, E.M.R., B.d.A., and F.B.; Validation, E.M.R., B.d.A., F.B., and J.P.D.; Formal analysis, E.M.R., B.d.A., and F.B.; Investigation, E.M.R., B.d.A., F.B., and J.P.D.; Resources, E.M.R., B.d.A., F.B., and J.P.D.; Data curation, E.M.R., B.d.A., and F.B.; Writing—original draft preparation, F.B.; Writing—review and editing, E.M.R., B.d.A., F.B., and J.P.D.; Visualization, E.M.R., B.d.A., and F.B.; Supervision, E.M.R., B.d.A., and J.P.D.; Project administration, E.M.R. and B.d.A.; Funding acquisition, E.M.R., B.d.A., and F.B.

Funding: This research was funded by the Ministry of Economy and Competitiveness and the Industrial Engineering School-UNED (DPI2014-58007-R; Ref: 2019-ICF03 and Ref: 2019-ICF05).

Acknowledgments: The authors thank the Research Group of the UNED "Industrial Production and Manufacturing Engineering (IPME)" for the support given during the development of this work and to the Grupo Antolín for donating the materials used in this work.

Conflicts of Interest: The authors declare no conflict of interest.

References

1. Rubio, E.M.; Villeta, M.; Valencia, J.L.; Sáenz de Pipaón, J.M. Experimental Study for Improving the Repair of Magnesium–Aluminium Hybrid Parts by Turning Processes. *Metals* **2018**, *8*, 59.
2. Kainer, K. Challenges for Implementation of Magnesium into More Applications. In *Magnesium Technology 2016*; Springer International Publishing: Basel, Switzerland, 2016; pp. 5–6.
3. Bhowmick, S.; Lukitsch, M.J.; Alpas, A.T. Dry and minimum quantity lubrication drilling of cast magnesium alloy (AM60). *Int. J. Mach. Tools Manuf.* **2010**, *50*, 444–457. [CrossRef]
4. Pantelakis, S.G.; Alexopoulos, N.; Chamos, A. Mechanical performance evaluation of cast magnesium alloys for automotive and aeronautical applications. *J. Eng. Mater. Technol.* **2007**, *129*, 422–430. [CrossRef]
5. Mouritz, A.P. *Introduction to Aerospace Materials*; Elsevier: Amsterdam, The Netherlands, 2012; ISBN 978-0-85709-515-2.
6. Fores, F.H.; Eliezer, D.; Aghion, E. How to Increase the use of Magnesium in Aerospace Applications. In Proceedings of the 2nd Israeli International Conference on Magnesium Science and Technology, Dead Sea, Israel, 1 January 2000.
7. Fleming, S. An Overview of Magnesium based Alloys for Aerospace and Automotive Applications. Available online: https://pdfs.semanticscholar.org/5596/35646ba740d5fcb9a308e384a1b6cef46f2e.pdf (accessed on 20 June 2019).
8. Gwynne, B. Magnesium Alloys in Aerospace Applications, Past Concerns, Current Solutions. Available online: https://www.fire.tc.faa.gov/2007conference/files/Materials_Fire_Safety/WedAM/GwynneMagnesium/GwynneMagnesiumPres.pdf (accessed on 20 June 2019).
9. Sáenz de Pipaón, J.M. Diseño y Fabricación de Probetas de Componentes Híbridos con Aleaciones de Magnesio para Ensayos de Mecanizado. Ph.D. Thesis, Universidad Nacional de Educación a Distancia, Madrid, Spain, December 2013.
10. Carou, D.; Rubio, E.M.; Davim, J.P. Machinability of Magnesium and Its Alloys: A Review. In *Traditional Machining Processes: Research Advances*; Davim, J.P., Ed.; Materials Forming, Machining and Tribology; Springer Berlin Heidelberg: Berlin, Heidelberg, 2015; pp. 133–152. ISBN 978-3-662-45088-8.
11. Joksch, S. *Safe and Economically Efficient Use of Coolants in Mechanical Processing of Magnesium Alloys.*; Wiley Online Library: Hokoben, NJ, USA, 2004; pp. 888–894.
12. Villeta, M.; Rubio, E.M.; De Pipaón, J.S.; Sebastián, M.A. Surface finish optimization of magnesium pieces obtained by dry turning based on Taguchi techniques and statistical tests. *Mater. Manuf. Process.* **2011**, *26*, 1503–1510. [CrossRef]
13. Weinert, K.; Liedschulte, M.; Opalla, D.; Schroer, M. High-Tech Machining of Magnesium and Magnesium Composites. *Magnes.-Alloys Technol.* **2004**, 130–151.
14. Froes, F.; Eliezer, D.; Aghion, E. The science, technology, and applications of magnesium. *JOM J. Miner. Met. Mater. Soc.* **1998**, *50*, 30–34. [CrossRef]
15. Tikal, F.; Schmier, M.; Vollmer, C. High-speed-drilling in AZ91 D without Lubricoolants. *Magnes. Alloys Their Appl.* **2000**, 371–379.
16. Tomac, N.; Tønnessen, K.; Rasch, F. Safe machining of magnesium. In *Advanced Manufacturing Systems and Technology*; Springer: Vienna, Austria, 1996; pp. 177–184.
17. Habashi, F. *Alloys: Preparation, Properties, Applications*; John Wiley & Sons: Hoboken, NJ, USA, 2008; ISBN 978-3-527-61192-8.
18. Akyuz, B. Machinability of magnesium and its alloys. *TOJSAT Online J. Sci. Technol.* **2011**, *1*, 31–38.
19. Rubio, E.M.; Valencia, J.L.; Saá, A.J.; Carou, D. Experimental study of the dry facing of magnesium pieces based on the surface roughness. *Int. J. Precis. Eng. Manuf.* **2013**, *14*, 995–1001. [CrossRef]

20. Rubio, E.M.; Villeta, M.; Carou, D.; Saá, A. Comparative analysis of sustainable cooling systems in intermittent turning of magnesium pieces. *Int. J. Precis. Eng. Manuf.* **2014**, *15*, 929–940. [CrossRef]
21. Rubio, E.M.; Valencia, J.L.; de Agustina, B.; Saá, A.J. Tool selection based on surface roughness in dry facing repair operations of magnesium pieces. *Int. J. Mater. Prod. Technol.* **2014**, *48*, 116–134. [CrossRef]
22. Carou, D.; Rubio, E.M.; Lauro, C.H.; Davim, J.P. Experimental investigation on finish intermittent turning of UNS M11917 magnesium alloy under dry machining. *Int. J. Adv. Manuf. Technol.* **2014**, *75*, 1417–1429. [CrossRef]
23. Gariboldi, E. Drilling a magnesium alloy using PVD coated twist drills. *J. Mater. Process. Technol.* **2003**, *134*, 287–295. [CrossRef]
24. Bhowmick, S.; Alpas, A.T. The role of diamond-like carbon coated drills on minimum quantity lubrication drilling of magnesium alloys. *Surf. Coat. Technol.* **2011**, *205*, 5302–5311. [CrossRef]
25. Wang, J.; Liu, Y.B.; An, J.; Wang, L.M. Wear mechanism map of uncoated HSS tools during drilling die-cast magnesium alloy. *Wear* **2008**, *265*, 685–691. [CrossRef]
26. Berzosa, F.; de Agustina, B.; Rubio, E.M. Tool Selection in Drilling of Magnesium UNSM11917 Pieces under Dry and MQL Conditions Based on Surface Roughness. *Adv. Mater. Process. Technol. Conf.* **2017**, *184*, 117–127. [CrossRef]
27. Bagci, E.; Ozcelik, B. Effects of different cooling conditions on twist drill temperature. *Int. J. Adv. Manuf. Technol.* **2007**, *34*, 867–877. [CrossRef]
28. Weinert, K.; Lange, M. Machining of magnesium matrix composites. *Adv. Eng. Mater.* **2001**, *3*, 975–979. [CrossRef]
29. Carou, D.; Rubio, E.M.; Davim, J. Analysis of ignition risk in intermittent turning of UNS M11917 magnesium alloy at low cutting speeds based on the chip morphology. *Proc. Inst. Mech. Eng. Part B J. Eng. Manuf.* **2014**, *229*, 365–371. [CrossRef]
30. Rubio, E.; Villeta, M.; Valencia, J.; Sáenz de Pipaón, J. Cutting Parameter Selection for Efficient and Sustainable Repair of Holes Made in Hybrid Mg–Ti–Mg Component Stacks by Dry Drilling Operations. *Materials* **2018**, *11*, 1369. [CrossRef]
31. Villeta, M.; de Agustina, B.; Sáenz de Pipaón, J.M.; Rubio, E.M. Efficient optimisation of machining processes based on technical specifications for surface roughness: application to magnesium pieces in the aerospace industry. *Int. J. Adv. Manuf. Technol.* **2011**, *60*, 1237–1246. [CrossRef]
32. American Society of Mechanical Engineers. *ANSI/ASME B46.1-2009. Surface Texture (Surface Roughness, Waviness, and Lay)*; American Society of Mechanical Engineers: New York, NY, USA, 2010.
33. Luan, B.L.; Gray, J.; Yang, L.X.; Cheong, W.J.; Shoesmith, D. Surface modification of AZ91 magnesium alloy. *Trans Tech Publ.* **2007**, *546*, 513–518.
34. Deutsches Institut für Normung e.V. *DIN 1897; Parallel Shank Twist Drills, Stub Series*; Deutsches Institut Fur Normung E.V.: Berlin, Germany, 2006.
35. Deutsches Institut für Normung e.V. *DIN 1836; Groups of Tool Application for Chip Removal es of Points*; Deutsches Institut für Normung, e.V.: Berlin, Germany, 1984.
36. Deutsches Institut für Normung e.V. *DIN 1412; Twist Drills Made of High-Speed Steel - Shapes of Points*; Deutsches Institut für Normung e.V.: Berlin, Germany, 2001.
37. International Organization for Standardization. *ISO 4287. Geometrical Product Specifications (GPS). Surface Texture: Profile Method. Terms, Definitions and Surface Texture Parameters*; International Organization for Standardization: Geneva, Switzerland, 1997.
38. International Organization for Standardization. *ISO 4288. Geometrical Product Specifications (GPS). Surface Texture: Profile Method. Rules and Procedures for the Assessment of Surface Texture*; International Organization for Standardization: Geneva, Switzerland, 1996.
39. Astakhov, V.P. *Drills: Science and Technology of Advanced Operations*; CRC Press: Boca Raton, FL, USA, 2014; ISBN 978-1-4665-8435-8.

Article

Parametric Analysis of Macro-Geometrical Deviations in Dry Turning of UNS A97075 (Al-Zn) Alloy

Sergio Martín Béjar, Francisco Javier Trujillo Vilches *, Carolina Bermudo Gamboa and Lorenzo Sevilla Hurtado

Civil, Material and Manufacturing Engineering Department, EII, University of Malaga, 29071 Malaga, Spain; smartinb@uma.es (S.M.B.); bgamboa@uma.es (C.B.G.); lsevilla@uma.es (L.S.H.)

* Correspondence: trujillov@uma.es; Tel.: +34-951-953-245

Received: 28 September 2019; Accepted: 23 October 2019; Published: 24 October 2019

Abstract: Macro-geometrical deviations play a very important role in the functionality and reliability of structural parts for aircraft. The use of environmentally friendly techniques, such as dry machining, may negatively affect these deviations. Despite its importance, there is a lack of research that analyzes them as a function of the cutting parameters in the case of aluminum alloys for aeronautical purpose. In this work, the cutting speed and feed influence on several macro-geometrical deviations (parallelism, straightness, circular run-out, roundness, concentricity, total circular run-out and cylindricity) in dry turning of UNS A97075 alloy was analyzed. The main novelty of this work lies in the use of high slenderness parts used in further fatigue tests. The results showed that feed seems to be the most influential parameter in most of the deviations studied. In addition, the parts with lower rigidity exhibited higher sensitivity to change with the cutting parameters. Finally, different parametric models were proposed to obtain the geometrical deviations as a function of the cutting parameters.

Keywords: UNS A97075; dry turning; surface integrity; straightness; parallelism; roundness; concentricity; circular run-out; total run-out; cylindricity

---

1. Introduction

One of the most appreciated quality requirements in machining processes is related to the surface integrity (SI) concept [1]. Field et al. [2,3] defined the SI as the inherent or enhanced condition of a surface produced in machining operations or other surface generation processes. Griffiths [4] defined it as the topological, mechanical, chemical and metallurgical worth of a manufactured surface and its relationship with its functional performance. A new definition of this concept is developed by Astakhov [5], who defined it as a set of properties (both, superficial and in-depth) of an engineering surface that affect its service behavior. These properties include geometrical, physical-chemical and biological parameters.

In more recent works, Gómez-Parra et al [6,7] provided an expanded view of the SI concept, defining the SI as a set of properties that the material surface exhibits, acquires or becomes modified during a forming process. These properties can be analyzed from three points of view connected to each other: micro-geometrical (surface roughness, micro and macro cracks, waviness, particle adhesion), macro-geometrical (cylindricity, concentricity, straightness) and physical-chemical properties (micro hardness, residual stress, stress corrosion, tensile strength, fatigue behavior). The authors highlight that these properties may not only improve the functional performance of the part, but also worsen it.

The macro-geometrical deviations or material mechanical properties inclusion within the SI definition presents great controversy. On one hand, the macro-geometrical deviations and the mechanical properties affect not only the part surface, but also the bulk. On the other hand, the mechanical properties are included within the surface physical-chemical properties. The surface is considered as a whole, and not only from the micro-geometrical and physical-chemical approach at a point on the surface. In this way, the macro-geometrical deviations may affect negatively the mechanical behavior or increase the appearance of micro-geometrical or physical-chemical defects, and vice versa. Thereby, these three points of view are interconnected and may not be considered in isolation. Regardless of their inclusion or not within the SI definition, the macro-geometrical deviations play a fundamental role in mechanical properties [8,9], such as fatigue behavior [10]. In that way, fatigue test standards are very demanding regarding the tolerances of these geometrical deviations [11].

Fatigue behavior is one of the most important properties to take into account in the behavior of aircraft structural parts in service [12]. The quality requirements of these components are highly demanding because they are placed in critical areas. Consequently, geometrical tolerances (at macro and micro scale) are usually very narrow in order to make their assembly easier and to improve their functionality, reliability and longevity [13,14]. However, these high-quality demands result in higher costs. Hence, one of the most important challenges is to balance the manufacturing process performance of these components, from four different points of view: functional, economic, environmental and energetic [15–17].

Light alloys (mainly Al and Ti alloys) are widely used in the manufacturing of structural parts for aircrafts, individually or combined with composites (such as carbon fiber reinforced polymers, CFRP) to form fiber metal laminates structures. Specifically, aluminum alloys series 2000 (Al-Cu) and 7000 (Al-Zn) are used in the components under fatigue load in service, such as the pressurized cabins fuselage, ribs, spars and wings upper/lower skins [18–20]. Machining (mainly milling, drilling and turning) is frequently used to manufacture these structures [21,22]. The current trend in machining of these alloys moves towards reducing or eliminating the use of cutting fluids (dry machining), due to environmental and occupational health reasons [23–25]. The machining process based on the minimum quantity of cooling lubrication (MQCL) or the minimum quantity of lubrication (MQL) are good alternatives [26]. However, the performance balance of dry machining is currently a challenge. On one hand, the CFRP/Al structures show a bad behavior under wet machining conditions and the mixture of aluminum chip, CFRP and cutting fluids is complex and expensive to recycle [27]. On the other hand, dry machining results in very aggressive cutting conditions, which gives rise to a fast temperature increase in the cutting area and fast tool wear [28]. This fact may negatively affect the surface integrity of the machined parts. Hence, the environmental component of the process performance is improved, whereas the economical and functional components may be reduced.

Within this context, the cutting parameters (cutting speed, feed and cutting depth) play a very important role [29]. A large amount of research can be found in the literature analyzing the cutting parameters influence on the SI micro-geometrical aspects of dry machined wrought aluminum alloys [28,30–33] taking in to account the influence of the tool wear, chip geometry or axial machining length. Usually, the mean average roughness (*Ra*) is the selected parameter to evaluate these deviations. Most of the research agree that feed is the most influential cutting parameter on *Ra*, showing *Ra* a general trend to increase with feed [34–36]. Some of these works develop parametric models that allow predicting the *Ra* evolution as a function of the cutting parameters. Usually, these models adopt a potential form [35–38] due to the simplicity of the model, where the exponent of each variable represent the influence in the general term, and to the good fit that the models usually show. In addition, some authors show a relationship between *Ra* and the cutting forces [8,39]. Therefore, an online monitoring of micro-geometrical deviations was carried out to analyze this relationship.

Nevertheless, there is a lack of research focused on the analysis of the cutting parameters influence on macro-geometrical deviations for these alloys, despite their importance and influence on the functional behavior of these parts [9]. Clares et al. [7] proposed an experimental methodology to evaluate the SI in dry turning of aerospace alloys from the three aforementioned points of view (geometrical, at micro and macro scale and physical-chemical features). Sánchez-sola et al. [40] studied the cutting speed (v_c = 40–170 m/min) and feed (f = 0.05–0.30 mm/r) influence on straightness, parallelism and roundness of the UNS A92024 alloy. The cylindrical bars (200 mm length, 80–120 mm diameter) were dry turned. The worst results were obtained when the highest f and lowest v_c values were used. A general trend to increase the straightness and parallelism with f was found for low v_c. An opposite trend was found for high v_c, with a strong dispersion in the results for the highest f values. Regarding the roundness, its value tends to increase with f, regardless v_c. No clear trend was found with v_c. On the one hand, Sánchez-sola et al. explain that straightness and parallelism are measured along the whole part and, therefore, they are more influenced by the roughness profile, chip generation, built-up edge (BUE) detaching and, as a result, by the cutting parameters. On the other hand, roundness is measured from the transversal section. Hence, it is less influenced by these parameters. In addition, exponential parametric models were obtained. These models showed a good fit for high f and low v_c.

Trujillo et al. [9] analyzed the cutting parameters influence (v_c = 40–200 m/min; f = 0.05–0.20 mm/r) on roundness, circular run-out, straightness and parallelism of dry turned UNS 97075 alloy cylindrical bars (150 mm length, 30–60 mm diameter). The cutting depth (a_p) remained constant (1 mm). These geometrical deviations showed low sensitivity to change with the cutting parameters, unlike what was observed for micro-geometrical deviations, strongly influenced by f. However, v_c exhibited higher influence on straightness and parallelism, whereas f showed higher influence on roundness and circular run-out. These results can be explained in a similar way that were exposed in [40]. The exponential parametric models were also developed for each macro-geometrical deviation. Parallelism and straightness models exhibited a good fit for low v_c. However, the roundness and circular run-out showed a good fit for all v_c tested. Finally, Trujillo et al. highlighted that these models should be tested under different conditions (cutting parameters range, specimens' geometry) to check their generality.

It is necessary to point out that these previous studies have been carried out on low slenderness parts. Nevertheless, the structural parts for the aircraft usually show high slenderness rates [41]. Therefore, extended research on slender parts should be developed in order to analyze the cutting parameters' influence on the macro-geometrical deviations and their influence on mechanical properties, such as fatigue behavior. Given the aforementioned, the main novelty of the study, this work focusses on the analysis of the cutting speed and feed influence on several geometrical deviations (straightness, parallelism, circular run-out, roundness, concentricity and cylindricity) of dry turned UNS A97075 alloy. For this purpose, the specimens with high slenderness were used. These specimens were designed to be used in further fatigue tests. Finally, the different experimental parametric models were developed. These models allow predicting some geometrical deviations as a function of the cutting parameters within the studied range.

2. Materials and Methods

Several dry turning tests were carried out on UNS A97075-T6 alloy specimens in order to evaluate the cutting parameters' influence on different macro-geometrical deviations. This material is widely used in the manufacturing of aeronautical structural parts that work under compressive and fatigue loads [42]. Arc atomic emission spectroscopy (AES) was used to obtain the tested alloy composition (% mass). The results are shown in Table 1.

Table 1. Tested alloy composition (% mass).

Zn	Mg	Cu	Cr	Si	Mn	Al
6.01	2.61	1.88	0.19	0.08	0.07	Rest

The final specimens' geometry obtained from cylindrical bars (D = 20 mm) was selected according to the rotating bar bending fatigue test standard, ISO 1143:2010 [11]. Among the different geometries proposed in this standard, the cylindrical smooth geometry was selected. On the one hand, this geometry is less rigid that those used in previous research [9,40], being this one of the main novelties of this work. On the other hand, these specimens are being used in current works regarding the cutting parameters influence on fatigue behavior. The specimens' shape and geometrical dimensions are shown in Figure 1.

In addition, it is necessary to highlight that the standard quality requirements regarding the cylindricity and concentricity are strongly demanding. Therefore, the specimen slenderness was taken into account and calculated as the relation between the lengths (L = 167 mm) and the lower diameter (d = 7.6 mm), being this value, 22.37.

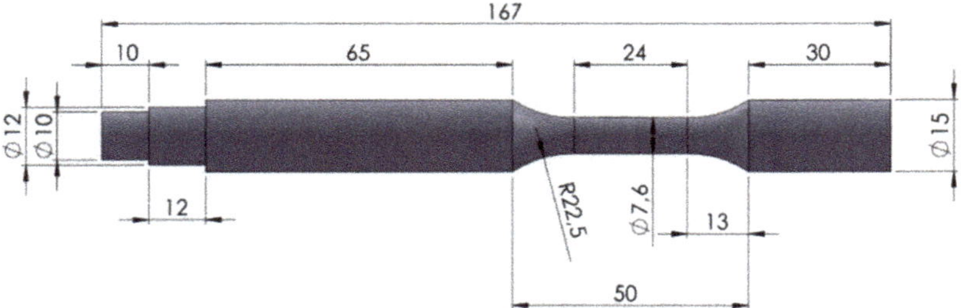

Figure 1. Specimens geometrical dimensions (mm).

The machining tests were conducted in a Computer Numerical Control (CNC) turning center. The different combinations of cutting speed (v_c) and feed (f) were used (Table 2). The cutting depth (a_p) remained constant. Every test was performed under dry conditions, in order to use environmentally friendly techniques. It must be pointed that although low cutting speeds are not recommended for machining aluminum alloys, these alloys are often hybridized with other materials in which these low cutting speeds are required, such as fiber metal laminates, FML (CFRP +Al + Ti). In addition, this fact allows the comparison with previous studies on the geometrical deviations performed in the same cutting parameters range [9,40]. Every test was performed under dry conditions in order to use environmentally friendly techniques.

Table 2. Cutting parameters.

v_c (m/min)	f (mm/r)	a_p (mm)
40 60 80	0.05 0.10 0.15 0.20	1.0

The tool (Figure 2) was an uncoated WC-Co insert (ISO DCMT 11T308-14 IC20) and each test was carried out using a new tool to ensure identical initial conditions. The cutting angles setup can be observed in Table 3. The tool geometry and position respect the specimen allow the machining of the entire specimen profile with a single operation. Additionally, these geometrical cutting conditions reduce the deflection effect on slenderness parts due to the predominance of the axial cutting forces over the radial cutting forces. Due to the cutting tool geometry (a major cutting-edge angle of 66.5°), the force axial component is dominant over the radial component. Therefore, the bending effect on sample was limited.

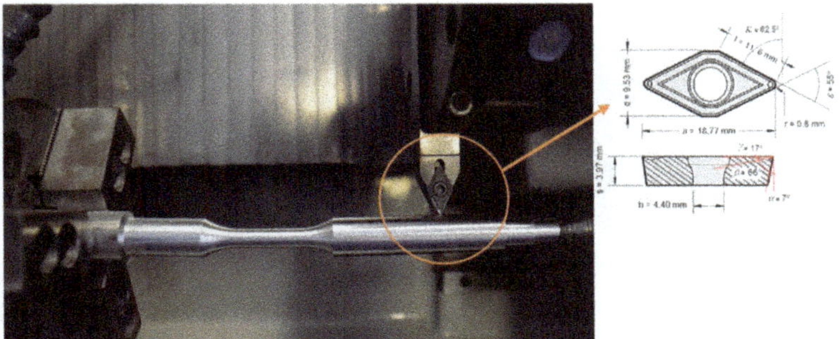

Figure 2. Specimen positioning and turning operation tool.

Table 3. Cutting angles setup.

Cutting Angles	Value
Relief angle (α)	7°
Cutting edge angle (β)	66°
Rake angle (γ)	17°
Major cutting edge angle (κ_r)	62.5°
Insert included angle (ε)	55°

Once the specimens were machined, different macro-geometrical deviations were controlled in the calibrated area (parallelism, straightness, circular run-out, roundness, concentricity, total run out and cylindricity). The experimental setup is shown in Figure 3. A millesimal dial gauge with a measuring span of 12.5 mm, scale division of 0.001 mm and a maximum permissible error (MPE) of 4 μm, was used to control these deviations. This device was placed on the tool carriage to avoid removing the specimen from the turning center and achieve a faster process. Previously, the setup rigidity and the run-out of the spindle were controlled in order to assess their contribution to the part run-out. This contribution was found to be negligible. In addition, some of the specimens were off-line measured in a geometrical deviation measurement machine (Figure 4) in order to validate the experimental setup. The differences found did not exceed 10%.

Figure 3. Geometrical control setup on the specimen calibrated area.

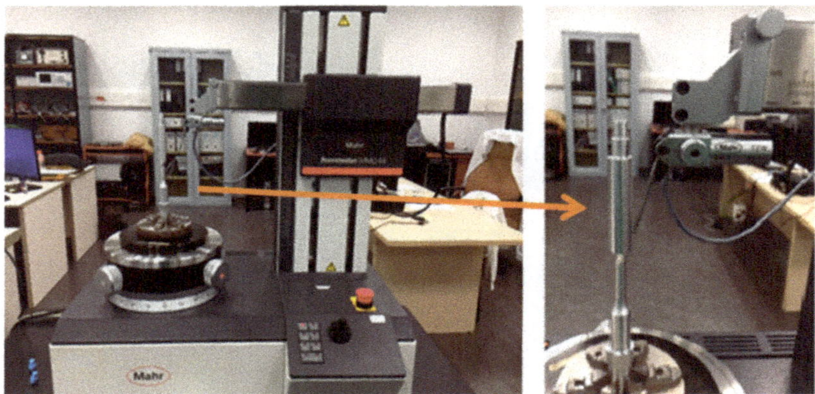

Figure 4. Form measuring system.

The circular run-out (CRO) was measured along six sections (S1 to S6, separated 4 mm from each other) in the calibrated area (Figure 5a). For each section, twelve measurements were performed at 30° each. The parallelism (PAR) was controlled along twelve generatrix (G1–G12), separated 30° from each other, Figure 5b.

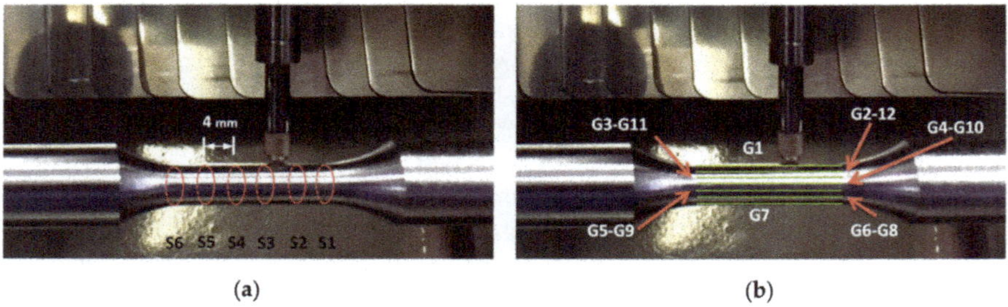

(**a**) (**b**)

Figure 5. (**a**) The roundness (RON), circular run-out (CRO) and concentricity (CON) measured sections; (**b**) straightness (STR) and parallelism (PAR) measured generatrix.

The circular run-out (CRO) was obtained as the difference between the maximum and the minimum profile radius ($R_{\max} - R_{\min}$), as in Figure 6a [42]. The roundness (RON) and concentricity (CON) were calculated from the CRO experimental results (Figure 6a). Among the different mathematical methods available, the least squares circles method [43] was applied in this work. To evaluate the least square circumference center, a nonlinear iterative mathematical model was considered, minimizing the function error (Equation (1)) and taking the seed as the rotation center.

$$\mathrm{SSE}(a,b) = \sum_{i=1}^{n}\left(R - \sqrt{(x_i - a)^2 - (y_i - b)^2}\right)^2 \quad (1)$$

where different variables corresponding with:

- R: Radius of the profile
- x_i: x coordinate of profile point
- y_i: y coordinate of profile point
- a: x coordinate of the center of the least square circumference
- b: y coordinate of the center of the least square circumference

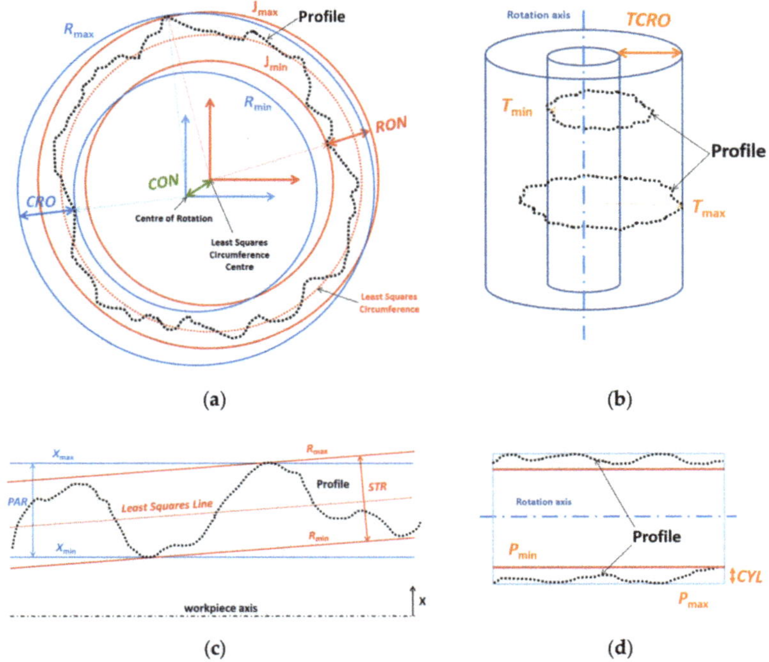

Figure 6. Macrogeometrical deviations: (**a**) RON, CON and CRO, (**b**) total circular run-out (TCRO), (**c**) STR and PAR and (**d**) cylindricity (CYL).

Once the center of the least square circumference has been calculated, *RON* is obtained as the difference between the radiuses of two concentric circumferences ($J_{max} - J_{min}$) which delimit the area containing all points of the profile, as in Figure 6a. The CON is the distance between the rotation center (0, 0) and the least squares circumference center previously obtained (a, b) (Equation (2)).

$$\mathrm{CON} = \sqrt{a^2 + b^2} \tag{2}$$

The parallelism (PAR) was obtained as the distance between two parallel lines ($X_{max} - X_{min}$), which delimit the area containing all the profile points and are parallel to the work piece axis (Figure 6c). The straightness (STR) was calculated as the distance between two parallel lines ($D_{max} - D_{min}$) which delimit the area containing all the profile points and are parallel to the least squares regression line (Figure 6c).

Taking into account the total calibrated area volume, the total circular run-out (TCRO) was calculated as the difference between the maximum and minimum of every radio sections measured ($T_{max} - T_{min}$) in the calibrated area (Figure 6b). Finally, the cylindricity (CYL) was obtained as the difference between two co-axial cylinders, such that their radial difference is at a minimum ($P_{max} - P_{min}$), as in Figure 6d.

The theorical PAR respect CYL was taken in to account. Considering the tool tip radius (r = 0.8 mm), the depth of cut (a_p = 1 mm) and the worst feed rate condition (f = 0.20 mm/r), the scallop calculated is 6.27 µm. The standard 1143:2010 allows a maximum of 20 µm. Therefore, the scallop value for each feed rate implemented in this work is considered within the standard values.

3. Results and Discussion

3.1. Parallelism and Straightness

Figure 7 shows STR and PAR experimental results as a function of v_c and f. In Figure 7a, the PAR shows a general trend to slightly increase at low f (0.05–0.10 mm/r), regardless v_c. From f = 0.10 mm/r to 0.20 mm/r, PAR tends to decrease, whereas it remains more or less constant for v_c = 60 and 80 m/min. Notwithstanding this, a higher dispersion can be observed. From f = 0.15 to 0.20 mm/r, PAR remains constant for v_c = 40 and 60 m/min. Nevertheless, PAR tends to increase for v_c = 80 m/min for this f range. In fact, the worst results (and the highest dispersion) are obtained when the highest f and v_c are combined.

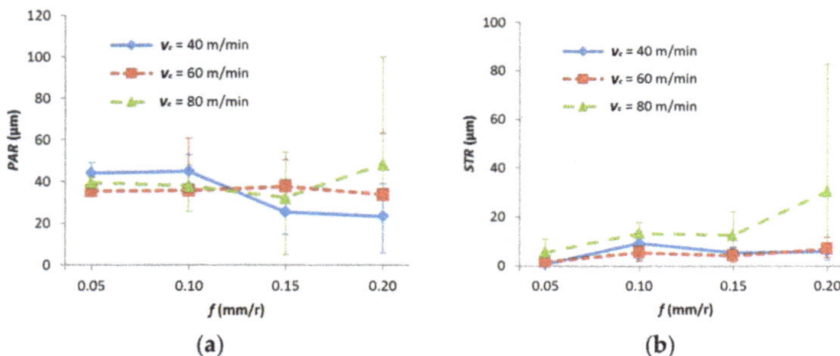

Figure 7. (**a**) Parallelism and (**b**) straightness deviation as a function of cutting speed (v_c) and feed (f).

Figure 7b shows that STR presents a higher dependence on the cutting parameters studied than PAR. On one hand, the highest values are always obtained for v_c = 80 m/min, regardless f. In addition, a general trend to increase STR with f is observed, mainly from f = 0.05 to 0.10 mm/r and from f = 0.15 to 0.20 mm/r. On the other hand, for the low range of f studied (0.05 to 0.10 mm/r), STR tends to increase with f, regardless v_c. Nevertheless, in general terms, v_c seems to be the most influent parameter. Moreover, its effect increased when it is combined with the highest v_c (80 m/min). In addition, these results show higher deviations than other research with a less slender workpiece [9,40].

These results can be explained taking into account how PAR and STR are measured along the machining length. This fact results in a higher dependence of those geometrical deviations on the built-up edge (BUE) formation and detaching vibrations and/or deflections of the specimen. Figure 8 shows the tool rake face for v_c = 40 m/min, for f 0.05 and 0.10 mm/r. These images show that the indirect adhesion wear phenomenon is higher for 0.10 mm/r. A similar trend was found for v_c = 60 and 80 m/min. Therefore, STR is more sensitive to the BUE formation at low f. For higher f values (0.15–0.20 mm/r), STR becomes less sensitive to BUE (with similar intensity at high f values) and more sensitive to vibrations, more noticeable for v_c = 80 m/min (Figure 9).

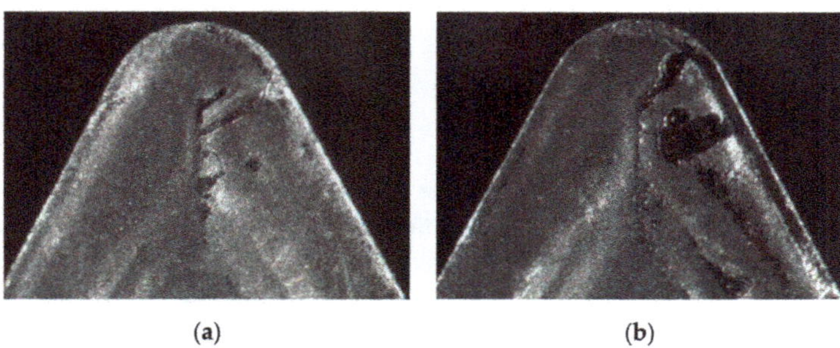

(**a**) (**b**)

Figure 8. Stereoscopic optical microscopy (SOM) of the tool rake face (40×) after the tests performed for (**a**) f = 0.05 mm/r and (**b**) f = 0.10 mm/r, for v_c = 40 m/min.

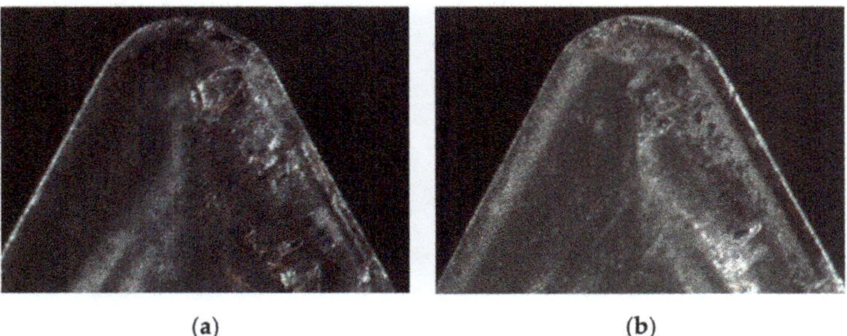

(a) (b)

Figure 9. Stereoscopic optical microscopy (SOM) of the tool rake face (40×) after the tests performed for (**a**) f = 0.15 mm/r and (**b**) f = 0.20 mm/r, for v_c = 80 m/min.

These results have revealed two important differences with previous research on similar alloys, in which more rigid specimens were used [9,40]. First, both PAR and STR have shown higher values. As a result, these macro-geometrical deviations tend to increase with the specimen slenderness. In addition, previous work did not reveal a clear influence of the cutting parameters on PAR and STR. However, only v_c seemed to be slightly more influential at high values. However, in this work, PAR and STR behavior was more sensitive with f, and its effect was maximized at a high cutting speed.

3.2. Circular Run-out, Roundness and Concentricity

Figure 10 shows CRO experimental results as a function of the cutting parameters for each section. With the different sections (1–6) in mind, the specimen slenderness and the experimental setup rigidity play a special role. Further sections from the chuck (1–3) have exhibited higher CRO deviations, whereas closer sections (4–6) have shown lower values. This general trend is more noticeable for v_c = 40 and 60 m/min, regardless of f, whereas it is less evident for v_c = 80 m/min. Therefore, the specimen slenderness and rigidity are more influential at low v_c. Nevertheless, vibrations become more relevant at high v_c.

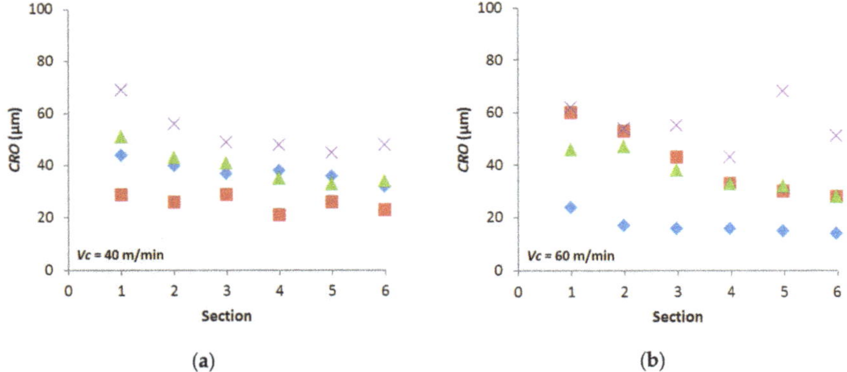

(a) (b)

Figure 10. *Cont.*

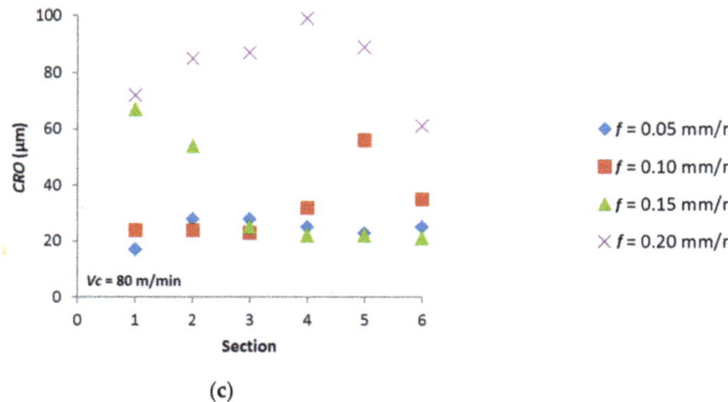

(c)

Figure 10. Circular run-out deviations for (**a**) v_c = 40 m/min, (**b**) v_c = 60 m/min and (**c**) v_c = 80 m/min.

In addition, Figure 11 plots the average CRO (taking all sections into account) as a function of f and v_c. In general, f seems to be the most influential parameter. The circular run-out (CRO) exhibits a general trend to increase with f, regardless v_c. This fact becomes more noticeable at high cutting speed values. The general trend is less clear with regard to v_c. Between f = 0.05 and 0.10 mm/r, the highest CRO value is obtained for v_c = 40 and 60 m/min, respectively. However, for f = 0.20 mm/r, the worst result is found for v_c = 80 m/min. As a result, the combination of the highest f and v_c values result in the highest CRO deviations. The high vibration levels, produced when the high values of cutting speed and feed are used, should explain this behavior.

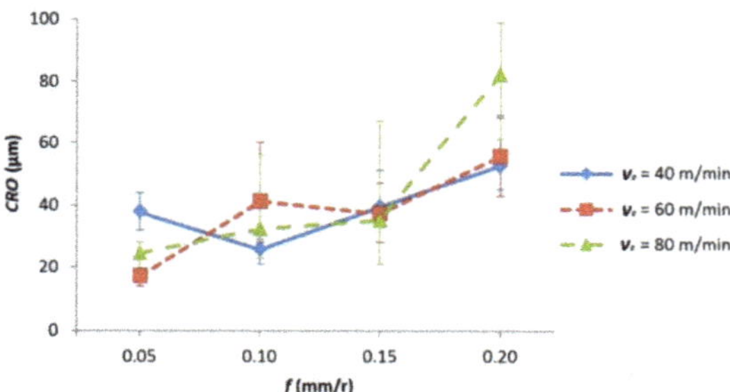

Figure 11. Circular run-out mean values as a function of cutting speed (v_c) and feed (f).

The RON and CON deviations can be calculated from CRO experimental results and Equations (1) and (2).

Figure 12 plots the RON experimental results in the function of the cutting parameters, for each section. A general trend to increase RON deviations in the function of f can be observed, regardless of v_c. However, a strong increment in the function of v_c can be noticed for f = 0.20 mm/r. Trivial variations of RON in the function of the section relativity position can be observed, except for v_c = 80 m/min and f = 0.20 mm/r, where severe cutting conditions show an increase on the central sections (2–5). Taking into account that feed is the most influenced parameter in cutting forces, regardless of v_c [31,44], this fact can be explained due to an increment of deflections, especially in a slender specimen, where the cutting forces are increased in the higher range of f. Notably, the CON evolution for the different sections measured can be considered as a function or the results shown for CRO (Figure 10) and RON (Figure 12).

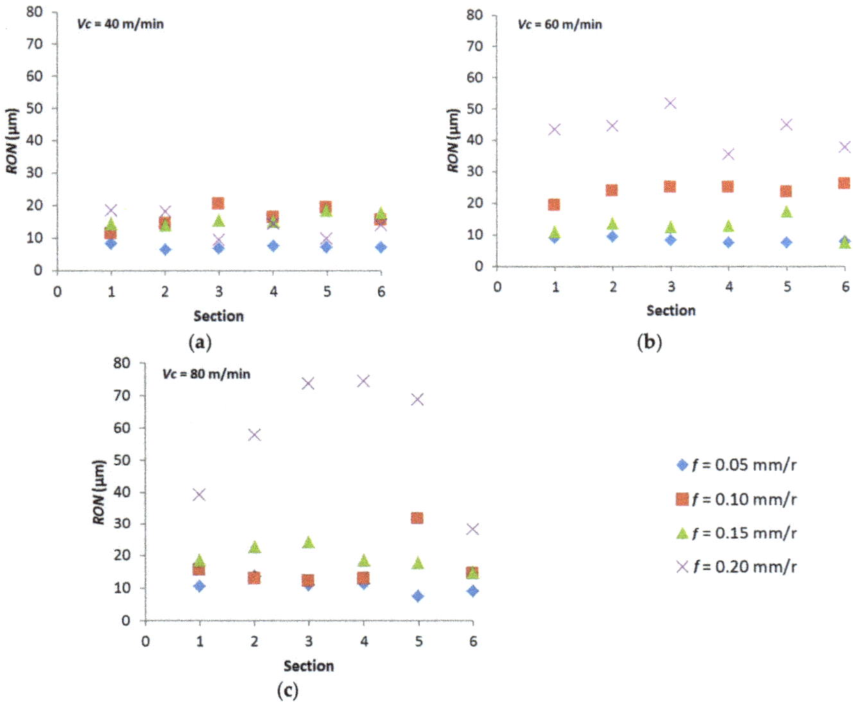

Figure 12. Roundness deviations for (**a**) v_c = 40 m/min, (**b**) v_c = 60 m/min and (**c**) v_c = 80 m/min.

Figure 13 plots the RON and CON mean values (considering all the sections) as a function of f and v_c. A general trend to increase RON mean values with f is observed, regardless of v_c, as in Figure 13a. This increment is more evident from f = 0.15 to 0.20 mm/r and softer for v_c = 40 m/min. Regarding v_c, the general trend is less clear. Only between f = 0.15 and 0.20 mm/r there is a trend to increase *RD* with v_c. The worst results are obtained when the highest values of f (0.20 mm/r) and v_c (60 and 80 m/min) are tested. Hence, the trend is very similar as that observed for CRO. Nevertheless, the cutting parameters influence is more evident on RON than on CRO. This fact is a consequence of deleting the concentricity effect, which shows a less clear trend with the cutting parameters, as in Figure 13b. The CON average values (Figure 13b) show a general trend to increase with f, but softer than RON. Regarding v_c, the CON values are more scattered, showing different behavior at a low and high feed.

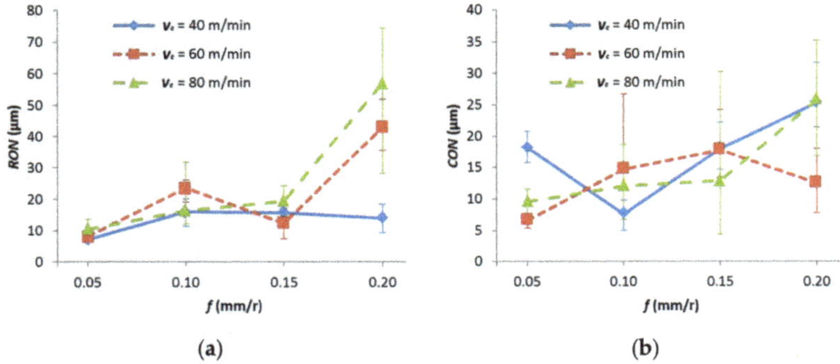

Figure 13. Mean deviations values in function of v_c and f for (**a**) roughness and (**b**) concentricity.

Considering that CRO is related with RON and CON, it is necessary to point out that CON takes more importance in CRO for low v_c and f values, whereas RON influence is more noticeable for higher values.

The found trend is slightly different from those obtained in previous research for UNS A97075 and UNS A92024 alloys [9,40]. CRO and RON sensitivity to change with the cutting parameters was very low (in specimens with higher rigidity). Nevertheless, this sensitivity increases when the specimen geometry is less rigid, especially with f. In addition, the deviations values are significantly higher for high slenderness specimens.

3.3. Total Circular Run-out and Cylindricity

Figure 14a shows the TCRO values as a function of v_c and f. No clear trend can be observed as a function of v_c or f for this deviation. From f = 0.05 to 0.10 mm/r, TCRO tends to increase for v_c = 60 and 80 m/min, whereas it tends to decrease for v_c = 40 m/min. From f = 0.10 to 0.20 mm/r, its value remains more or less constant, for v_c = 40 and 60 m/min, regardless of f. However, TCRO only shows a general trend to increase with f for v_c = 80 m/min. This fact is more noticeable for f = 0.20 mm/r, where the worst result is obtained. Therefore, TCRO exhibits less sensitivity to change with the cutting parameters than CRO, RON or STR, and similar to CON. The value of TCRO strongly depends on the maximum value of CRO in each section, but also on the angular position where that maximum value is obtained. Therefore, the maximum value for CRO in one section may occur in a different angular position than another one and, as a result, the effect of the cutting parameters may be less evident.

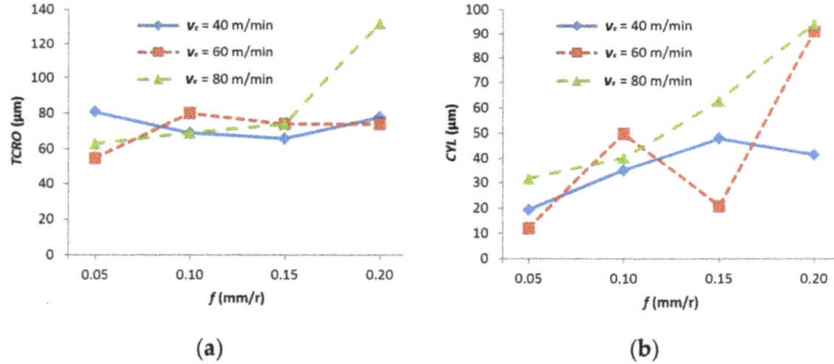

Figure 14. Mean deviations values as a function of v_c and f for the (**a**) total circular run-out, and (**b**) cylindricity.

Finally, Figure 14b plots the CYL values as a function of f and v_c. For this deviation, the influence of f is more evident. A general trend to increase CYL with f can be observed in a wide range of v_c studied. This effect is more noticeable at v_c = 80 m/min, mainly at higher f values (0.15–0.20 mm/r). The worst results appear when the highest f (0.20 mm/r) is combined with high v_c (60 and 80 m/min). For v_c = 60 m/min, this trend is fulfilled from 0.05 to 0.10 mm/r and 0.20 mm/r. Nevertheless, for f = 0.15 mm/r, a significant decrease is observed. Therefore, in a similar way than CRO, RON or STR, CYL exhibits a higher sensitivity to change with f than v_c. This behavior was to be expected, taking into account that CYL combines the effect of the profile deviations along the specimen length, as well as along its section.

3.4. Geometrical Tolerance in Rotating Bar Bending Specimens

ISO 1143:2010 standard establishes the different geometrical tolerances for the rotating bar bending specimen, in order to minimize their effect in the fatigue behavior. In this sense, in the expected fatigue fracture zone, the standard indicates a CYL and CON maximum tolerance deviation. These requirements are 20 µm for CYL and 15 µm for CON, between sections S1 and S6 [11]. Nevertheless, the geometrical deviations in a manufacturing process are usually far from the standard requirements. Therefore, it may be interesting to compare these requirements with those obtained in the manufacturing process under different conditions.

On the one hand, the CON values were below the required standard limit, in general. Only higher values of f (0.15 and 0.20 mm/r), in combination with high v_c (60 and 80 m/min) exceed that limit. Therefore, the influence of CON on fatigue behavior may not be neglected under these cutting conditions. On the other hand, only for low f and v_c value combinations (f = 0.05 mm/r, v_c = 40 and 60 m/min), the CYL results were below the standard requirements (Figure 12b). This is due to the stronger influence of the cutting parameters on this geometrical deviation. As a result, there is a wide range of cutting conditions where the CYL deviations should be considered in the fatigue behavior analysis.

3.5. Parametric Models for Macro-Geometrical Deviations

The experimental results suggest the possibility of obtaining parametric models that allow relating some of the analyzed geometrical deviations with the cutting parameters. These experimental models may be useful to predict these deviations before machining [9,40]. These models were developed for those deviations (GD) that have shown a greater dependence on the cutting parameters (STR, CRO and RON). Different models were tested. The best fit was obtained for a potential model, as shown in Equation (3).

$$\mathrm{GD} = C \cdot v_c^x f^y \quad (3)$$

where C, x and y are constant. Table 4 shows the results obtained for these constants. It is necessary to point out that these models have shown, in general, a reasonable fit (coefficient of determination, $R^2 \approx 0.6$–0.7).

Table 4. Model coefficients.

GD	C	x	y
STR	0.120	1.574	1.123
CRO	119.720	0.028	0.584
RON	5.297	0.736	0.832

Figure 15 plots the experimental data versus the model results, for STR, CRO and RON respectively. Regarding STR, both cutting parameters (f and v_c) show a strong influence on the model. Nevertheless, the higher value of the x exponent indicates a higher influence of the cutting speed. This is in good agreement with the experimental results (previously discussed). Only for f = 0.20 mm/r and v_c = 80 m/min, the model shows a worse fit (Figure 15a).

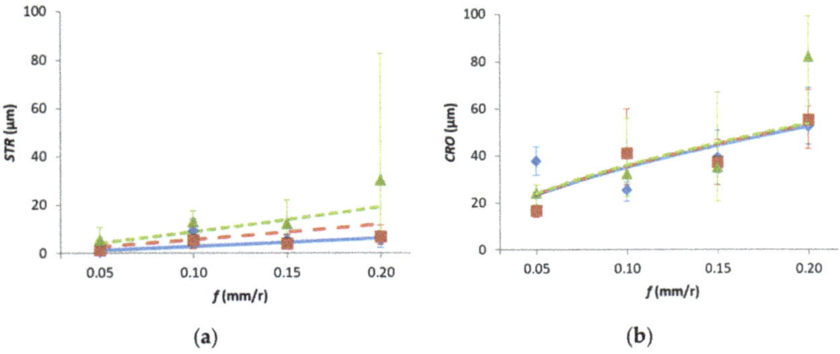

Figure 15. *Cont.*

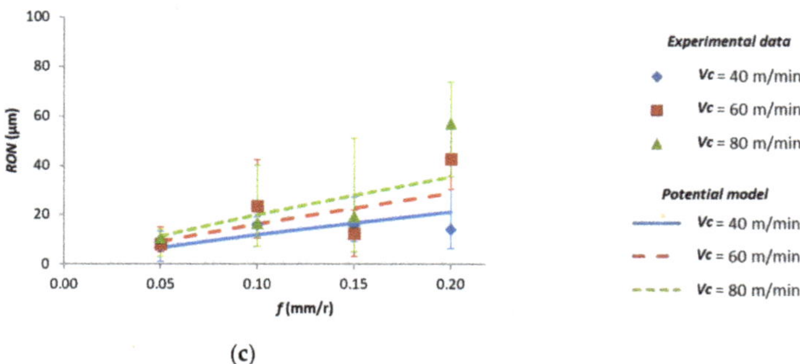

(c)

Figure 15. Potential models as a function of v_c and f (**a**) straightness, (**b**) circular run-out and (**c**) roundness.

This fact may be a consequence of the higher dispersion in the experimental data. This can be considered as normal, taking into account the higher vibration levels obtained for this cutting parameter combination, as previously discussed.

With regard to CRO, the x low value indicates a negligible influence of v_c. This fact can be also observed in Figure 15b. Therefore, f is the most influential parameter. This is in good agreement with the experimental observations, as previously commented. However, this is the model that showed a worse fit. This may be explained due to the fact that CRO includes the effect of RON and CON. The effect of the cutting parameters on CON was not clear. As a result, the CRO experimental results exhibited higher dispersion and, therefore, the model for CRO showed a lower fit (Figure 15b).

Finally, the model for RON shows a similar value for x and y. Hence, the influence of both cutting parameters (v_c and f) is similar, as previously commented. In addition, the model for RON shows a better fit than for CRO. However, as it happens for STR, the model shows a lower fit for the highest values of v_c and f. As a result, it is useful only for the low range of v_c and f studied.

Additionally, Figure 16 plots these parametric potential models in 3D, where each geometrical deviation is represented by a surface. As previously commented, Figure 16a,c (STR and RON, respectively) show a strong dependence of v_c and f, whereas Figure 16b exhibits a very low influence with v_c. In addition, an increment in these deviations can be observed when the cutting parameters are increased.

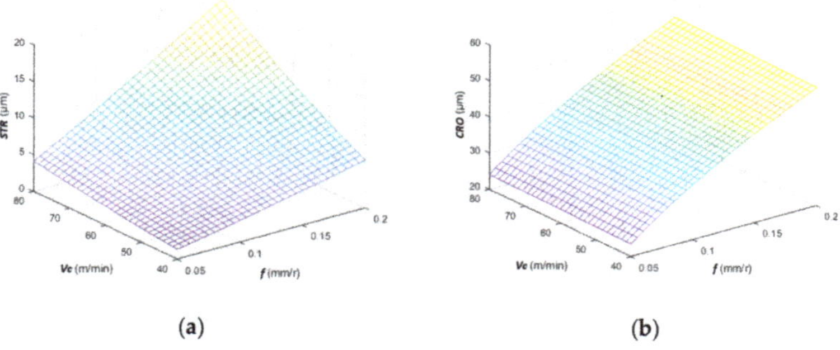

(a) (b)

Figure 16. *Cont.*

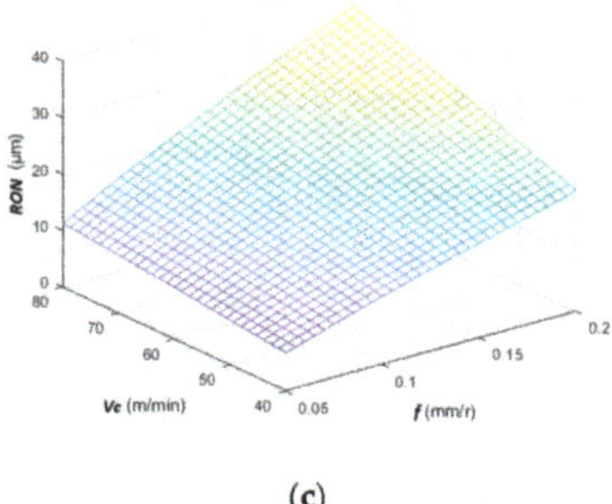

(c)

Figure 16. Potential models for (**a**) STR = $g(v_c, f)$, (**b**) CRO = $h(v_c, f)$ and (**c**) RON = $j(v_c, f)$.

It is necessary to point out that similar models have been obtained in previous research, for specimens with low slenderness (UNS A97075 and UNS A92024 alloys) [9,40]. Those models presented an exponential form for these macro-geometrical deviations under similar cutting conditions. Notwithstanding this, they exhibited a lower sensitivity to change with the cutting parameters due to the higher rigidity of the specimens' geometry. Therefore, the potential models presented in this work are more suitable for specimens with higher slenderness. Finally, it is necessary to highlight that these models are valid within the range of the tested cutting parameters and under the cutting conditions exposed. The general validity of these models should be contrasted in further works, under different conditions. However, they are useful for a better understanding of the experimental results obtained.

4. Conclusions

In this work, the influence of the cutting speed and feed on several macro-geometrical deviations (parallelism, straightness, circular run-out, roundness, concentricity, total circular run-out and cylindricity) in dry turning of UNS A97075 (Al-Zn) alloy was analyzed. The analysis was performed using high slenderness parts in order to compare the experimental results with previous research carried out with more rigid parts.

Straightness has exhibited a general trend to increase with the feed, regardless the cutting speed. This trend was more evident at low feed values. The worst results were obtained when the highest cutting speed and feed rate were combined. These results should be explained taking into account that this deviation is measured along the specimen length. Therefore, it is influenced by the BUL and BUE formation and detaching, the process rigidity, the part slenderness and vibrations. On one hand, at low feed, the BUE formation and detaching becomes more relevant and the feed rate influence is more noticeable. On the other hand, at high cutting speed, vibrations become more relevant and the sensitivity to change with the cutting speed is higher.

Regarding parallelism, the influence was less evident and no clear trend was found as a function of the feed rate or cutting speed. Notwithstanding this, the obtained results for both deviations were higher than those obtained in previous works with more rigid parts under similar cutting conditions.

The circular run-out and roundness showed a general trend to increase with the feed rate. This dependence was more evident for the roundness. The cutting speed influence was lower in both cases. The feed rate effect worsened when it was combined with high cutting speeds. With regard to the evolution of those deviations along the specimen section, further sections from the chuck have shown higher deviations, whereas closer sections have exhibited lower values. This general trend was more noticeable at a low cutting speed, regardless of f. In addition, an increased f influence was observed in the sections more distant from the chuck. Hence, vibrations became more relevant at a high spindle rotational speed, whereas the specimen slenderness and rigidity were more influential at a low cutting speed.

Regarding concentricity, no clear trend was found. Therefore, its sensitivity to change with the cutting parameters was lower. Usually, this deviation is more influenced by the rigidity setup and spindle concentricity than by the cutting parameters.

Furthermore, no clear trend was observed as a function of the cutting parameters for the total circular run-out. Therefore, this deviation exhibited lower sensitivity to change with the cutting parameters. This fact may be explained taking into account that this deviation strongly depends on the maximum value of the circular run-out in each section, but also on the angular position for this value.

With regard to cylindricity, this deviation showed a general trend to increase with the feed rate in a wide range of cutting speeds studied. Its sensitivity to change with the feed was higher than with the cutting speed, in a similar way than the circular run-out, the roundness or the straightness. This is due to the fact that cylindricity combines the effect of the profile deviations along the specimen length, as well as along its section.

Therefore, the experimental results revealed higher sensitivity to change with the cutting parameters than the results obtained in previous research with lower slenderness parts, for most macro-geometrical deviations. In addition, the feed rate seems to be the most influential parameter whereas the cutting speed has shown less influence. On the other hand, the obtained deviations have been noticeably higher in parts with lower slenderness, compared with those obtained in previous research with more rigid parts of UNS A92024 and UNS A97075 alloys.

Additionally, the experimental results for CON and CYL have been compared with the quality requirements of the specimens used in the rotating bar bending fatigue test standard (ISO 1143:2010). The results revealed that there is a wide range of cutting conditions where both geometrical deviations should be considered for the fatigue behavior analysis.

Finally, a set of potential parametric models were proposed for STR (v_c, f), CRO (v_c, f) and RON (v_c, f). These models exhibited a reasonable fit for STR and RON, whereas the fit for CRO was lower. These models may be useful to analyze the influence of cutting conditions (v_c, f) in these deviations before machining. It is necessary to point out that these models are useful in the range of cutting conditions evaluated and can be considered as a first step to obtain more complex models. In addition, these models were compared with other models obtained from the tests with less slender specimens (20 times less slenderness). In spite of this, the results dispersion is of the same order.

It is necessary to point out that although the study was carried out on the parts with a geometry different from that used for the manufacture of aircraft structural parts, this work revealed the importance of this kind of analysis in further works. In addition, it can be considered as the starting point to analyze the influence of the geometrical deviations on mechanical properties, such as fatigue behavior.

Author Contributions: S.M.B. and F.J.T.V. conceived and designed the experiments; S.M.B. performed the experiments; S.M.B., F.J.T.V., C.B.G. and L.S.H. analyzed the data; S.M.B. and F.J.T.V. wrote the paper; L.S.H. and C.B.G. revised the paper.

Funding: This research received no external funding.

Acknowledgments: The authors thank the University of Cádiz for its form measuring equipment and the University of Malaga-Andalucia Tech Campus of international Excellence for its economic contribution to this paper.

Conflicts of Interest: The authors declare no conflict of interest.

References

1. Grzesik, W. Introduction. In *Advanced Machining Processes of Metallic Materials*; Elsevier: Amsterdam, The Netherlands, 2017; pp. 1–5.
2. Field, M.; Kahles, J.F. Review of surface integrity of machined components. *Ann. CIRP* **1971**, *20*, 153–163.
3. Field, M.; Kahles, J.F.; Cammett, J.T. A review of measuring methods for surface integrity. *CIRP Ann. Manuf. Technol.* **1972**, *21*, 219–238.

4. Griffiths, B. *Manufacturing Surface Technology: Surface Integrity & Functional Performance*; Butterworth-Heinemann: Oxford, UK, 2001.
5. Astakhov, V.P. Surface Integrity—Definition and Importance in Functional Performance. In *Surface Integrity in Machining*; Springer: London, UK, 2010; pp. 1–35.
6. Gómez Parra, A. Study of the influence of machining on the functional performance of aluminum alloys for strategic use in the aeronautical industry. Ph.D. Thesis, Universidad de Cádiz, Cadiz, Spain, 2016.
7. Clares Rodriguez, J.M.; Vazquez Martinez, J.M.; Gomez-Parra, A.; Puerta Morales, F.J.; Marcos, M. Experimental Methodology for Evaluating Workpieces Surface Integrity in Dry Turning of Aerospace Alloys. In Proceedings of the Annals of DAAAM, the 26th International DAAAM Symposium, Zadar, Croatia, 18–25 October 2015; pp. 0849–0855.
8. Risbood, K.A.; Dixit, U.S.; Sahasrabudhe, A.D. Prediction of surface roughness and dimensional deviation by measuring cutting forces and vibrations in turning process. *J. Mater. Process. Technol.* **2003**, *132*, 203–214. [CrossRef]
9. Trujillo, F.J.; Sevilla, L.; Marcos, M. Cutting speed-feed coupled experimental model for geometric deviations in the dry turning of UNS A97075 Al-Zn alloys. *Adv. Mech. Eng.* **2014**, *2014*, 382435. [CrossRef]
10. Abroug, F.; Pessard, E.; Germain, G.; Morel, F.; Chové, E. The influence of machined topography on the HCF behaviour of the Al 7050 alloy. *Procedia Eng.* **2018**, *213*, 613–622. [CrossRef]
11. ISO 1143:2010 (E). Metallic materials: Rotating Bar Bending Fatigue Testing. Available online: https://www.iso.org/standard/41875.html (accessed on 28 September 2019).
12. Javidi, A.; Rieger, U.; Eichlseder, W. The effect of machining on the surface integrity and fatigue life. *Int. J. Fatigue* **2008**, *30*, 2050–2055. [CrossRef]
13. M'Saoubi, R.; Outeiro, J.C.; Chandrasekaran, H.; Dillon, O.W., Jr.; Jawahir, I.S. A review of surface integrity in machining and its impact on functional performance and life of machined products. *Int. J. Sustain. Manuf.* **2008**, *1*, 203.
14. Martín-Béjar, S.; Trujillo, F.J.; Herrera, M.; Sevilla, L.; Marcos, M. Experimental methodology design for fatigue behaviour analysis of turned aluminum alloys. *Procedia Manuf.* **2017**, *13*, 73–80. [CrossRef]
15. Apostolos, F.; Alexios, P.; Georgios, P.; Panagiotis, S.; George, C. Energy efficiency of manufacturing processes: A critical review. *Procedia CIRP* **2013**, *7*, 628–633. [CrossRef]
16. Gómez-Parra, A.; Puerta, F.J.; Rosales, E.I.; González-Madrigal, J.M.; Marcos, M. Study of the influence of cutting parameters on the Ultimate Tensile Strength (UTS) of UNS A92024 alloy dry turned bars. *Procedia Eng.* **2013**, *63*, 796–803. [CrossRef]
17. Trujillo, F.J.; Sevilla, L.; Marcos, M. Experimental parametric model for indirect adhesion wear measurement in the dry turning of UNS A97075 (Al-Zn) alloy. *Materials* **2017**, *10*, 152. [CrossRef] [PubMed]
18. Starke, E.A.; Staley, J.T. Application of modern aluminium alloys to aircraft. *Fundam. Alum. Metall.* **2011**, 747–783. [CrossRef]
19. Santos, M.C.; Machado, A.R.; Sales, W.R.; Barrozo, M.A.S.; Ezugwu, E.O. Machining of aluminum alloys: a review. *Int. J. Adv. Manuf. Technol.* **2016**, *86*, 3067–3080. [CrossRef]
20. Polmear, I.J. 3-Wrought aluminium alloys. In *Light Alloys*, 4th ed.; Butterworth-Heinemann: Oxford, UK, 2005; pp. 97–204.
21. Trent, E.M.; Wright, J.R. *Metal Cutting*, 4th ed.; Butterworth-Heinemann: Oxford, UK, 2000.
22. Campbell, F.C. *Manufacturing Technology for Aerospace Structural Materials*; Elsevier Science: Amsterdam, The Netherlands, 2006.
23. Sugihara, T.; Nishimoto, Y.; Enomoto, T. On-machine tool resharpening for dry machining of aluminum alloys. *Procedia CIRP* **2014**, *24*, 68–73. [CrossRef]
24. Goindi, G.S.; Sarkar, P. Dry machining: A step towards sustainable machining—Challenges and future directions. *J. Clean. Prod.* **2017**, *165*, 1557–1571. [CrossRef]
25. Krolczyk, G.M.; Maruda, R.W.; Krolczyk, J.B.; Wojciechowski, S.; Mia, M.; Nieslony, P.; Bukzik, G. Ecological trends in machining as a key factor in sustainable production—A review. *J. Clean. Prod.* **2019**, *218*, 601–615. [CrossRef]
26. Maruda, R.W.; Krolczyk, G.M.; Wojciechowski, S.; Zak, K.; Habrat, W.; Nieslony, P. Effects of extreme pressure and anti-wear additives on surface topography and tool wear during MQCL turning of AISI 1045 steel. *J. Mech. Sci. Technol.* **2018**, *32*, 1585–1591. [CrossRef]

27. Chen, L.; Hsieh, C.-C.; Wetherbee, J.; Yang, C.-L. Characteristics and treatability of oil-bearing wastes from aluminum alloy machining operations. *J. Hazard. Mater.* **2008**, *152*, 1220–1228. [CrossRef]
28. García-Jurado, D.; Vazquez-Martinez, J.M.; Gámez, A.J.; Batista, M.; Puerta, F.J.; Marcos, M. FVM based study of the influence of secondary adhesion tool wear on surface roughness of dry turned Al-Cu aerospace alloy. *Procedia Eng.* **2015**, *132*, 600–607. [CrossRef]
29. Trujillo, F.J.; Sevilla, L.; Martín, F.; Bermudo, C. Analysis of the chip geometry in dry machining of aeronautical aluminum alloys. *Appl. Sci.* **2017**, *7*, 132.
30. Trujillo, F.J.; Sevilla, L.; Marcos, M. Influence of the axial machining length on microgeometrical deviations of horizontally dry-turned UNS A97075 Al-Zn alloy. *Procedia Eng.* **2013**, *63*, 405–412. [CrossRef]
31. Gökkaya, H. The effects of machining parameters on cutting forces, surface roughness, Built-Up Edge (BUE) and Built-Up Layer (BUL) during machining AA2014 (T4) Alloy. *J. Mech. Eng.* **2010**, *56*, 584–593.
32. Rubio, E.M.; Camacho, A.M.; Sánchez-Sola, J.M.; Marcos, M. Surface roughness of AA7050 alloy turned bars: Analysis of the influence of the length of machining. *J. Mater. Process. Technol.* **2005**, *162–163*, 682–689. [CrossRef]
33. Horváth, R.; Drégelyi-Kiss, A. Analysis of surface roughness of aluminum alloys fine turned: United phenomenological models and multi-performance optimization. *Meas. J. Int. Meas. Confed.* **2015**, *65*, 181–192. [CrossRef]
34. Jomaa, W.; Songmene, V.; Bocher, P. Surface finish and residual stresses induced by orthogonal dry machining of AA7075-T651. *Materials* **2014**, *7*, 1603. [CrossRef]
35. De Lacerda, J.C.; Martins, G.D.; Signoretti, V.T.; Teixeira, R.L.P. Evolution of the surface roughness of a low carbon steel subjected to fatigue. *Int. J. Fatigue* **2017**, *102*, 143–148. [CrossRef]
36. Novovic, D.; Dewes, R.C.; Aspinwall, D.K.; Voice, W.; Bowen, P. The effect of machined topography and integrity on fatigue life. *Int. J. Mach. Tools Manuf.* **2004**, *44*, 125–134. [CrossRef]
37. Trujillo, F.J.; Sevilla, L.; Salguero, J.; Batista, M.; Marcos, M. Parametric potential model for determining the microgeometrical deviations of horizontally dry-turned UNS A97075 (Al–Zn) alloy. *Adv. Sci. Lett.* **2013**, *19*, 731–735. [CrossRef]
38. Trujillo, F.J.; Marcos, M.; Sevilla, L. Experimental prediction model for roughness in the turning of UNS A97075 alloys. *Mater. Sci. Forum* **2014**, *797*, 59–64. [CrossRef]
39. Salguero, J.; Puerta, F.J.; Gomez-Parra, A.; Trujillo, F.J.; Sevilla, L.; Marcos, M. An analysis of geometrical models for evaluating the influence of feed rate on the roughness of dry turned UNS A92050 (Al-Cu-Li) alloy. *Adv. Mater. Process. Technol.* **2016**, *2*, 578–589. [CrossRef]
40. Sánchez Sola, J.M.; Batista, M.; Salguero, J.; Gómez, A.; Marcos, M. Cutting speed-feed based parametric model for macro-geometrical deviations in the dry turning of UNS A92024 Al-Cu alloys. *Key Eng. Mater.* **2012**, *504–506*, 1311–1316. [CrossRef]
41. Benchergui, D.; Svoboda, C. Aircraft design. *Aerosp. America* **2012**, *50*, 28.
42. ISO 1101:2017 (E). Geometrical Product Specifications (GPS)—Geometrical tolerancing—Tolerances of Form, Orientation, Location and Run-Out. Available online: https://www.iso.org/obp/ui/#iso:std:iso:1101:ed-4:v1:en (accessed on 28 September 2019).
43. Sui, W.; Zhang, D. Four methods for roundness evaluation. *Phys. Procedia* **2012**, *24*, 2159–2164. [CrossRef]
44. Wojciechowski, S. Machined surface roughness including cutter displacements in milling of hardened steel. *Metrol. Meas. Syst.* **2011**, *18*, 429–440. [CrossRef]

Article

Effects of Tool Edge Geometry on Chip Segmentation and Exit Burr: A Finite Element Approach

Muhammad Asad

Mechanical Engineering Department, Prince Mohammad Bin Fahd University, AL-Khobar 31952, Saudi Arabia; masad@pmu.edu.sa

Received: 18 October 2019; Accepted: 13 November 2019; Published: 18 November 2019

Abstract: The effects of different tool edge geometries (hone and chamfer (T-land)) on quantitative measurement of end (exit) burr and chip segmentation (frequency and degree) in machining of AA2024-T351 are presented in this work. The finite element (FE) approach is adopted to perform cutting simulations for various combinations of cutting speed, feed, and tool edge geometries. Results show an increasing trend in degree of chip segmentation and end burr as hone edge tool radius or chamfer tool geometry macro parameters concerning chamfer length and chamfer angle increase. Conversely, the least effects for chip segmentation frequency have been figured out. Statistical optimization techniques, such as response surface methodology, Taguchi's design of experiment, and analysis of variance (ANOVA), are applied to present predictive models, figure out optimum cutting parameters, and their significance and relative contributions to results of end burr and chip segmentation. Various numerical findings are successfully compared with experimental data. The ultimate goal is to help optimize tool edge design and select optimum cutting parameters for improved productivity.

Keywords: tool edge preparation; segmented chip; machining simulation; burr; optimization

---

1. Introduction

Aluminum alloys are widely used in the aerospace industry due to their excellent strength-to-weight ratios and thermal properties. Aluminum alloys are categorized as easy to machine materials and are ideal candidates to subject to dry high-speed machining. However, certain complex combinations of tool materials, tool cutting angles (mainly rake angles), tool edge geometry (hone edge and chamfer edge), chip breaker profiles, cutting process parameters, machine dynamics, among others, greatly influence high-speed cutting processes and may result in high cutting temperatures and intense localized deformations, as reported in numerous experimental and numerical studies performed on aluminum alloys, such as AA2024-T351, AA7010-T7451, and AA7050-T7451. The severe cutting conditions lead to highly segmented chip morphology (higher "chip segmentation frequency" and higher "degree of chip segmentation"), poor surface finish, compromised surface integrity, along with high residual stresses and early failure of tools [1–5].

Furthermore, burr formation is another unlikely phenomenon associated with machining processes. Burr (the undesired and detrimental sharp material formed on workpiece edges) is formed during machining of metallic materials and composite/metal stacks in all sorts of machining processes, such as drilling, milling, turning, and broaching. However, ductile of machining materials generally results in pronounced burr lengths [6,7]. Deburring or burr removal is a necessary process before the component is ready for its functional life, providing the required surface quality and allowing integration into product assembly. Various mechanical, thermal, electrical, or chemical deburring processes employed in industry are costly, require technical expertise, and are quite time consuming [6,7]. These non-value-added post-machining deburring processes undermine the benefits of high-speed machining of aluminum alloys. All of this necessitates the optimization of cutting parameters, tool materials, and angles and edge geometries to improve machined component quality, improve tool life, and eventually increase productivity. Worthy analytical, experimental, and numerical efforts have been carried out in this context to comprehend the chip formation process [8–12] and optimize cutting parameters to control surface quality and residual stresses [13–15]. Most recently, an integrated finite element and finite volume numerical model was presented by Hegab et al. [16] to analyze nano-additive-based minimum quantity lubrication (MQL) effects on machining forces, temperatures, and residual stresses. A considerable decrease in cutting temperatures and residual stress was reported using nano-additive-based MQL. This ultimately will help to increase tool life and improve surface integrity. Furthermore, physical comprehension of burr formation mechanisms and burr control through parametric optimization and tool and workpiece geometry optimization have also been widely discussed in literature [3,4,6,7,17–19].

The present work aims to examine the effect of tool edge geometry design (hone (round) edge and chamfer (T-land) edge), also called "tool edge preparation", on chip formation, chip segmentation frequency, degree of chip segmentation, and exit burr formation processes. Various combinations of two macro-level parameters of chamfer edge geometry, namely chamfer length (l_β) and chamfer angle (γ_β), and the macro geometry of the hone edge radius (r_β) are investigated (Figure 1, Section 2.1). Micro-level cutting edge geometry segments such as "cutting edge segment on flank face" and "cutting edge segment on rake face", as discussed by Denkena et al. [19], are not considered, as feed values taken in the current study are higher than the equivalent edge radii (Table 1, Section 2.1). Additionally, the workpiece material in the vicinity of the stagnation point (around which micro cutting geometry is defined by Denkena et al. [19]) is extremely deformed during machining and is removed during simulation after attaining the defined damage criteria (described later in Section 2.2).

To simulate chip segmentation and exit burr formation processes for orthogonal cutting of AA2024-T351 finite element analyses using various combinations of tool edge geometry, cutting speed and feed tests were performed. Higher values for the tool edge chamfer length (l_β), chamfer angle (γ_β), and hone edge radius (r_β) will certainly increase the negative rake angle in the vicinity of the stagnation point, and the increased workpiece area will experience high thermo-mechanical load. This will largely influence the primary shear zone, negative shear zone (responsible for exit burr formation), and material degradation, in turn reducing the augmentation of chip segmentation and leading to longer burr lengths. Chip segmentation and exit burr formation processes are the main focus of the present work due to their direct and indirect effects on machined surface quality and tool life. For example, chip segmentation frequency and degree of chip segmentation directly dictate residual stress patterns, intensity, and depth on machined surfaces [11,20]. The chip segmentation phenomenon also causes fluctuating cutting forces and harmful chatter vibration affecting machined surface and tool life [21–23], whereas burr not only influences machined surface quality but also influences the fatigue life of machined parts [4,6,7]. A phenomenal shift from "thermal softening" to "crack initiation and propagation" has also been highlighted [12,21,24], causing formation of segmented chips using varying tool edge geometries, cutting speeds, and feeds. This paper also provides more comprehensive information on burr formation ("negative burrs" at the exit end of workpiece), crack propagation at the front of the tool edge, formation of negative shear zones and pivot point locations, boot-type chip formation, and associated burr generation phenomena. The eventual aim of the presented work is to provide further insight into chip and burr formation in machining of AA2024-T351 and to optimize cutting parameters and tool edge design for improved productivity, employing a finite element (FE)-based design and analysis approach. Numerically computed results of chip morphology, cutting forces, and chip segmentation frequency are compared with the ones obtained previously by performing orthogonal cutting experimental investigations on AA2024-T351 under similar cutting conditions [11].

A full factorial Taguchi's design of experiment (DOE) technique is employed to determine optimum combinations of tool edge geometry, cutting speed, and cutting feed to curtail burr lengths, chip segmentation frequency, and degree of chip segmentation. Analysis of variance (ANOVA) is performed to determine the percentage influence of these factors on exit burr lengths, segmentation frequency, and degree of segmentation. Response surface methodology (RSM)-based quadratic predictive models are also proposed.

2. Finite Element Based Orthogonal Cutting Model

2.1. Geometrical Model, Mesh, Constraints, and Hypothesis

Figure 1 shows workpiece and tool geometrical models for orthogonal cutting cases, conceived in Abaqus explicit software (Abaqus, 6.16, Dassault Systemes, Johnston, RI, USA, 2016). For the present work, six different cutting edge geometries are considered: two hone edge (r_β = 5 μm and 20 μm) and four chamfer edge (chamfer length (l_β) = 0.1 mm, chamfer angle (γ_β) = 15°; l_β = 0.1 mm, γ_β = 25°; l_β = 0.2 mm, γ_β = 15°; l_β = 0.2 mm, γ_β = 25°) technologies. In the current work, chip separation is based on ductile damage of a predefined sacrificial material layer approach [11], named the "chip separation zone" in Figure 1. The width of the "chip separation zone" is kept to the order of the tool hone edge radius (r_β), as per experimental evidence [25]. For hone edge radii of 5 μm and 20 μm, the "chip separation zone" width is taken as 20 μm, while for chamfer edge geometries, the "chip separation zone" is taken as the "equivalent radius (r_{eq})" of chamfer edge geometries, as shown in Figure 1 and summarized in Table 1.

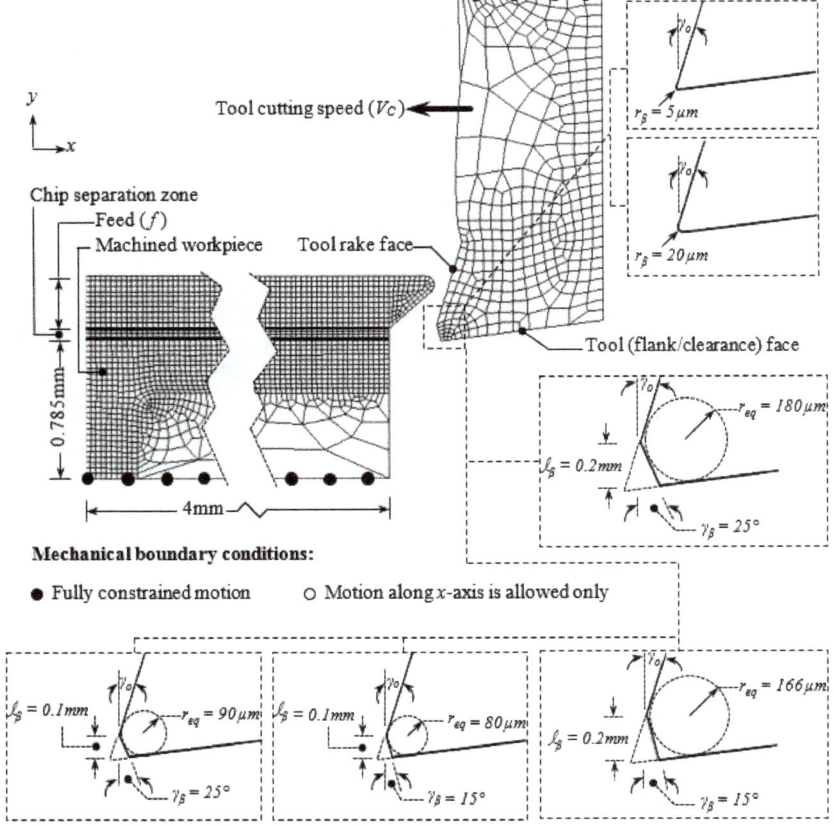

Figure 1. Orthogonal geometrical model and constraints.

Table 1. The "chip separation zone" width for various tool edge geometries.

Tool Edge Geometry	Equivalent Radius, r_{eq} (μm)	"Chip Separation Zone" Width (μm)
Hone edge (r_β = 5 μm)	5	20
Hone edge (r_β = 20 μm)	20	20
Chamfer length (l_β) = 0.1 mm, chamfer angle (γ_β) = 15°	80	80
Chamfer length (l_β) = 0.1 mm, chamfer angle (γ_β) = 25°	90	80
Chamfer length (l_β) = 0.2 mm, chamfer angle (γ_β) = 15°	166	170
Chamfer length (l_β) = 0.2 mm, chamfer angle (γ_β) = 25°	180	170

In the FE model, the tool rake angle = 17.5° and the clearance angle = 7°, and the profile of insert chip breaker geometry are obtained using scanning electron microscope (SEM: Zeiss SUPRA 55-VP FEGSEM, Oberkochen, Germany) and are similar to that of Sandvik's "uncoated carbide insert: CCGX 12 04 08-AL 93 H10 (Sandvik Coromant Sandviken, Sweden)" geometry used in experimental work [11]. The workpiece geometry is modeled initially in three parts: the "machined workpiece", "chip separation zone", and the chip (with specific feed, f). Later on, parts are assembled, as per Figure 1, with the Abaqus built-in tie constraint algorithm, which ensures that all parts behave as a single entity during simulation. The objective for generating distinct parts (the "machined workpiece", "chip separation zone", and chip) lies in the ease of defining different material behaviors and governing equations in different sections of the workpiece.

During machining, heat is generated due to plastic work and friction at the tool and workpiece interface; therefore, to perform coupled temperature–displacement simulations, both the tool and workpiece are meshed with four-node, bilinear, quadrilateral continuum, displacement and temperature, reduced integration elements (CPE4RT), using the plane strain hypothesis. In these elements, along with displacement, temperature is also a nodal variable. Selection of an optimum mesh density in metal machining simulation producing physical results is quite challenging because of the non-availability of a specifically defined criterion in the literature. However, as a general rule, the finer the mesh, the higher the cutting force due to the size effect phenomenon [2]. A mesh sensitivity analysis for various mesh densities (Figure 2) was performed for f = 0.4 mm/rev and V_C = 100 m/min. The increase in cutting forces as a function of mesh density can be figured out. An asymptotic value of mesh size of approximately 25 μm was achieved. Any further decrease in mesh density will not change cutting forces considerably, however, it will attract a time penalty in numerical simulation. A mesh density in the order of 20 μm is chosen in the "chip separation zone", chip, and upper layer (~0.3 mm) of the machined workpiece. The workpiece is fully constrained, while the tool advances with defined cutting speed in the x-direction during simulation, as shown in Figure 1. Cutting simulations were performed with twenty-four various combinations of cutting speed (V_C), feed (f), and tool edge geometries (Table 2).

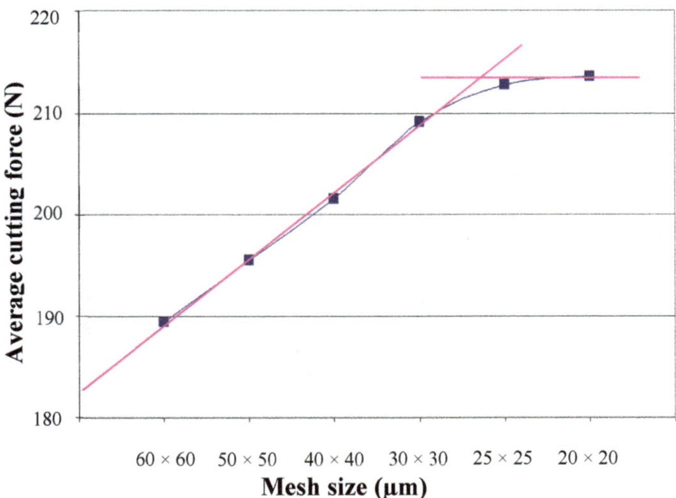

Figure 2. Average cutting force (N) for various mesh densities (μm) for plain strain conditions.

Table 2. Levels of cutting parameters.

Level	Factors		
	Tool Edge Equivalent Radius, r_{eq} (μm)	Cutting Speed, V_C (m/min)	Feed Rate, f (mm/rev)
1	5	800	0.3
2	5	400	0.3
3	20	800	0.3
4	20	400	0.3
5	80	800	0.3
6	80	400	0.3
7	90	800	0.3
8	90	400	0.3
9	166	800	0.3
10	166	400	0.3
11	180	800	0.3
12	180	400	0.3
13	5	800	0.4
14	5	400	0.4
15	20	800	0.4
16	20	400	0.4
17	80	800	0.4
18	80	400	0.4
19	90	800	0.4
20	90	400	0.4
21	166	800	0.4
22	166	400	0.4
23	180	800	0.4
24	180	400	0.4

2.2. Material Behavior, Chip Separation, Friction, and Thermal Models

The workpiece material's behavior is defined by the Johnson–Cook thermo-elasto-visco-plastic constitutive model (Equation (1)). This law adequately defines material behavior in high-speed metal deformation applications. Chip formation and separation are based on the evolution of ductile fracture [5]. The Johnson–Cook shear damage model (Equation (2)) is used to simulate ductile damage.

Initially, Equation (3) is used to calculate scalar damage initiation. Then, modeling of damage evolution is based on Equation (4), representing the linear evolution of scalar damage evolution parameter (*D*), and Equation (5), representing the exponential evolution of scalar damage evolution parameter (*D*). Equations (4) and (5) are used in chip separation and chip regions, respectively. In the latter equation, G_f, represents the fracture energy required to open the unit area of a crack, as per Hillerborg et al.'s fracture energy proposal [26], and is considered a material property. As per the approach, the material softening response after damage initiation is characterized by a stress–displacement response rather than a stress–strain response, and fracture energy is then given as Equation (6). In the present work, G_f is taken as an input material parameter calculated by Equation (7). Finally, Equation (8) is used to calculate the equivalent plastic displacement at failure.

$$\overline{\sigma}_{JC} = \underbrace{\left(A + B\overline{\varepsilon}^{n}\right)}_{Elasto-plastic\ term} \underbrace{\left[1 + Cln\left(\frac{\dot{\overline{\varepsilon}}}{\dot{\overline{\varepsilon}}_0}\right)\right]}_{Viscosity\ term} \underbrace{\left[1 - \left(\frac{T - T_r}{T_m - T_r}\right)^{m}\right]}_{Softening\ term} \tag{1}$$

$$\overline{\varepsilon}_{0i} = \left[D_1 + D_2 exp\left(D_3 \frac{P}{\overline{\sigma}}\right)\right]\left[1 + D_4 ln\left(\frac{\dot{\overline{\varepsilon}}}{\dot{\overline{\varepsilon}}_0}\right)\right]\left[1 + D_5\left(\frac{T - T_r}{T_m - T_r}\right)\right] \tag{2}$$

$$\omega = \sum \frac{\Delta\overline{\varepsilon}}{\overline{\varepsilon}_{0i}} \tag{3}$$

$$D = \frac{\overline{u}}{\overline{u}_f} \tag{4}$$

$$D = 1 - exp\left(-\int_0^{\overline{u}} \frac{\overline{\sigma}}{G_f} d\,\overline{u}\right) \tag{5}$$

$$G_f = \int_0^{\overline{u}_f} \sigma_y d\,\overline{u} \tag{6}$$

$$\left(G_f\right)_{I,II} = \frac{1 - \nu^2}{E}\left(K_C^2\right)_{I,II} \tag{7}$$

$$\overline{u}_f = \frac{2G_f}{\sigma_y} \tag{8}$$

During the progression of material damage, as the damage evolution parameter (*D*) approaches a value of one, it is assumed that the element's stiffness is fully degraded and that it can be removed from the mesh. Hence, chip separation from the workpiece body is realized. The tool (tungsten carbide) is modeled as a purely elastic body in the present work. Tool and workpiece material properties and model equation parameters are shown in Tables 3 and 4, respectively.

Table 3. Physical properties of tool and workpiece materials [11].

Parameters	Workpiece (AA2024-T351)	Insert (Tungsten Carbide)
Density, ρ	2700	11,900
Young's modulus, E	73,000	534,000
Poisson's ratio, ν	0.33	0.22
Fracture energy, G_f	20×10^3	X
Specific heat, C_p	$0.557\ T + 877.6$	400
Expansion coefficient, α_d	$8.91^{-3}\ T + 22.2$	X
Thermal conductivity, λ	$25 \le T \le 300$: $\lambda = 0.247T + 114.4$ $300 \le T \le T_m$: $\lambda = -0.125T + 226$	50
Meting temperature, T_m	520	X
Room temperature, T_r	25	25
Fracture toughness (K_{IC} and K_{IIC})	26 and 37	X

Table 4. Johnson–Cook model parameters for AA2024-T351 [11].

A	B	n	C	m	D_1	D_2	D_3	D_4	D_5
352	440	0.42	0.0083	1	0.13	0.13	−1.5	0.011	0

During the machining process, heat is produced due to friction and plastic work. Conduction is the only mode of heat transfer considered in the present work, while the definition of contact conductance between the tool and workpiece ensures thermal conduction between them. Heat generation due to plastic work is modeled via Equation (9).

$$\dot{q}_p = \eta_p \overline{\sigma}.\dot{\overline{\varepsilon}} \tag{9}$$

where $\dot{q}_p$ is the heat generation rate due to plastic deformation and η_p is the plastic (inelastic) heat fraction, taken as equal to 0.9. The heat generation rate due to friction is calculated by employing Equation (10).

$$\dot{q}_f = \rho C_p \frac{\Delta T_f}{\Delta t} = \eta_f\, J\, \tau_f\, \dot{\gamma} \tag{10}$$

An amount of heat J (from the fraction of dissipated energy η_f caused by friction) remains in the chip $(1 - J)$ and is conducted to the tool. The fraction of heat J is a function of conductivities and diffusivities of tool and workpiece materials [27]. These thermal properties are temperature-dependent (Table 3) and vary with tool and workpiece contact during highly dynamic cutting processes. All of this makes it quite challenging to consider an accurate value of J for tool–workpiece contact. Therefore, in the present work the Abaqus default value of $J = 0.5$ is taken. The steady state, two-dimensional form of the energy equation is given by Equation (11).

$$\lambda\left(\frac{\partial^2 T}{\partial x^2} + \frac{\partial^2 T}{\partial y^2}\right) - \rho C_p\left(u_x \frac{\partial T}{\partial x} + u_y \frac{\partial T}{\partial y}\right)\dot{q}_f + \dot{q}_p = 0 \tag{11}$$

Accurate and precise definition of friction characteristics between the tool and workpiece is important as well as challenging, since it depends on tool and workpiece material properties and geometries, cutting temperature, cutting speed, contact pressure, cutting forces, and contact length, among others [28,29]. Valuable research studies have been dedicated to this important aspect of metal machining to develop a more precise and realistic friction model under variable cutting conditions, owing to its importance in affecting the chip geometry, built-up edge formation, cutting temperature, tool wear, and surface integrity, among others. Application of these friction models in finite-element-based machining models can be taken into account when numerical models are based on the Eulerian formulation; nevertheless, it is still challenging when numerical models are based on the Lagrangian formulation. In the finite element cutting models based on the latter formulation, the workpiece mesh experiences high deformation in the vicinity of the tool–workpiece interaction. Simultaneously, when damage and fracture energy approaches are used in constitutive models, the contact conditions become highly dynamic and complex. As the present work is based on the Lagrangian formulation, to avoid complexities in simulation, a basic Coulomb's fiction law has been adopted.

3. Results and Analysis

3.1. Finite Element Analysis and Discussion

Coupled temperature displacement cutting simulations for 24 combinations of feed, cutting speed, and tool edge geometries were performed, as per Table 2. Computational results concerning cutting forces, chip segmentation frequency, chip segmentation intensity, temperature distribution in the workpiece and tool, and end (exit) burr are calculated. Results of average cutting forces, chip morphology, and chip segmentation frequency (with tool edge equivalent radius, r_{eq} = 20 μm) are compared with the related available results of the experimental work [11]. Numerical results of cutting forces are found to have good correlation with the related experimental ones, as shown in Table 5. The results of chip segmentation frequencies for levels 15 and 16 (V_C = 800 m/min, f = 0.4 mm/rev, r_{eq} = 20 μm and V_C = 400 m/min, f = 0.4 mm/rev, r_{eq} = 20 μm) adequately correspond to their experimental counterparts. However, chip segmentation frequencies for levels 3 and 4 (V_C = 800 m/min, f = 0.3 mm/rev, r_{eq} = 20 μm and V_C = 400 mm/min, f = 0.3 mm/rev, r_{eq} = 20 μm) do not correspond well. The latter is due to the fact that at lower cutting feeds, segmentation intensity decreases (i.e., more uniform chip or less intense segmented chip morphology results). A more refined mesh would be required to obtain more accurate "segmentation frequency" results at lower cutting feeds, which would attract a greater time penalty in numerical simulations. Numerical findings (as presented in Table 5 and Figure 3a) only at levels 3, 4, 15, and 16 are compared with available experimental data results [11]. This comparison is made to validate the numerical model, whereas the rest of the numerical simulations made with various combinations of speed, feed, and tool edge geometry (levels 1, 2, 5–14, and 17–24) are merely exploitation of the validated numerical model (with no experimental results found in the literature). Numerically simulated and experimentally acquired chip morphologies (level 15 only) are compared in Figure 3.

Table 5. Numerical and experimental [11] comparison of mean cutting forces (at constant cutting depth, a_P = 4 mm) and chip segmentation frequencies.

Levels	Numerically Computed Cutting Forces (N)	Experimentally Registered Cutting Forces (N)	Numerically Computed Chip Segmentation Frequency (kHz)	Experimentally Registered Chip Segmentation Frequency (kHz)
3	669	769	53	90
4	657	769	26	37
15	840	976	63	65
16	833	978	31	32

3.1.1. Cutting Parameters and Tool Geometry Effects on Chip Segmentation Frequency and Segmentation Intensity

In almost all parametric combinations of cutting speed, feed, and tool edge radius, a slightly segmented to highly segmented chip morphology is reported. This shows the high plasticity properties of the alloy. Segmented chips (with high segmentation frequency and segmentation intensity) negatively affect machined surface integrity in terms of the quality of the surface profile, residual stress patterns, and the intensity of residual stresses. In the literature, these chips were also reported to produce periodic fluctuations in cutting forces and tool vibrations, which eventually effect tool life. The mechanism of formation of segmented chips is still not well understood, owing to the complex nature of the machining process, which is greatly influenced by the material properties and microstructure, tool geometries, cutting parameters, machine tool dynamics, and friction, among others [12,30]. However, there are mainly two theories explaining the phenomenon of chip segmentation in most of the ductile materials: (a) thermoplastic deformation and formation of adiabatic shear bands because of thermal softening; and (b) fracture, where cracks initiate and propagate in the primary shear zone [12]. In the present work, both phenomena have been witnessed.

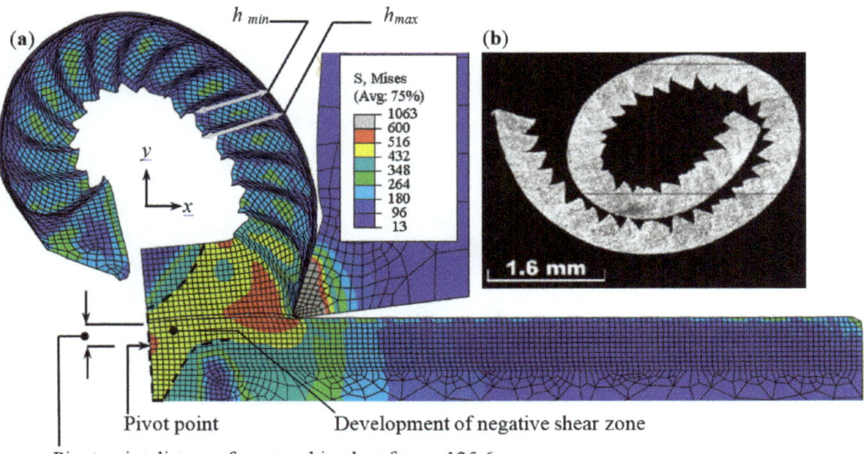

Figure 3. Chip morphology for cutting speed V_C = 800 m/min, f = 0.4 mm/rev, r_{eq} = 20 μm (level 15): (**a**) numerically simulated; (**b**) experimentally generated.

At high cutting speed, frictional resistance causes an increase in cutting temperatures at the tool–workpiece interface, resulting in thermal softening (Figure 4). The thermal softening phenomenon dominates strain hardening, the material stiffness degrades (lower stresses in the vicinity of the tool edge; Figure 3a), and the material flows in the primary shear zones with ease, leading to generation of adiabatic shear bands. Apart from obvious results of higher cutting temperatures due to higher cutting speed, it can also be seen from Figure 4 that an increase in tool edge radius (especially tools with chamfer geometry) results in lower cutting temperatures. Similar trends have also been reported by Ozel [31] for cutting of AISI H-13 with cubic boron nitride (CBN) cutting inserts. This phenomenon is due to the size effect (i.e., more specific cutting energy is required as the tool radius increases in comparison to uncut chip thickness). A wider area now experiences plastic deformation, which requires more energy, and more heat is generated. However, the heat due to inelastic work is more easily dispersed over a large surface area with a larger equivalent edge radius, and consequently maximum temperatures are lower. At higher feed, higher temperatures are produced due to larger amount of plastic work (Figure 4). However, the rate of increase of temperature is not high enough (for feed variation studied in this work this ranges from 0.3 to 0.4 mm/rev) to cause any considerable thermal softening. Furthermore, at higher feed values, due to length effect, longer segments of chips are generated (i.e., frequency of segments will decrease). This shows that higher cutting speeds supplemented with a lower feed rate and lower tool edge radius promote formation of more adiabatic shear bands (high frequency of segmented chip morphology), mainly due to thermal softening. Segmentation frequency is greatly influenced by variation of cutting speed, while segmentation intensity or degree of chip segmentation, calculated by "$(h_{\max} - h_{\min})/h_{\max}$", seems to be least effected by speed variation, as can be seen in Figure 5.

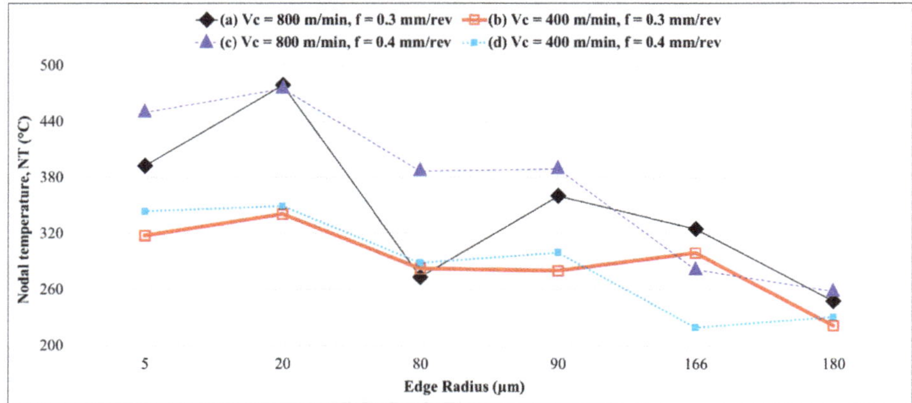

Figure 4. Maximum nodal temperature evolution for cutting speeds, feed, and tool edge radius variations.

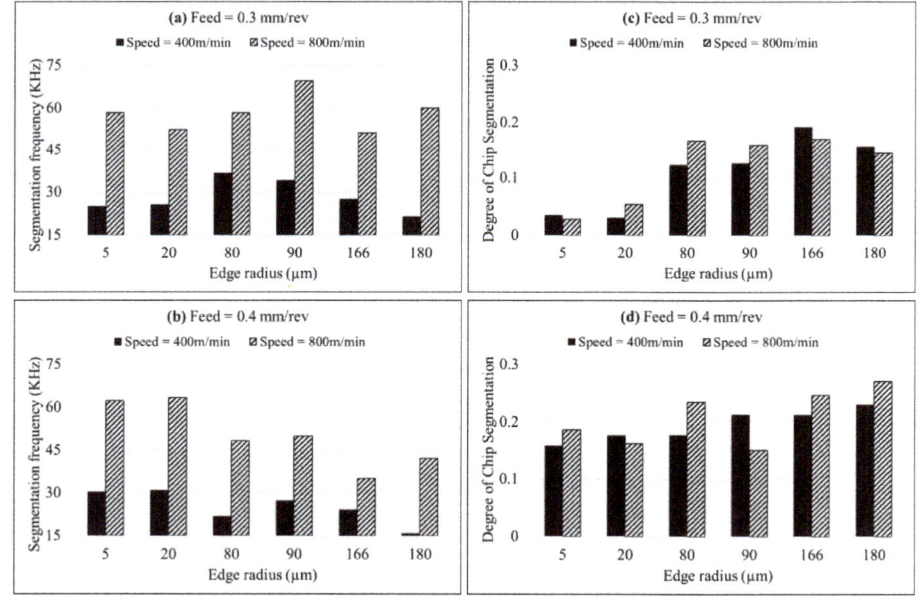

Figure 5. Cutting speed effect on segmentation frequency and degree of chip segmentation: (**a**) Segmentation frequency for f = 0.3 mm/rev; (**b**) Segmentation frequency for f = 0.4 mm/rev; (**c**) Degree of chip segmentation for f = 0.3 mm/rev; (**d**) Degree of chip segmentation for f = 0.4 mm/rev.

Figure 6 shows that an increase in cutting edge radius rarely influences the segmentation frequency, which largely influences the degree or intensity of segmentation. Indeed, as the chamfer tool angle (γ_β) increases, the effective rake angle in the vicinity of the stagnation point becomes more negative, and as the chamfer tool length (l_β) or hone edge radius (r_β) increase, the workpiece area experiences high thermo-mechanical load, leading to initiation and propagation of fracture in the primary shear zone. Furthermore, it can be noticed that chamfer tool length (l_β) contributes more than chamfer tool angle (γ_β) in intensifying the degree of segmentation and the equivalent edge radius (Table 1). On the other hand, as discussed previously and depicted in Figure 4, the increase in cutting edge radius results in decreasing temperature; hence, thermal softening is not the dominant or responsible mechanism for chip segmentation at higher values of tool cutting edge radii.

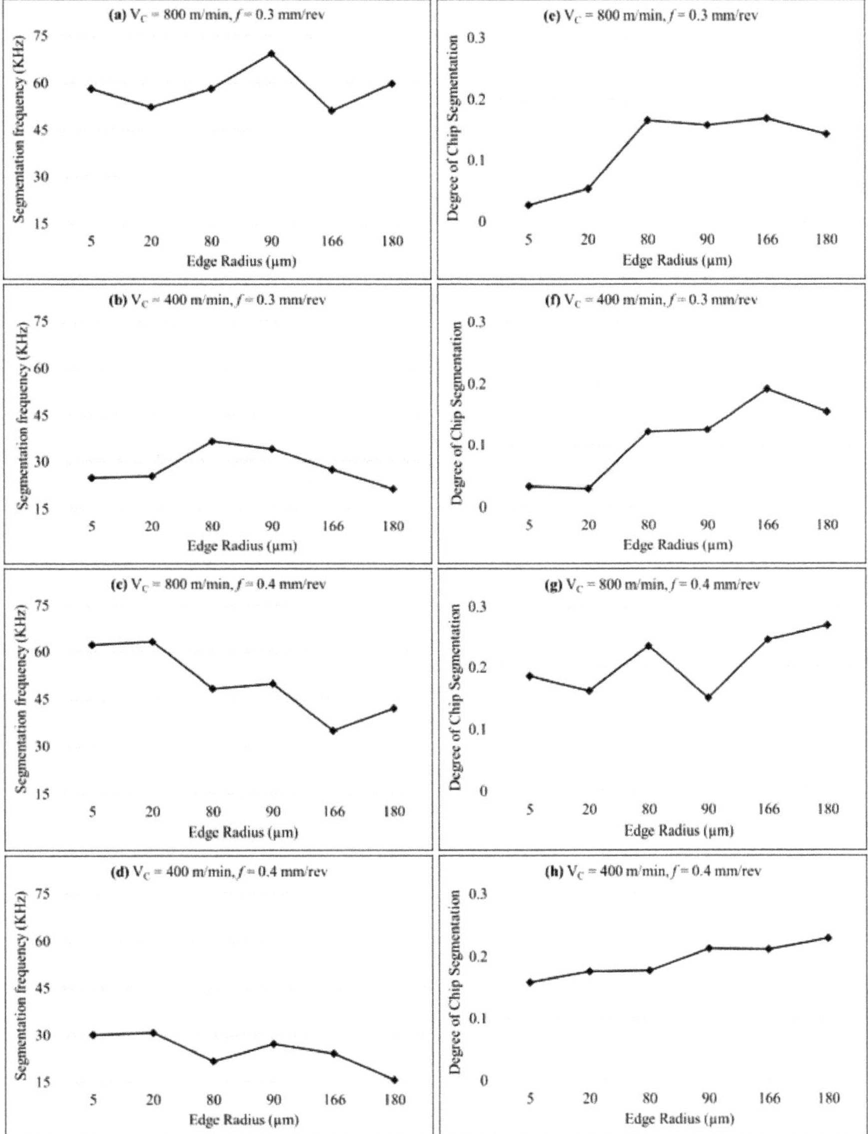

Figure 6. Edge radius effect on segmentation frequency and degree of chip segmentation: (**a**) Segmentation frequency for V_c = 800 m/min, f = 0.3 mm/rev; (**b**) Segmentation frequency for V_c = 400 m/min, f = 0.3 mm/rev; (**c**) Segmentation frequency for V_c = 800 m/min, f = 0.4 mm/rev; (**d**) Segmentation frequency for V_c = 400 m/min, f = 0.4 mm/rev; (**e**) Degree of chip segmentation for V_c = 800 m/min, f = 0.3 mm/rev; (**f**) Degree of chip segmentation for V_c = 400 m/min, f = 0.3 mm/rev; (**g**) Degree of chip segmentation for V_c = 800 m/min, f = 0.4 mm/rev; (**h**) Degree of chip segmentation for V_c = 400 m/min, f = 0.4 mm/rev.

Figure 7 shows a highly segmented chip morphology (with higher degree of chip segmentation) generated for V_C = 800 m/min, f = 0.4 mm/rev, r_{eq} = 180 μm (level 23). In shear bands, the stiffness is fully degraded, with almost zero value for stresses. This shows the probability of fracture in the primary shear zone. Similar trends can also be seen in Figure 8 with variation of feed. The degree of chip segmentation is highly influenced by the change in feed, although by decreasing feed, segmentation frequency increases (due to length effect, longer segments of chips are generated), but this effect is not as pronounced as can be seen for the degree of segmentation.

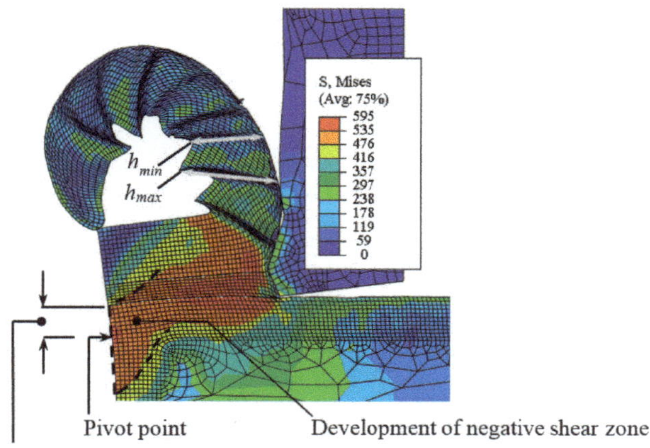

Figure 7. Chip morphology for V_C = 800 m/min, f = 0.4 mm/rev, r_{eq} = 180 μm (level 23).

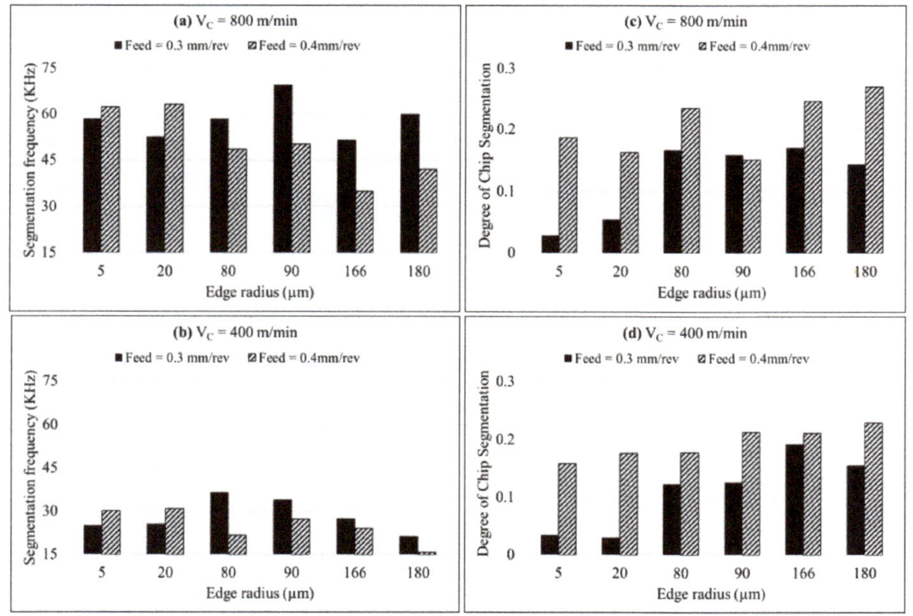

Figure 8. Feed rate effect on segmentation frequency and degree of chip segmentation: (**a**) Segmentation frequency for V_c = 800 m/min; (**b**) Segmentation frequency for V_c = 400 m/min; (**c**) Degree of chip segmentation for V_c = 800 m/min; (**d**) Degree of chip segmentation for V_c = 400 m/min.

Considering the above, it can be summarized that cutting speed greatly influences the chip segmentation frequency, while feed and tool edge radius largely effect the degree of chip segmentation. The thermal softening phenomenon plays a vital role in chip segmentation at higher cutting speeds, lower feed rates, and with smaller tool edge radius values (mainly in increasing segmentation frequency), while crack propagation in primary shear bands occurs at higher values of cutting edge radius and feed (largely influence segmentation degree). To predict optimal combinations of speed, feed, and tool edge radius to minimize the generation of segmented chip morphology (segmentation frequency and degree of chip segmentation), statistical analyses are performed in the next section.

3.1.2. Cutting Parameters and Tool Geometry Effects on End (Exit) Burr Formation

During the course of chip formation, as the tool keeps on advancing in the cutting direction towards the end of the workpiece, a negative shear zone starts to grow from the workpiece free end (exit end) towards the primary shear zone (Figures 3 and 7). The formation of the negative shear zone is specifically due to the bending load experienced by the workpiece free end during tool advancement in the cutting direction. As the tool advances further, the bending load keeps on increasing, the material experiences higher stresses in this deformation zone, and a pivot point (high stressed point) appears on the exit edge of the workpiece (Figures 3 and 7). The location of the "pivot point" is measured from the machined surface along the y-axis. The distance of the "pivot point" has a direct relationship with burr lengths (produced at the exit end)—longer distances represent longer burr lengths. The pivot point distance highly depends on the cutting parameters, materials, and tool geometry. During the course of cutting, the negative shear zone expands further around the pivot point and reaches the tool edge. Higher stresses far ahead of the tool tip position (due to the negative shear zone) promote the material's ductile failure and initiation of cracks in the chip separation zone far ahead of the tool tip (Figure 9). The material deviates from the actual cutting phenomenon, the chip formation process ceases, the tool pushes away the boot-type chip (combination of chip and uncut material), and the end burr (workpiece's deformed exit edge) appears at the end of the workpiece. Figure 9 shows early and advanced failure of chip separation zone material with formation of cracks and generation of an end burr for V_C = 800 m/min, f = 0.4 mm/rev, r_{eq} = 180 μm (level 23).

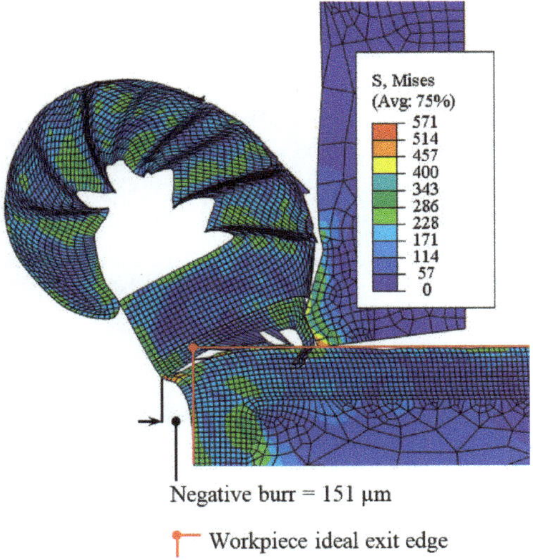

Figure 9. The chip separation zone's material advance failure and formation of negative burr for V_C = 800 m/min, f = 0.4 mm/rev, r_{eq} = 180 μm (level 23).

During machining of aluminum alloys, for various combinations of cutting parameters, both negative and positive burrs at the end of the workpiece have been reported in the literature [3]. Positive burrs (without considerable damage to workpiece edge) are normally generated at lower feed values, and vice versa [3]. In the present work, for AA2024-T351, with investigated combinations of cutting speed, feed, and tool edge geometry, only negative burrs (with edge breakout) were formed. It is found that machining performed with higher feeds along with larger tool edge radii produces highly stressed and more widened shear zones (both primary and negative), and the pivot point location is further away from the machined surface, generating longer burrs than for machining performed at lower feed rates and with smaller tool edge radii. Figures 10 and 11 quantify and produce a trend for exit burrs as a function of the feed and tool edge radius. On the other hand, speed variation was been found to have non-noticeable effects in changing exit burr lengths (Figure 12). The results, in general, are consistent with the findings of experimental burr formation studies performed on aluminum alloys [3,32]. Table 6 details numerically computed exit burr lengths for twenty-four various combinations (defined in Table 2) of cutting speed (V_C), feed (f), and tool edge geometries.

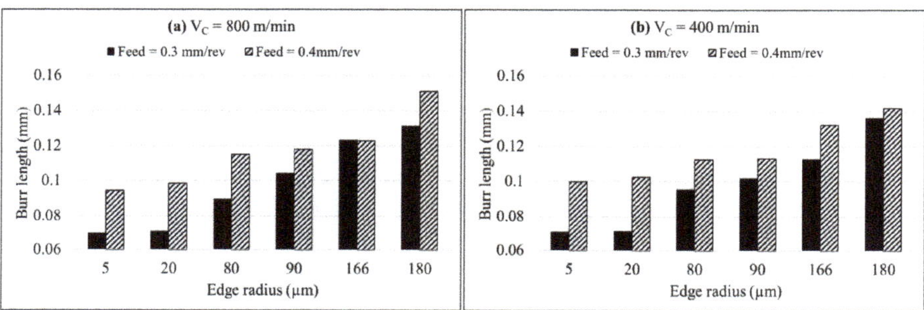

Figure 10. Feed rate variation effects on exit burr lengths: (**a**) Burr length at V_c = 800 m/min; (**b**) Burr length at V_c = 400 m/min.

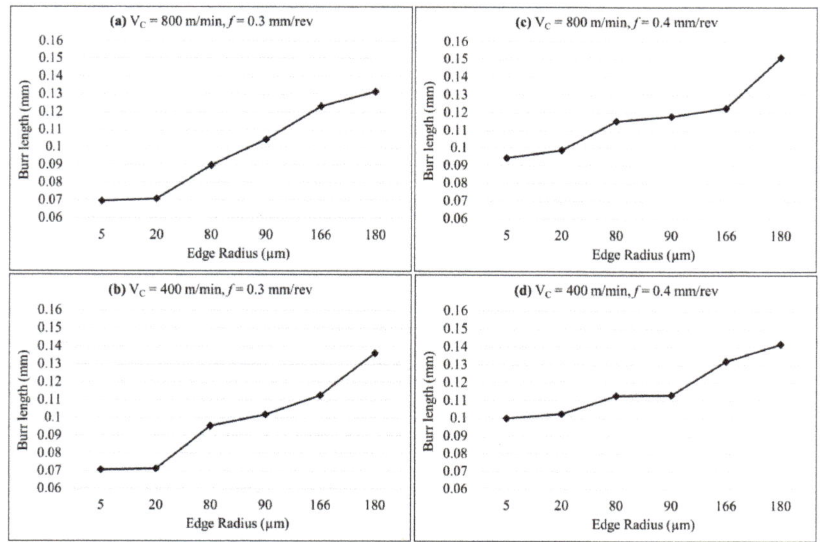

Figure 11. Edge radius variation effects on exit burr lengths: (**a**) Burr length for V_c = 800 m/min, f = 0.3 mm/rev; (**b**) Burr length for V_c = 400 m/min, f = 0.3 mm/rev; (**c**) Burr length for V_c = 800 m/min, f = 0.4 mm/rev; (**d**) Burr length for V_c = 400 m/min, f = 0.4 mm/rev.

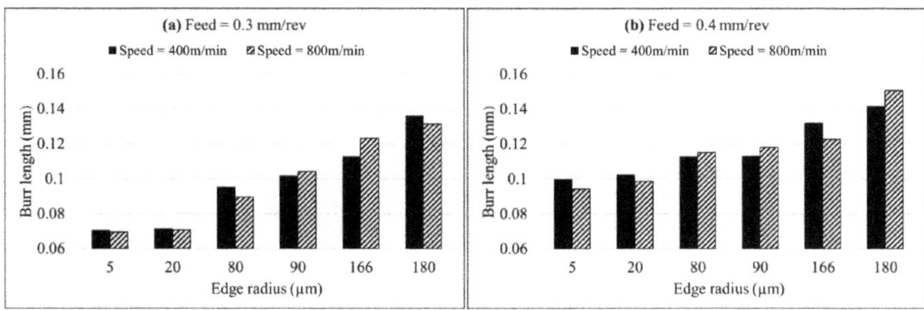

Figure 12. Cutting speed variation effects on exit burr lengths: (**a**) Burr length for f = 0.3 mm/rev; (**b**) Burr length for f = 0.4 mm/rev.

Table 6. Numerically computed exit burr lengths.

Level	Factors			Numerically Computed Exit Burr Lengths (μm)
	Tool Edge Equivalent Radius, r_{eq} (μm)	Cutting Speed, V_C (m/min)	Feed Rate, f (mm/rev)	
1	5	800	0.3	69.5
2	5	400	0.3	70.6
3	20	800	0.3	70.9
4	20	400	0.3	71.2
5	80	800	0.3	89.6
6	80	400	0.3	95.4
7	90	800	0.3	104.3
8	90	400	0.3	101.8
9	166	800	0.3	123.2
10	166	400	0.3	112.6
11	180	800	0.3	131.5
12	180	400	0.3	136.2
13	5	800	0.4	94.3
14	5	400	0.4	99.8
15	20	800	0.4	98.6
16	20	400	0.4	102.3
17	80	800	0.4	115.2
18	80	400	0.4	112.6
19	90	800	0.4	118
20	90	400	0.4	113
21	166	800	0.4	122.7
22	166	400	0.4	132.1
23	180	800	0.4	150.9
24	180	400	0.4	141.8

3.2. Statistical Analysis and Optimization

In the preceding section, finite element method (FEM) approach was employed to predict the likelihood of chip segmentation features (segmentation frequency and degree of chip segmentation) and exit burr formation under various combinations of speed, feed, and tool edge radius. Various associated phenomena such as maximum nodal temperature, material stiffness degradation, early fracture of material in the tool's advancement direction, and location of the pivot point are also discussed. Interesting conclusions can be drawn for optimizing the machining of AA2024-T351 using tungsten carbide inserts. Nevertheless, further investigations are required to predict optimum combinations of speed, feed, and tool edge radius to minimize the generation of segmented chip morphology (segmentation frequency and degree of chip segmentation) and reduce burr formation. The relative significance of each cutting parameter on the latter phenomenon would also be interesting from a production engineer's perspective. Predictive models of chip segmentation features (segmentation frequency and degree of chip segmentation) and exit burr lengths would be advantageous to minimize the cutting trials to optimize the cutting. In this framework, the present section exploits statistical analysis tools, such as Taguchi's design of experiment (DOE), analysis of variance (ANOVA), and response surface methodology (RSM).

3.2.1. Statistical Analyses on Burr Optimization

To determine the optimum combination of cutting parameters (speed, feed, and tool edge radius) for minimum end burr lengths, Taguchi's DOE is employed. The quality criterion approach "the-smaller-the-better" is used for the data (exit burr lengths computed over twenty-four tests via finite element analysis (FEA) and Equation (12) is used to determine the signal-to-noise (S/N) ratio.

$$\frac{S}{N} = -10log\left(\sum\left(\frac{y_i^2}{n}\right)\right) \quad (12)$$

In the relationship, "y_i" represents the response value of the ith test and "n" is the number of test repetitions (taken as one). Feed and speed have two levels of variations (f = 0.3 and 0.4 mm/rev and V_C = 800 and 400 m/min), while tool edge radius has six levels of variations (r_{eq} = 5, 20, 80, 90, 166, and 180 μm). The parametric combination V_C = 800 m/min, f = 0.3 mm/rev, r_{eq} = 5 μm (Level 1, Table 2) represents the optimum combination for generation of minimum burr, as can be figured out by the plots of main effects of S/N ratio (Figure 13a) and data mean (Figure 13b). Table 7 results show that the edge radius is the most influential factor and speed is the least influential factor in burr formation. Results show a good match with the experimental findings of Niknam and Songmene [32].

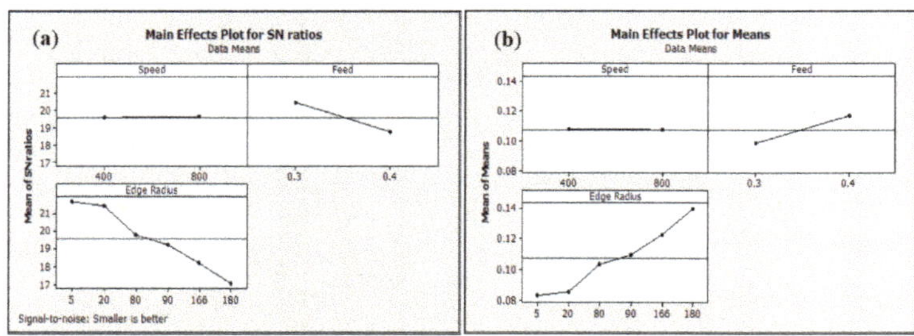

Figure 13. Exit burr lengths as a function of cutting parameters: (**a**) variation of signal to noise (S/N) ratios and (**b**) data means.

Table 7. Response table for data means.

Level	Factor (Speed)	Factor (Feed)	Factor (Edge Radius)
1	0.107385925	0.098054667	0.083556375
2	0.107456283	0.116787542	0.08575725
3	-	-	0.10319225
4	-	-	0.1092785
5	-	-	0.122657
6	-	-	0.14008525
Difference	7.03583×10^{-5}	0.018732875	0.056528875
Rank	3	2	1

Next, to establish a relationship between exit burr lengths and machining parameters, a second order multiple regression model (Equation (13)) based on RSM is used. The developed regression model (Equation (14) using Minitab software (Minitab, 16.2, Minitab-LLC, State College, PA, USA, 2010). The predicted value for burr length (for optimal cutting parameters: V_C = 800 m/min, f = 0.3

mm/rev, r_{eq} = 5 μm) to generate minimum burr using Equation (14) matches the value acquired through finite element simulation (Table 6).

$$y = \beta_0 + \sum_{i=1}^{3} \beta_i x_i + \sum_{i=1}^{3} \beta_{ij} x_i x_j + \sum_{i=1}^{3} \beta_{ii} {x_i}^2 + \varepsilon \tag{13}$$

$$\begin{aligned} &Burr\ length = -0.0160032 - 1.41323e^{-6}(Speed) + 0.290049\ (Feed) - 0.000581417\ (Edge\ radius) \\ &-1.34362e^{-5}(Speed \times Feed) + 6.58783e^{-8}(Speed \times Edge\ radius) - 0.00104982\ (Feed \times Edge\ radius) \\ &+2.17843e^{-7}(Edge\ radius \times Edge\ radius) \end{aligned} \tag{14}$$

In Equation (14), quadratic terms of speed and feed are been included as they are insignificant. Finally, to determine the significance of the regression model and relative contribution of each of the machining parameters, analysis of variance (ANOVA) is performed. Terms used in ANOVA Table 8 are defined in Equations (15)–(18).

$$Sum\ of\ Squares\ (SS) = \frac{N}{nf} \sum_{i=1}^{nf} (\overline{y}_i - \overline{y})^2 \tag{15}$$

where N is the total number of tests, nf represents the level of each factor, $\overline{y}$ is the mean of the response, and $\overline{y}_i$ is the mean of the response at each level of the respective factor.

$$Mean\ square(Variance) : MS_i = \frac{SS}{DF_i} \tag{16}$$

$$Fisher\ Coefficient(F\text{-}value) = \frac{MS_i}{MS_{Error}} \tag{17}$$

$$Percent\ Contributi\ on : PP(\%) = \left(\frac{SS}{SS_T}\right) \times 100 \tag{18}$$

Table 8. Analysis of variance (ANOVA) results for exit (end) burr.

Source	DF	SS	MS	*F*-Value	*P*-Value	PP%	Remarks
Regression	7	0.011474	0.001639	41.56	0	-	Significant
Speed	1	0	0	0.001	0.997	0	Insignificant
Feed	1	0.002106	0.002048	51.94	0.000002	17.39776952	Significant
Edge Radius	1	0.009051	0.00898	227.7	0	74.77075589	Significant
Edge Radius × Edge Radius	1	0.000011	0.000011	0.29	0.598	0.090871541	Insignificant
Speed × Feed	1	0	0	0.01	0.918	0	Insignificant
Speed × Edge Radius	1	0.000018	0.000018	0.46	0.508	0.148698885	Insignificant
Feed × Edge Radius	1	0.000288	0.000288	7.29	0.016	2.379182156	Significant
Error	16	0.000631	0.000039	-	-	5.212722016	-
Total	23	0.012105	-	-	-	100	-

In the ANOVA table, significance or insignificance is attributed to each of the source factors based on the Fisher coefficient value (*F*-value). ANOVA for significance level = 5% (95% confidence level) was performed. The probability values (*P*-values) of the regression model, feed, and edge radius are <0.05. This shows the significance of the regression model and the factors that contribute the most: feed and edge radius. Speed, "quadratic terms", and "interactive terms" have the least effect on burr formation. Table 8 also shows that the edge radius has the highest contribution in producing burr at 74.77%, the feed contribution is 17.39%, while speed variation has the least effect in exit burr formation. This hierarchy of contribution also confirms the findings of Taguchi's DOE methodology (Table 7). It is interesting to note that ANOVA produced for "pivot point location" (considering it as target function, Table 9) has similar trends in term of % contribution of machining parameters in producing burr (Table 8). This helps to conclude that a distant pivot point location (for larger edge radius and higher feed values) is a strong sign that longer burr will be produced.

Table 9. ANOVA results for pivot point location.

Source	DF	SS	MS	*F*-Value	*P*-Value	PP %	Remarks
Regression	7	0.049389	0.007056	12.91	0	-	Significant
Speed	1	0.00037	0.000374	0.68	0.42	0.636493437	Significant
Feed	1	0.004971	0.004976	9.11	0.008	8.551375342	Insignificant
Edge Radius	1	0.04362	0.043299	79.25	0	75.03741549	Insignificant
Edge Radius × Edge Radius	1	0.000044	0.000044	0.08	0.781	0.075691111	Insignificant
Speed × Feed	1	0.00037	0.00037	0.68	0.423	0.636493437	Insignificant
Speed × Edge Radius	1	0.00001	0.00001	0.02	0.894	0.017202525	Insignificant
Feed × Edge Radius	1	0.000005	0.000005	0.01	0.924	0.008601263	Insignificant
Error	16	0.008742	0.000546	-	-	15.03844764	-
Total	23	0.058131	-	-	-	100	-

3.2.2. Statistical Analyses of Segmented Chip Morphology (Segmentation Frequency and Degree of Chip Segmentation)

Figure 14a,b presents the plots of the main effects of S/N ratios and data means on segmentation frequency, respectively. Analyses of plots show that segmentation frequency increases as speed increases, while higher feed and larger edge radii suppress the segmentation phenomenon, though their effect is negligible. The parametric combination V_C = 400 m/min, f = 0.4 mm/rev, r_{eq} = 180 μm (level 24, Table 2) represents the optimum combination for generating the least amount of segmentation frequency.

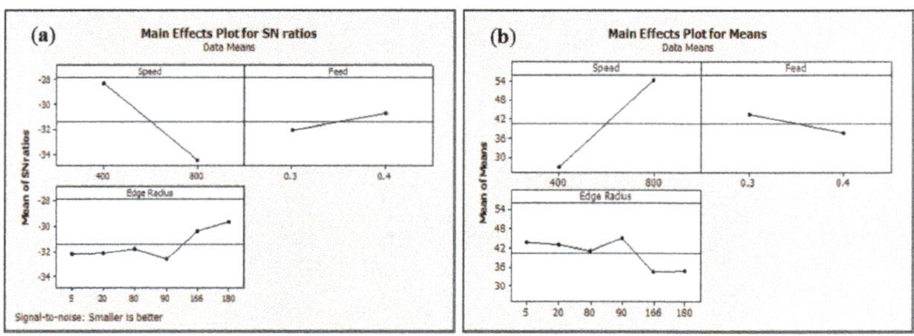

Figure 14. Segmentation frequency as a function of cutting parameters: (**a**) variation of signal-to-noise (S/N) ratios and (**b**) data means.

A second order multiple regression model based on RSM is presented in Equation (19) to define the relationship of segmentation frequency as a function of cutting parameters. In the model, quadratic terms of speed and feed are not included as they are insignificant.

$$\begin{aligned} Segmentation\ Frequency = & -39.8568 + 0.11879\ (Speed) + 105.777\ (Feed) + 0.44496\ (Edge\ radius) \\ & -0.120458\ (Speed \times Feed) - 8.47311e^{-5}(Speed \times Edge\ radius) - 1.00696\ (Feed \times Edge\ radius) \\ & -5.16853e^{-4}(Edge\ radius \times Edge\ radius) \end{aligned} \quad (19)$$

To outline the significance of the model and relative contribution of each of the cutting parameters on segmentation frequency, analysis of variance (ANOVA) is performed and results are summarized in Table 10. Results show that speed has the highest contribution in producing segmentation frequency at 76.63%, edge radius contributes 5.37%, while feed variation has the least effect in generating chips with high segmentation frequencies.

Table 10. ANOVA results for segmentation frequency.

Source	DF	SS	MS	*F* Value	*P*-Value	PP %	Remarks
Regression	7	5480.35	782.91	25.99	0	-	Significant
Speed	1	4569.18	4537.35	150.65	0	76.6351629	Significant
Feed	1	196.94	213.16	7.08	0.017	3.303115435	Insignificant
Edge Radius	1	320.62	302.58	10.05	0.006	5.377500105	Insignificant
Edge Radius × Edge Radius	1	64.15	64.15	2.13	0.164	1.075936098	Insignificant
Speed × Feed	1	34.82	34.82	1.16	0.298	0.584007715	Insignificant
Speed × Edge Radius	1	29.98	29.98	1	0.333	0.502830307	Insignificant
Feed × Edge Radius	1	264.65	264.65	8.79	0.009	4.438760535	Insignificant
Error	16	481.92	30.12	-	-	8.082854627	-
Total	23	5962.25	-	-	-	100	-

As discussed in Section 3.1.1, machining performed at higher speeds generates higher cutting temperatures (Figure 3), leading to thermal softening and generation of adiabatic shear bands (segmented chips). In this context, ANOVA analysis is performed for "maximum nodal temperature" (considering it as target function, Table 11) to figure out the % contribution of machining parameters (speed, feed, and tool edge radius) in influencing the temperature rise. It can be seen (Table 11) that edge radius has the highest contribution to temperature variation; indeed, temperature decreases as the edge radius increases (Figure 4), whereas speed is the second highest contributor in effecting the temperature; temperature increases as speed increases (Figure 4). Feed has been found to have the least effect on maximum temperature variations. Further analyses of Tables 10 and 11 help to conclude that higher temperatures produced at higher cutting speeds promote thermal softening and generation of more frequent adiabatic shear bands (higher segmentation frequency), whereas higher feed and larger edge radii reduce segmentation frequencies, though their effects are minimal.

Table 11. ANOVA results for maximum nodal temperature.

Source	DF	SS	MS	*F* Value	*P*-Value	PP %	Remarks
Regression	7	108782	15540.3	12.2	0	-	Significant
Speed	1	29698	28782.6	22.59	0	22.99140667	Significant
Feed	1	992	882	0.69	0.418	0.767980181	Insignificant
Edge Radius	1	68989	68900.8	54.07	0	53.4094604	Insignificant
Edge Radius × Edge Radius	1	25	25.5	0.02	0.889	0.019354339	Insignificant
Speed × Feed	1	1247	1246.9	0.98	0.337	0.965394441	Insignificant
Speed × Edge Radius	1	5282	5282.4	4.15	0.059	4.089184795	Insignificant
Feed × Edge Radius	1	2548	2547.6	2	0.177	1.972594256	Insignificant
Error	16	20387	1274.2	-	-	15.78307657	-
Total	23	129170	-	-	-	100	-

Figure 15a,b presents the plots of the main effects of S/N ratios and data means on "degree of chip segmentation", respectively. Analyses of plots show that all cutting parameters promote segmentation degree, though speed's effect seems negligible. The parametric combination V_C = 400 m/min, f = 0.3 mm/rev, r_{eq} = 5 µm (level 2, Table 2) represents the optimum combination for generation of chips with the least degree of segmentation.

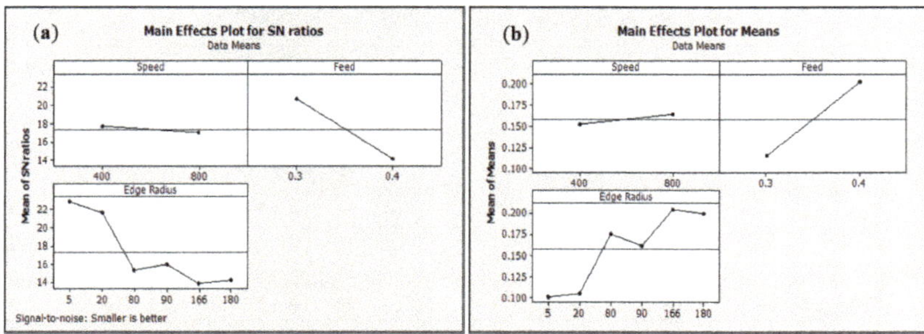

Figure 15. Degree of chip segmentation as a function of cutting parameters: variation of signal to noise (S/N) ratios (**a**) and (**b**) data means.

A second order multiple regression model based on RSM is presented in Equation (20) to define the relationship of the degree of chip segmentation as a function of cutting parameters. In the model, quadratic terms of speed and feed are been included as they are insignificant.

$$\begin{aligned} Degree\ of\ Segmentation &= -0.317506 - 5.7063\,(Speed) + 1.1144\,(Feed) + 0.22237096\,(Edge\ radius) \\ &+0.000102409\,(Speed \times Feed) - 1.36297e^{-9}(Speed \times Edge\ radius) - 0.00342995\,(Feed \times Edge\ radius) \\ &-3.03896e^{-6}(Edge\ radius \times Edge\ radius) \end{aligned} \quad (20)$$

To outline the significance of the model and relative contribution of each of the cutting parameters on the degree of chip segmentation, analysis of variance (ANOVA) is performed and results are summarized in Table 12. Results show that speed has the least contribution to producing highly segmented chips (with high degree of chip segmentation), while feed (43.895%) and edge radii (36.46%) significantly affect the production of highly segmented chips. Finite element analyses provide explicit explanation in this context (Section 3.1.1). Larger material area experiences severe plastic deformation when cutting is performed at higher feed rates supplemented with larger tool radii. Material stiffness degrades, leading to crack initiation and propagation in primary shear bands, resulting in highly segmented chips.

Table 12. ANOVA results for degree of chip segmentation.

Source	DF	SS	MS	*F* Value	*P*-Value	PP %	Remarks
Regression	7	0.08866	0.012666	14.49	0	-	Significant
Speed	1	0.000865	0.000864	0.99	0.335	0.842710312	Significant
Feed	1	0.045057	0.044174	50.54	0	43.89595207	Insignificant
Edge Radius	1	0.037425	0.038387	43.92	0	36.46061669	Insignificant
Edge Radius × Edge Radius	1	0.002218	0.002218	2.54	0.131	2.160845633	Insignificant
Speed × Feed	1	0.000025	0.000025	0.03	0.867	0.024355789	Insignificant
Speed × Edge Radius	1	0	0	0	0.998	0	Insignificant
Feed × Edge Radius	1	0.003071	0.003071	3.51	0.079	2.991865166	Insignificant
Error	16	0.013984	0.000874	-	-	13.62365434	-
Total	23	0.102645	-	-	-	100	-

4. Conclusions

The paper provides a staggered comprehension-to-optimization approach for chip segmentation and end burr (exit burr) formation phenomena in machining of an aerospace-grade aluminum alloy AA2024-T351. These phenomena effect tool life, workpiece machined surface quality and integrity, and hence the overall productivity. Primarily, a finite-element-based cutting model has been established and used to simulate orthogonal machining and chip formation processes for multiple parametric combinations of cutting speed, feed, and tool edge geometry. Results concerning chip segmentation (segmentation frequency and degree of segmentation) and end burr are numerically computed and comprehensively analyzed. To validate the numerical machining model, cutting forces, chip segmentation frequency, and chip morphology results are adequately compared with their experimental counterparts. Then, statistical optimization techniques such as Taguchi's DOE and ANOVA are employed to identify optimum cutting parameters and their % influence in effecting chip segmentation and end burr formation processes. Lastly, RSM-based quadratic predictive models for the aforementioned phenomena are presented.

The results presented in the current work are equally interesting for designers and researchers, providing further insight into machining and related phenomena. From a production engineering perspective, they provide optimum cutting conditions to enhance productivity through optimum selection of tool geometry and cutting parameters. Important findings of the present work are listed below.

- Machining operations performed with chamfer (T-land) edges can be represented with equivalent hone (round) edge radii.
- Only negative burr with a boot-type chip was witnessed for all investigated parametric cutting combinations of speed, feed, and tool edge geometry in machining of AA2024-T351.
- The negative shear zone is wider for cutting performed at higher cutting feed accompanied with larger tool edge radii. This promotes the material's early ductile failure, initiation, and progression of fracture in the chip separation zone far ahead of the tool tip location. Consequently, the material escapes the cutting process and the tool pushes away the boot-type chip (combination of chip and uncut workpiece material), and a longer negative end burr (deformed workpiece exit edge) appears at the exit edge of the workpiece. Statistical analyses show that tool edge radius is the major contributor (74%), while feed rate contributes up to 17.4% in generating burr. Cutting speed variation has been found to have negligible effects on burr quantification.
- Pivot point (the highly stressed point in the negative shear zone) location on the exit edge of the workpiece shows a direct relation in quantifying burr lengths. The distant location of the pivot point from the machined surface results in longer burr lengths, and vice versa.
- Higher cutting speeds enhance thermal softening and more frequent generation of chip shear bands (high frequency of chip segmentation). Finite-element-based parametric analyses and subsequent application of statistical optimization approaches show that speed is the highest contributor (76%) among the cutting parameters in generating highly segmented chips, significantly more so than feed and tool edge radii. Any variation in the latter parameters were found to have insignificant effects in this area.
- Wider workpiece materials undergo severe plastic deformation when machining is performed at higher cutting feeds complemented with larger tool edge radii. The material stiffness degrades easily, leading to crack initiation and propagation in primary and secondary shear zones, resulting in highly segmented chips (chips with a higher degree of chip segmentation). However, cutting speeds, on the other hand, did not been noticeably effect the degree of chip segmentation. Statistical analyses show that feed and tool edge radius both dominantly effect the phenomena, with contributions of 43.9% and 36.4%, respectively.
- Optimum cutting parametric combinations of feed, speed, and tool edge radius to minimize chip segmentation and exit burr formation have been presented. Furthermore, quadratic regression models have been proposed to quantify segmentation frequencies, degree of segmentation, and exit burr lengths as functions of cutting speed, feed, and tool edge radius.

In future studies, a more realistic friction model along with the most accurate heat fraction coefficient, J, will be incorporated into the finite element model to present more realistic results of industrial interest. Furthermore, the study will be extended for other materials and processes, such as drilling.

Funding: This research received no external funding.

Acknowledgments: Technical support provided by Francois Girardin of Laboratoire Vibrations Acoustique, INSA de Lyon, France is highly appreciated.

Conflicts of Interest: The authors declare no conflict of interest.

Abbreviations

A	Initial yield stress (MPa)
a_P	Cutting depth (mm)
B	Hardening modulus (MPa)
C	Strain rate dependency coefficient
Cp	Specific heat ($\text{Jkg}^{-1}\ {}^\circ\text{C}^{-1}$)
D	Damage evolution parameter
$D1\ D5$	Coefficients of Johnson–Cook material shear failure initiation criterion
E	Young's modulus (MPa)
f	Feed rate (mm/rev)
G_f	Fracture energy (N/m)
$KC_{I,II}$	Fracture toughness ($MPa\sqrt{m}$) for failure mode I and mode II
l_β	Chamfer length
m	Thermal softening coefficient
n	Work-hardening exponent
P	Hydrostatic pressure (MPa)
$\dot{q}_p$	Heat generation rate due to plastic deformation W/m^3
$\dot{q}_f$	Heat generation rate due to friction W/m^3
r_β	Hone edge radius
r_{eq}	Equivalent edge radius
T	Temperature at a given calculation instant (°C)
T_m	Melting temperature (°C)
T_r	Room temperature (°C)
$\overline{u}$	Equivalent plastic displacement (mm)
$\overline{u}_f$	Equivalent plastic displacement at failure (mm)
V_C	Cutting speed (m/min)
$\overline{\sigma}$	Stress, MPa
$\overline{\sigma}_{JC}$	Johnson-Cook equivalent stress (MPa)
σ_y	Yield stress (MPa)
T_f	Frictional shear stress, MPa
$\frac{p}{\overline{\sigma}}$	Stress triaxiality
$\overline{\varepsilon}$	Equivalent plastic strain
$\dot{\overline{\varepsilon}}$	Plastic strain rate (s^{-1})
$\dot{\overline{\varepsilon}}_0$	Reference strain rate ($10^{-3}\ \text{s}^{-1}$)
$\overline{\varepsilon}_f$	Equivalent plastic strain at failure
$\Delta\overline{\varepsilon}$	Equivalent plastic strain increment
$\overline{\varepsilon}_{0i}$	Plastic strain at damage initiation
η_p	Inelastic heat fraction
η_f	Frictional work conversion factor
ω	Damage initiation criterion
v	Poisson's ratio
α_d	Expansion coefficient ($\mu\text{m m}^{-1}\ {}^\circ\text{C}^{-1}$)
λ	Thermal conductivity ($\text{W m}^{-1}\ \text{C}^{-1}$)
ρ	Density (kg/m^3)
γ_o	Rake angle (degrees)

γ_β	Chamfer angle (degrees)
β_0	Constant of regression model equation
β_i and β_j	Coefficient of linear terms of regression model equation
β_{ij}	Coefficient of quadratic terms of regression model equation
x_i and x_j	Explicative variables (control factors)
ε	Random error
ANOVA	Analysis of variance
DOE	Design of experiment
RSM	Response surface methodology
DF	Degrees of freedom
MS	Mean squares (variance)
SS	Sum of squares
S/N ratio	Signal-to-noise ratio
PP	Percent contribution
P-value	Probability of significance
F-value	Fisher coefficient (variance ratio)

References

1. Xiuli, F.; Wenxing, L.; Yongzhi, P.; Wentao, L. Morphology evolution and micro-mechanism of chip formation during high speed machining. *Int. J. Adv. Manuf. Technol.* **2018**, *98*, 165–175. [CrossRef]
2. Ijaz, H.; Zain-ul-abdein, M.; Saleem, W.; Asad, M.; Mabrouki, T. Numerical simulation of the effects of elastic anisotropy and grain size upon the machining of AA2024. *Mach. Sci. Technol.* **2018**, *22*, 522–542. [CrossRef]
3. Régnier, T.; Fromentin, G.; Marcon, B.; Outeiro, J.; D'Acunto, A.; Crolet, A.; Grunder, T. Fundamental study of exit burr formation mechanisms during orthogonal cutting of AlSi aluminium alloy. *J. Mater. Process. Technol.* **2018**, *257*, 112–122. [CrossRef]
4. Abdelhafeez, A.M.; Soo, S.L.; Aspinwall, D.K.; Dowson, A.; Arnold, D. The influence of burr formation and feed rate on the fatigue life of drilled titanium and aluminium alloys used in aircraft manufacture. *CIRP Ann. Manuf. Tech.* **2018**, *67*, 103–108. [CrossRef]
5. Mabrouki, T.; Courbon, C.; Zhang, Y.; Rech, J.D.; Asad, M.; Hamdi, H.; Belhadi, S.; Salvatore, F. Some insights on the modelling of chip formation and its morphology during metal cutting operations. *C. R. Mec.* **2016**, *344*, 335–354. [CrossRef]
6. Aurich, J.C.; Dornfeld, D.A.; Arrazola, P.J.; Frankea, V.; Leitza, L.; Min, S. Burrs—Analysis, control and removal. *CIRP Ann. Manuf. Tech.* **2009**, *58*, 519–542. [CrossRef]
7. Niknam, S.A.; Davoodi, B.; Davim, J.P.; Victor, S. Mechanical deburring and edge-finishing processes for aluminum parts—A review. *Int. J. Adv. Manuf. Technol.* **2018**, *95*, 1101–1125. [CrossRef]
8. Uysal, A.; Jawahir, I.S. A slip-line model for serrated chip formation in machining of stainless steel and validation. *Int. J. Adv. Manuf. Technol.* **2019**, *101*, 2449–2464. [CrossRef]
9. Jafarian, F.; Ciaran, M.I.; Umbrello, D.; Arrazola, P.J.; Filice, L.; Amirabadi, H. Finite element simulation of machining Inconel alloy including microstructure changes. *Int. J. Mech. Sci.* **2014**, *88*, 110–121. [CrossRef]
10. Harzallah, M.; Pottier, T.; Senatore, J.; Mousseigne, M.; Germain, G.; Landon, Y. Numerical and experimental investigations of Ti-6Al-4V chip generation and thermo-mechanical couplings in orthogonal cutting. *Int. J. Mech. Sci.* **2017**, *134*, 189–202. [CrossRef]
11. Mabrouki, T.; Girardin, F.; Asad, M.; Rigal, J.F. Numerical and experimental study of dry cutting for an aeronautic aluminium alloy (A2024-T351). *Int. J. Mach. Tools Manuf.* **2008**, *481*, 187–197. [CrossRef]
12. Hernández, Y.S.; Vilches, F.J.T.; Gamboa, C.B.; Hurtado, L.S. Experimental Parametric Relationships for Chip Geometry in Dry Machining of the Ti6Al4V Alloy. *Materials* **2018**, *11*, 1260. [CrossRef] [PubMed]
13. Saleem, W.; Asad, M.; Zain-ul-abdein, M.; Ijaz, H.; Mabrouki, T. Numerical investigations of optimum turning parameters—AA2024-T351 aluminum alloy. *Mach. Sci. Technol.* **2016**, *20*, 634–654. [CrossRef]
14. Saleem, W.; Zain-ul-abdein, M.; Ijaz, H.; Mahfouz, A.S.B.; Ahmed, A.; Asad, M.; Mabrouki, T. Computational analysis and artificial neural network optimization of dry turning parameters—AA2024-T351. *Appl. Sci.* **2017**, *7*, 642. [CrossRef]

15. Hua, J.; Shivpuri, R.; Cheng, X.; Bedekar, V.; Matsumoto, Y.; Hashimoto, F.; Watkins, T.R. Effect of feed rate, workpiece hardness and cutting edge on subsurface residual stress in the hard turning of bearing steel using chamfer + hone cutting edge geometry. *Mater. Sci. Eng. A* **2005**, *394*, 238–248. [CrossRef]
16. Hegab, H.; Kishawy, H.A.; Umer, U.; Mohany, A. A model for machining with nano-additives based minimum quantity lubrication. *Int. J. Adv. Manuf. Technol.* **2019**, *102*, 2013–2028. [CrossRef]
17. Asad, M.; Ijaz, H.; Saleem, W.; Mahfouz, A.S.B.; Ahmad, Z.; Mabrouki, T. Finite Element Analysis and Statistical Optimization of End-Burr in Turning AA2024. *Metals* **2019**, *9*, 276. [CrossRef]
18. Wu, X.; Li, L.; He, N. Investigation on the burr formation mechanism in micro cutting. *Precis. Eng.* **2017**, *47*, 191–196. [CrossRef]
19. Denkena, B.; Koehler, J.; Rehe, M. Influnce of the honed cutting edge on tool wear and surface integrity in slot milling of 42CrMo4 steel. *Procedia CIRP* **2012**, *1*, 190–195. [CrossRef]
20. Zanger, F.; Kacaras, A.; Bächle, M.; Schwabe, M.; León, F.P.; Schulze, V. FEM simulation and acoustic emission based characterization of chip segmentation frequency in machining of Ti-6Al-4V. *Procedia CIRP* **2018**, *72*, 1421–1426. [CrossRef]
21. Mabrouki, T.; Rigal, J.F. A contribution to a qualitative understanding of thermo-mechanical effects during chip formation in hard turning. *J. Mater. Process. Technol.* **2006**, *176*, 214–221. [CrossRef]
22. Mahnama, M.; Movahhedy, M.R. Prediction of machining chatter based on FEM simulation of chip formation under dynamic conditions. *Int. J. Mach. Tools Manuf.* **2010**, *50*, 611–620. [CrossRef]
23. Asad, M.; Mabrouki, T.; Rigal, J.F. Finite-element-based hybrid dynamic cutting model for aluminium alloy milling. *Proc. IMechE Part. B J. Eng. Manuf.* **2010**, *224*, 1–13. [CrossRef]
24. Yameogo, D.; Haddag, B.; Makich, H.; Nouari, M. Prediction of the cutting forces and chip morphology when machining the Ti6Al4V alloy using a microstructural coupled model. *Procedia CIRP* **2017**, *58*, 335–340. [CrossRef]
25. Subbiah, S.; Melkote, S.N. Effect of finite edge radius on ductile fracture ahead of the cutting tool edge in micro-cutting of Al2024-T3. *Mater. Sci. Eng. A* **2008**, *474*, 283–300. [CrossRef]
26. Hillerborg, A.; Modéer, M.; Petersson, P.E. Analysis of crack formation and crack growth in concrete by means of fracture mechanics and finite elements. *Cem. Concr. Res.* **1976**, *6*, 773–782. [CrossRef]
27. Grzesik, W.; Nieslony, P. A computational approach to evaluate temperature and heat partition in machining with multilayer coated tools. *Int. J. Mach. Tools Manuf.* **2003**, *43*, 1311–1317. [CrossRef]
28. Özel, T. The influence of friction models on finite element simulations of machining. *Int. J. Mach. Tools Manuf.* **2006**, *46*, 518–530. [CrossRef]
29. Zhang, C.; Lu, J.; Zhang, F.; Butt, S.I. Identification of a new friction model at tool-chip interface in dry orthogonal cutting. *Int. J. Adv. Manuf. Technol.* **2017**, *89*, 921–932. [CrossRef]
30. Asad, M.; Mabrouki, T.; Rigal, J.F. On the tool vibration effects during down-cut peripheral milling process. *Int. J. Interact. Des. Manuf.* **2010**, *4*, 215–225. [CrossRef]
31. Ozel, T. Modeling of hard part machining: Effect of insert edge preparation in CBN cutting tools. *J. Mater. Process. Technol.* **2003**, *141*, 284–293. [CrossRef]
32. Niknam, S.A.; Songmene, V. Statistical investigation on burrs thickness during milling of 6061-T6 aluminium alloy. In Proceedings of the CIRP, 1st International Conference on Virtual Machining Process Technology, Montreal, QC, Canada, 28 May–1 June 2012.

Article

Predicting Continuous Chip to Segmented Chip Transition in Orthogonal Cutting of C45E Steel through Damage Modeling

Ashwin Moris Devotta [1,*], P. V. Sivaprasad [2], Tomas Beno [3] and Mahdi Eynian [3]

1 R&D Turning, Sandvik Coromant AB, 81181 Sandviken, Sweden
2 R&D, Sandvik Materials Technology AB, 81181 Sandviken, Sweden; palla.sivaprasad@sandvik.com
3 Department of Engineering Science, University West, SE-46132 Trollhättan, Sweden; tomas.beno@hv.se (T.B.); mahdi.eynian@hv.se (M.E.)
* Correspondence: ashwin.devotta@sandvik.com; Tel.: +46-706-163-722

Received: 9 March 2020; Accepted: 8 April 2020; Published: 17 April 2020

Abstract: Machining process modeling has been an active endeavor for more than a century and it has been reported to be able to predict industrially relevant process outcomes. Recent advances in the fundamental understanding of material behavior and material modeling aids in improving the sustainability of industrial machining process. In this work, the flow stress behavior of C45E steel is modeled by modifying the well-known Johnson-Cook model that incorporates the dynamic strain aging (DSA) influence. The modification is based on the Voyiadjis-Abed-Rusinek (VAR) material model approach. The modified JC model provides the possibility for the first time to include DSA influence in chip formation simulations. The transition from continuous to segmented chip for varying rake angle and feed at constant cutting velocity is predicted while using the ductile damage modeling approach with two different fracture initiation strain models (Autenrieth fracture initiation strain model and Karp fracture initiation strain model). The result shows that chip segmentation intensity and frequency is sensitive to fracture initiation strain models. The Autenrieth fracture initiation strain model can predict the transition from continuous to segmented chip qualitatively. The study shows the transition from continuous chip to segmented chip for varying feed rates and rake angles for the first time. The study highlights the need for material testing at strain, strain rate, and temperature prevalent in the machining process for the development of flow stress and fracture models.

Keywords: chip segmentation; damage modeling; dynamic strain aging

1. Introduction

The machining process remains one of the critical manufacturing processes in the 21st century and it has critical engineering applications [1]. Developments in the fields of plasticity and fracture mechanics are used to improve the understanding of the machining process [2]. This improved understanding has become even more necessary with the ever-increasing reliability need for engineered components [3].

Machining process modeling has been carried out while using different approaches, such as empirical, analytical and numerical methods [4]. The finite element (FE) method has been quite extensively used in the modeling of the machining process to model the chip formation process. FE modeling of the machining process provides the ability to incorporate a newer advanced understanding of material behavior. This improved understanding is usually obtained through other methods, such as material testing or more sophisticated modeling techniques. Within FE modeling of the machining process, workpiece material modeling requires the material response to large deformations (large plastic strains) at very high strain rates and very high temperatures.

The ferritic-pearlitic steel group is one of the essential engineering steels having a wide area of application [5]. Material deformation mechanisms, such as strain hardening, strain rate hardening, and thermal softening are vital and have been used in FE modeling of the machining process for more than two decades. Besides, there are deformation mechanisms that are material specific, such as dynamic strain aging (DSA). Previous studies carried out in understanding the behavior of ferritic-pearlitic steel have shown that DSA is a function of temperature and strain rate [6–9]. Earlier studies have been uncertain about the need for the incorporation of DSA in machining process modeling [10,11]. However, recent studies [9,12] have shown the need to incorporate DSA to improve the model's prediction accuracy. DSA has been shown to influence strain hardening behavior, thermal softening behavior, and strain rate hardening [13]. C45E steel falls within the ferritic-pearlitic dual-phase steel group. Flow stress experiments in C45E steel through compression testing at varying temperatures and strain rates have shown the presence of DSA [14]. A modified form of the Johnson–Cook (JC) model was developed using the regression modeling approach. In another recently reported work [8], a material model (Voyiadjis-Abed-Rusinek (VAR) model) was developed, where a phenomenological model is combined with a physics-based model to capture DSA in C45 steel and it is better suited for FE simulation implementation. Using the VAR approach, an attempt has been made to modify the JC model to incorporate DSA. Further, the modified JC (MJC) model is to be implemented to simulate chip formation in the machining process.

Predicting chip segmentation through FE simulations requires the modification of flow stress curves through strain-softening modifications [15] or damage modeling [16–20]. Direct strain-softening modifications of flow stress are based on the adiabatic shear theory [21]. They have been used primarily in chip segmentation prediction in the machining of Ti alloys. Damage modeling where the material failure due to the ductile shear failure has also been used in the chip segmentation of ductile material machining. Most of the studies have concentrated on improving chip segmentation prediction accuracy in terms of chip segmentation intensity [22]. Besides, from a machining process design perspective, the transition from continuous chip to segmented chip is necessary and it has been the focus of very few studies [22,23]. The continuous chip to segmented chip transition with increasing cutting velocity is attributed to the adiabatic shearing process [24]. Recently developed models [16] have explored this area where the influence of damage parameter on chip segmentation has been evaluated. The ability of finite element simulations to predict the continuous chip to segmented chip transition is scarce [23], and none exists for steel machining to the authors' knowledge.

The present work aims to modify the Johnson–Cook model for the incorporation of the DSA influence. The modified Johnson–Cook model, in combination with the damage model, needs to be implemented then within an FE framework for simulation of chip formation in machining. The fracture initiation strain model's influence within the ductile damage model approach is to be evaluated in the prediction of the continuous chip to segmented chip transition in the machining of C45E steel under orthogonal cutting conditions.

In this section, the development of the modified JC model for C45E steel based on the approach developed for the VAR model [8] is presented.

2. Modified Johnson–Cook Model Development Incorporating DSA

2.1. Modified Johnson–Cook Model

The JC model, which is one of the most used material models in the finite element simulation of the machining process, as described in Equation (1).The JC model is built in the multiplicative form in the form of strain hardening component, strain rate hardening component and thermal softening component [25]. The strain hardening component is defined by initial yield stress (A), strain hardening coefficient (B) and strain hardening exponent (n), respectively. The strain rate hardening and thermal softening terms are fitted using the parameters C and m, respectively.

The fitting of the strain hardening component is traditionally carried out using standard quasistatic testing using tensile or compressive loading. In the previously reported work, A and B were modeled using flow stresses at varying temperatures and strain rates using regression modeling approach. The strain hardening exponent, n, was fitted as a first-order function of temperature. The strain hardening behavior of the JC model itself is written in the form, as shown in Equation (2) to reduce the FE implementation complexity.

$$\bar{\sigma} = (A + B\bar{\varepsilon}^n)\left(1 + C\ln\left(\frac{\dot{\varepsilon}}{\dot{\varepsilon_0}}\right)\right)\left(1 - \left(\frac{T - T_o}{T_m - T_0}\right)^m\right) \quad (1)$$

$$A + B\varepsilon^n = A\left(1 + \frac{B}{A}\varepsilon^n\right) \quad (2)$$

$$A = A_0 + A_\Delta\left[1 + \tanh\left(\frac{T - T_l}{\xi}\right)\right]\left[1 - \tanh\left(\frac{T - T_h}{\xi}\right)\right] \quad (3)$$

$$A_\Delta = \Delta\sigma\left(1 + A_1 \log\left(\frac{\dot{\varepsilon}}{\dot{\varepsilon_0}}\right)\right) \quad (4)$$

$$T_l = T_{ll}\left(1.15 + t\log\left(\frac{\dot{\varepsilon}}{\dot{\varepsilon_0}}\right)\right) \quad (5)$$

$$T_h = T_{hh}\left(1.15 + t\log\left(\frac{\varepsilon .}{\dot{\varepsilon_0}}\right)\right) \quad (6)$$

The initial yield stress parameter A is modified similarly to the VAR model to incorporate the DSA influence, leading to yield stress increase with temperature increase at specific temperature ranges [8], as in Equations (3)–(6) and shown in Figure 1. Reported experimental investigations have shown that the dynamic strain aging temperature range is a function of strain rate [6,7,14]. The value of A_0 is obtained from the room temperature quasistatic compression test. The value of A_Δ corresponds to the peak initial yield stress increase in the DSA regime. The constants A_1 and t are fitted with flow stress curves that are obtained at high strain rate conditions from El Magd et al. [6] and Hokka et al. [7]. The temperatures T_l and T_h represent the start and end of DSA, the regime at the reference strain rate. T_l and T_h are modeled as a function of strain rate instead of a constant as in the VAR model. This methodology is necessary to accommodate the observation of dynamic strain aging temperature range being dependent on strain rate [6,7]. The model assumes the DSA regime to extend to higher strain rates at which experimental results are unavailable. Nevertheless, the DSA regime for machining process modeling has been incorporated by modifying the temperature component of the Power-law model while using a regression equation by Childs et al. [26], with the limitation of the temperature range to be a constant. The constants, B/A and n, are fitted for three different temperature ranges to accommodate the influence of temperature on strain hardening.

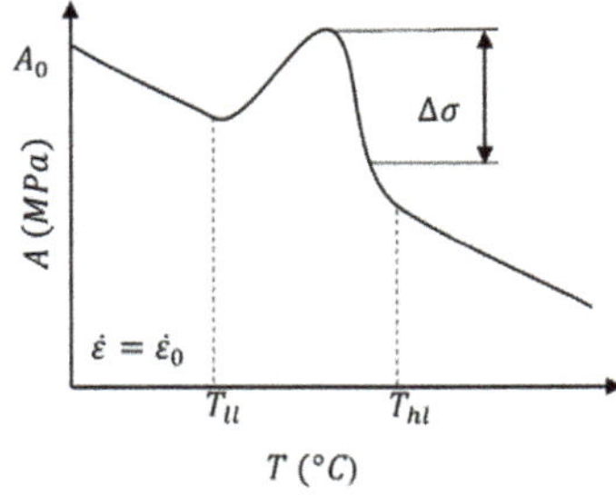

Figure 1. Schematic of the Voyiadjis-Abed-Rusinek (VAR) model-based modification of the JC model's initial yield stress parameter with corresponding mathematical equations with A0 representing the constant, A, of the JC model.

2.2. Fitting of Flow Stress Curves Using the Modified JC Model

The ability of the newly developed flow stress curves from the previous section to predict DSA has been validated using the Gleeble test data from previous work [14]. The flow stress curves are obtained for three different strain rates of 1 s^{-1}, 5 s^{-1}, and 60 s^{-1} with programmed temperatures between 200–700 °C. The temperature range at which DSA is exhibited is a function of the strain rate [6,7]. In the temperature-strain rate range where DSA is active, the initial yield stress increases with temperature and decreases with the strain rate. The presence of DSA has also been shown to lead to serrations in the flow stresses. The temperature at which DSA peaks has been identified to be around 650 °C at high strain rates present in machining conditions [12]. Table 1 presents the constants of the modified JC model. The parameter B and n in the modified Johnson-Cook model has been modified for three different strain hardening regimes. This approach is carried out to ease the implementation of the FE method. The constants are provided, as shown in Equation (7).

$$\begin{cases} B = 1.5A, & n = 0.5 & T < 400\,^{\circ}\mathrm{C} \\ B = 1.2A, & n = 0.3 & 400\,^{\circ}\mathrm{C} < T < 500\,^{\circ}\mathrm{C} \\ B = 1.0A, & n = 0.2 & T > 500\,^{\circ}\mathrm{C} \end{cases} \quad (7)$$

Table 1. Modified JC model parameters for normalized C45E steel.

A_0 (MPa)	$\Delta\sigma$ (MPa)	A_1 (-)	T_{ll} (°C)	T_{hh} (°C)	t (-)	ξ(-)	C(-)	m(-)
500.0	80.0	0.0001	200.0	500	0.1	100	0.0018	1.0

The strain rate hardening and thermal softening parameters of the original JC model for C45E steel are used in this study (C = 0.0018 and m = 1) [11]. With the parameters obtained from the MJC model, the flow stress curves at a strain of 0.1 are predicted under different temperatures and three different strain rates are shown, as in Figure 2. The flow stress prediction extrapolated to higher strain rates prevalent in machining conditions is also plotted to visualize the DSA regime being a function of strain rate.

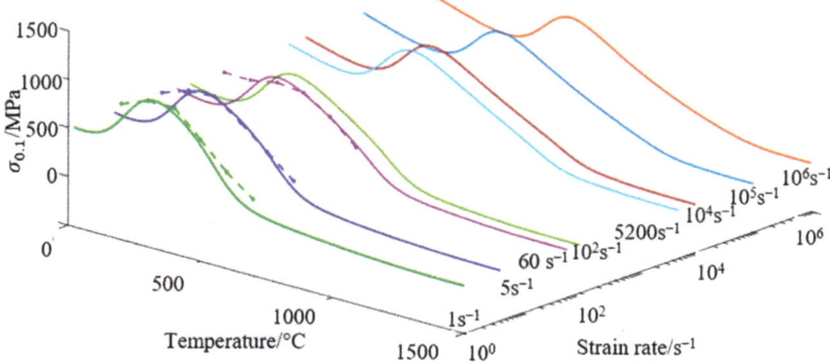

Figure 2. Initial yield stress ($\sigma_{0.1}$) for varying temperature and strain rate with the modified Johnson–Cook (JC) model extrapolated to high strain rates to accommodate machining conditions (Experimental data is plotted with dashed lines and model-predicted data is plotted with continuous lines).

The modification of the initial yield stress parameter of the JC model can capture the increase in initial yield stress with temperature increase. At the reference strain rate of 1 s^{-1}, the DSA regime is active in the temperature range of 200 °C to 500 °C. With a strain rate increase between 10^3 s^{-1} to

10^6 s^{-1}, the modeled DSA regime's initial yield stress moves to higher temperatures of 650 °C and it correlates with other published experimental results [6,7].

With the ability of the MJC model's initial yield stress parameter to capture the DSA regime in Figure 2, Figure 3 shows that the MJC model can capture the strain hardening behavior observed from Gleeble tests with reasonable accuracy. In the temperature ranges below 400 °C, the material strain hardens with increasing strain. At 500 °C, dynamic recovery influences strain hardening, leading to a lowering slope, and are well captured by the MJC model. Beyond 500 °C, at 600 °C and 700 °C, the flow stress curves exhibit constant flow stress with increasing strain.

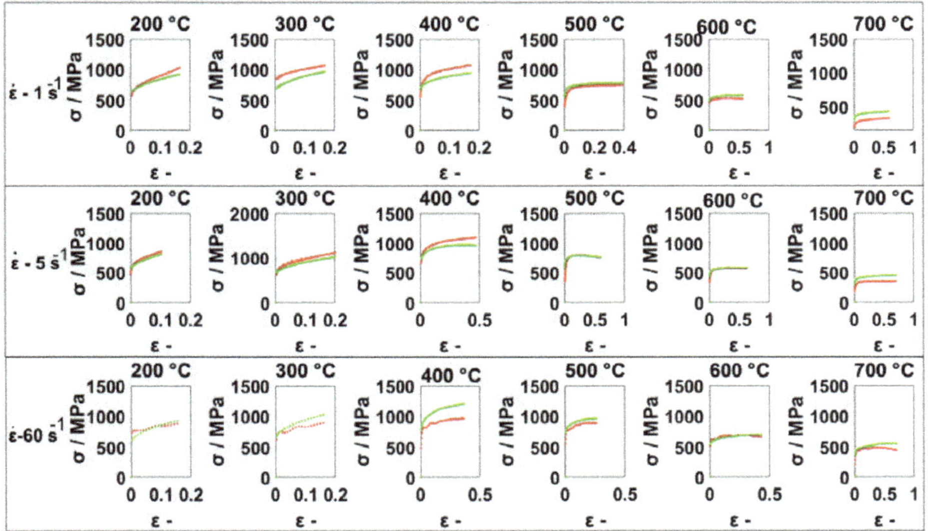

Figure 3. Flow stress curves predicted by the modified JC (MJC) model compared with the flow stress curves obtained using compression tests using Gleeble thermomechanical simulator at strain rates 1 s^{-1}, 5 s^{-1} and 60 s^{-1} (Model: Green color; Experimental: Red color).

MJC model predicted $\sigma_{0.1}$ at a $\dot{\varepsilon}$ of 0.52×10^4 s^{-1} is plotted in Figure 4 to evaluate the validity of extending the hypothesis of DSA presence at very high strain rates. The model can predict $\sigma_{0.1}$ with reasonable accuracy. The flow stress at a strain of 0.1 is chosen to avoid the transient conditions during the early stages of loading in compression testing.

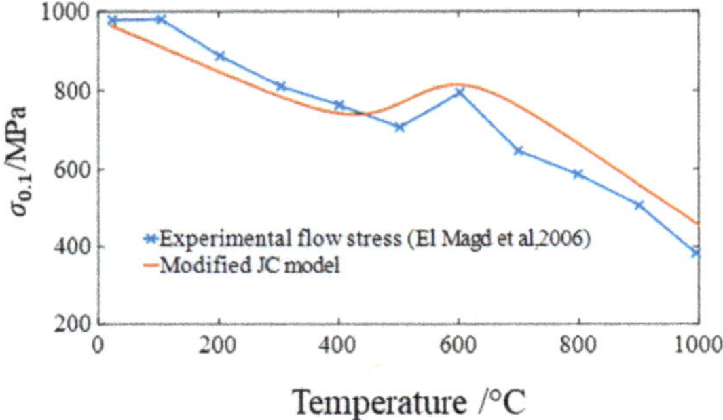

Figure 4. Experimental flow stress at 0.52 × 104 s^{-1} from El Magd et al. [6] fitted using the MJC Model.

3. Flow Stress Modification due to Damage

JC and MJC models both assume no change in material behavior with increasing strain. However, in reality, the flow stress is altered because of fracture. Therefore, to model chip segmentation in ductile materials at a lower cutting velocity where the possibility of adiabatic shear is less incorporation of fracture is essential. Figure 5a provides the schematic modification of the flow stress curves due to fracture from the works of Childs et al. [16,26]. The flow stress before the fracture is defined by Equation (8) and is marked by 'BEFORE FRACTURE' in Figure 5a. During plastic deformation, at the microstructure level, nucleation and growth of defects, such as micro-voids & micro-cracks and their coalescence into macro-cracks takes place leading to material damage [27]. The damage due to plastic deformation in the workpiece is accumulated through the damage factor (D) and is defined, as shown in Equations (8)–(12) and Figure 5a.

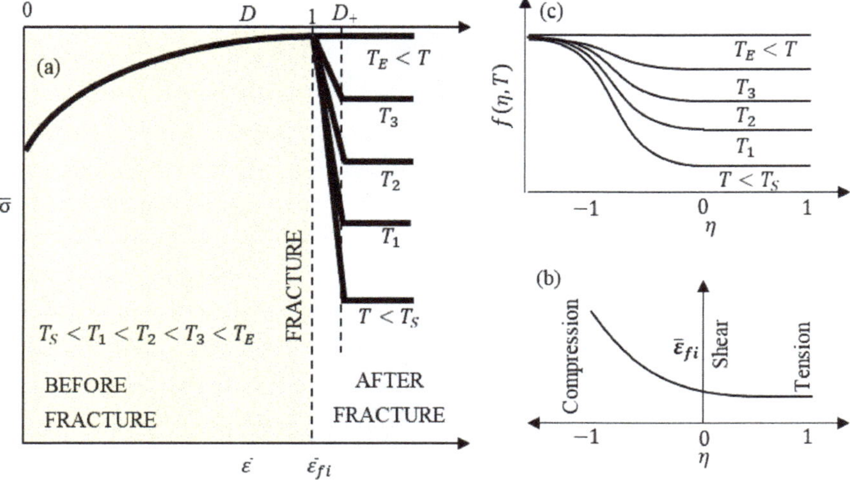

Figure 5. (**a**) Flow stress model for chip segmentation prediction with (**b**) flow stress modification factor as a function of stress triaxiality and temperature from Childs et al. using (**c**) the fracture initiation strain as a function of stress triaxiality.

The damage parameter is defined as the ratio of accumulated plastic strain to fracture initiation strain (ε_{fi}). The fracture initiation strain is presented in detail in the following section and it is the focus of the study. As the accumulated plastic strain equals fracture initiation strain, the damage parameter equals one and it is shown by "FRACTURE" in Figure 5a.

$$D = \int_0^{\overline{\varepsilon}} \frac{d\overline{\varepsilon}}{\varepsilon_{fi}} \tag{8}$$

$$\overline{\sigma}_D = \begin{cases} \overline{\sigma} & D < 1 \\ q(\eta, T,\ D,\ D_+)\cdot\overline{\sigma} & 1 < D < D_+ \\ f(\eta, T)\cdot\overline{\sigma} & D_+ < D \end{cases} \tag{9}$$

$$q(\eta, T,\ D,\ D_+) = \left(\frac{1 + f(\eta, T)}{2} - \frac{1 - f(\eta, T)}{2}\tanh\left(a\frac{2D - D_+ - 1}{D_+ - 1}\right)\right) f(\eta, T) \tag{10}$$

$$f(\eta, T) = \begin{cases} tanh\left(-\sqrt{3}\mu_i\eta\right) & T < T_S \\ p(\eta, T) & T_S < T < T_E \\ 1 & T_E < T \end{cases} \tag{11}$$

$$p(\eta, T) = \tanh\left(-\sqrt{3}\mu_i\eta\right) + \left[1 - \tanh\left(-\sqrt{3}\mu_i\eta\right)\right]\left(\frac{T - T_S}{T_E - T_S}\right) \tag{12}$$

The modeling of the flow stress behavior after the fracture point is defined by "AFTER FRACTURE" in Figure 5a. The flow stress curve is modified based on the loading conditions, η, and temperature, T, as shown in Equation (8) using the flow stress modification factor, $f(\eta, T)$. In an earlier development of Childs et al. [16], $f(\eta, T)$ was assumed to be a constant. With further development [26], $f(\eta, T)$ is defined as a function of stress triaxiality and temperature, as shown in Equation (8) and Figure 5c and presented in detail in the following sections.

3.1. Fracture Initiation Strain

The need for the fracture initiation strain and its use in the damage factor is presented in the previous section. The fracture initiation strain is modeled as a function of stress triaxiality (η), lode angle parameter $\left(\overline{\theta}\right)$, temperature (T), and strain rate $\left(\dot{\varepsilon}\right)$, as shown in Equation (13).

$$\varepsilon_{fi} = f\left(\eta, \overline{\theta}\right) \cdot g\left(\dot{\varepsilon}\right) \cdot h(T) \tag{13}$$

The function, $f\left(\eta, \overline{\theta}\right)$, is used to define the loading conditions. The stress triaxiality (η) defines the varying loading conditions in two-dimensional (2D) and the lode angle parameter, $\overline{\theta}$, extends the model to loading in three-dimensional (3D) space. In the case of the 2D cutting process (orthogonal cutting process), which falls under the plane strain condition, the lode angle parameter is zero. The stress triaxiality parameter defined by, η, as $\frac{\sigma_m}{\overline{\sigma}}$ defines the varying loading conditions from pure compression to combinations of shear/compression and shear/tension to tension, as shown in Figure 6.

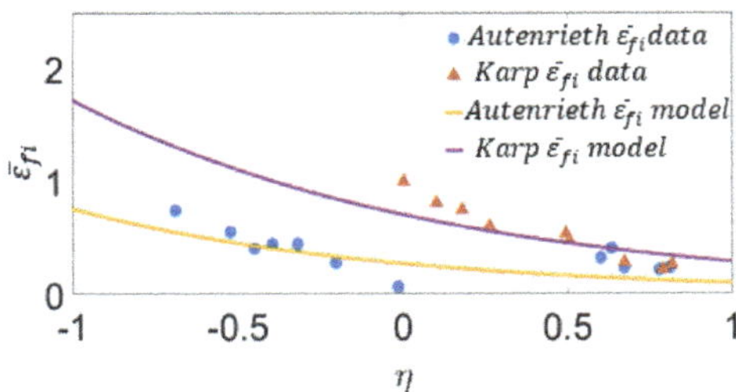

Figure 6. Fracture initiation strain predicted using Shear compression specimen under quasi-static loading, Shear specimens and compression testing as a function of stress triaxiality and fitted using an exponential function.

In order to obtain the fracture initiation strain under different conditions of η (−1 to 1), specimens of varying geometrical shapes are to be used [28]. Interestingly, the fracture initiation strain for variants of C45E steel has been studied for nearly 75 years [29]. With the advances in material testing, the variants of the C45E steel's fracture strain has been studied. Autenrieth et al. [30] obtained the fracture strain as a function of stress triaxiality using torsion and torsion-tension tests. These tests provide stress triaxiality from zero to positive stress triaxiality conditions. Recently, the same fracture strain as a function of stress triaxiality for conditions that range from negative to positive has been obtained using the Shear-compression disk specimen by Karp et al. [31]. In this study, the latter two fracture strain models identified as Autenrieth fracture strain data and Karp fracture strain data are used to evaluate their influence on chip segmentation prediction. The difference in the fracture strain between the two models can be attributed to the difference in processing history and the stress triaxiality evolution before fracture.

To implement the fracture strain in the finite element model, the exponential function used in the Bai–Wierzbicki damage model [32] is used and it is shown in Equation (14).

$$\varepsilon_{fi-X}(\eta) = D_{X1}e^{D_{X2}\eta} \quad (14)$$

The Autenrieth fracture strain data are fitted using the parameters (D_{X1} = 0.2339 and D_{X2} = −1.035) and the Karp fracture strain data are fitted using the parameters (D_{X1} = 0.7 and D_{X2} = −0.9).

The function $g(\dot{\varepsilon})$ is used to model the influence of strain rate. $h(T)$ is used to model the influence of temperature on the fracture initiation strain. In this study, $g(\dot{\varepsilon})$ and $h(T)$ are obtained from the JC damage fracture model, as shown in Equation (15) and Equation (16) and D_3 = 0.0018 and D_4 = 0.58 are obtained from the literature [33].

$$g(\dot{\varepsilon}) = \left(1 + D_3 \ln\left(\dot{\varepsilon}/\dot{\varepsilon_0}\right)\right) \quad (15)$$

$$h(T) = (1 + D_4T^*) \quad (16)$$

3.2. Flow Stress Modification Factor

As previously mentioned, the flow stress is modified, as shown in Figure 5a using the flow stress modification factor, $f(\eta, T)$ shown in Figure 5c. The flow stress in the 'AFTER FRACTURE' region is modeled while using a steady-state flow stress modification factor defined by Equation (9) and a transient flow stress modification factor. The steady-state is defined by $D > D+$ and the transient state is defined by $D+ > D > 1$. In this study, $D+$ is defined to be 1.25, as suggested by Childs et al. [26].

In the steady-state, the flow stress modification is a function of stress triaxiality and it is defined by Equation (12) below a certain temperature, T_S, as shown in Figure 5c. A negative stress triaxiality condition (e.g., $\eta = -1$) characterizing compression, the flow stress is modified only slightly due to incompressibility. A positive stress triaxiality condition, e.g., $\eta = 1$ characterizing tension, the flow stress is drastically reduced to zero, similar to necking in tensile testing. At temperatures above T_E, the material is assumed to be healed and continue to flow plastically with the flow stress defined by $\overline{\sigma}$, as shown in Equation (8). Between temperatures T_S and T_E, the flow stress curve is defined using the tanh function, as shown in Equation (11). The transient flow stress modification factor is defined as Equation (9) by multiplying the steady-state flow stress modification factor with a tanh function of D and $f(\eta, T)$.

The values for μ_i, T_S, and T_E used in this work are 1 °C, 600 °C, and 700 °C, respectively. The values for T_S and T_E based on the strain hardening transition behavior observed from the Gleeble tests.

4. Experimental Investigation of Chip Formation in Orthogonal Turning

The orthogonal turning process has been carried out in a normalized C45E steel tube with a diameter of 150 mm with a thickness of 3 mm, determining the uncut chip width. The material hardness is measured and an average hardness of 220 HV is recorded with an average grain size of 13.5 μm. The tubular workpiece reduces grain size variation influence. All of the experiments were carried out under dry cutting conditions. The cutting tool material is H13A grade carbide with TiCN coating. The cutting velocity is set at 150 m/min. The rake angle is varied as −5°, 5°, 10°, and 20°, and the feed is varied as 0.05 mm.rev^{-1}, 0.1 mm.rev^{-1}, 0.15 mm.rev^{-1}, 0.25 mm.rev^{-1}, 0.4 mm.rev^{-1}, and 0.6 mm.rev^{-1}. The cutting forces and feed forces are obtained through the Kistler dynamometer attached to the tool turret. The sample chips were collected for all cutting conditions and they have been used to quantify chip segmentation, as shown in Figure 7. The experimental investigation is described in detail in previous work [34]. The chips have to be plotted with varying magnification to capture the chip segmentation features at very low feed rate conditions (e.g., 0.05 mm.rev^{-1}) and similarly at very high feed rate conditions (e.g., 0.60 mm.rev^{-1}). This approach with varying magnification is carried out when the chips from simulations are also presented in the following sections.

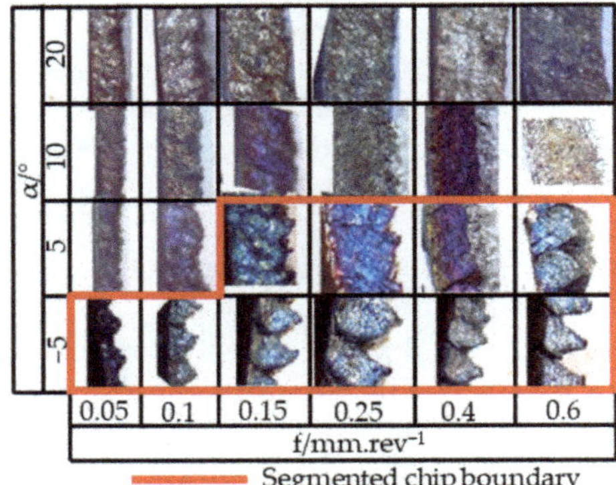

Figure 7. Chip morphology obtained through experimental investigation under varying rake angle and feed at a constant cutting speed of 150 m/min. with the boundary differentiating between continuous and segmented chip. Note: Image magnification not to scale.

5. FE Modeling of Chip Formation in Orthogonal Turning Process

The FE simulation of the cutting process was carried out using a commercial finite element software, Thirdwave Advantedge [35]. The software has been specifically built ground up for metal cutting simulations while using a dynamic explicit coupled Thermo-elastoplastic Lagrangian formulation [36].

The main requirement for metal cutting simulation is the ability for adaptive remeshing to account for the large plastic strains in the primary and secondary deformation zones. The simulations are carried out in 2D. The software provides the ability to customize the adaptive remeshing parameters. The adaptive remeshing parameters control the mesh transition in the tool vicinity (primary deformation zone) from coarse to fine and also the mesh transition from fine to coarse (chip mesh). The software through the mesh refinement and coarsening factors automatically control the conditions for adaptive remeshing. With predicting chip segmentation being the main aim of the study, mesh refinement factor of 6 (1-coarse → 8-fine) and mesh coarsening factor of 3 (1-fine → 8-coarse) is chosen.

A further increase in mesh refinement factors leads to computation time to increase by a factor of 3, making it practically unviable. Besides, the suggested minimum element size is set at 0.01 mm. The friction coefficient of 0.5 is used for all cutting conditions. The software provides the possibility to input the custom material model through a FORTRAN Subroutine, which and has been employed in the study to implement the modified JC model. Two different solution algorithms are supported by the software: Newton method and the secant method. At this stage, the secant method is used, as the method does not require the derivative implementation with the compromise of a slower convergence rate.

Simulations are carried out for all rake angles (−5°, 5°, 10° and 20°) and feed from 0.05 mm.rev^{-1} to 0.6 mm.rev^{-1}. The relief angle and the cutting-edge radius are kept constant with the values 7° and 30 µm, respectively, both in the experimental conditions and simulations. Simulations are run with the JC model and MJC model combined with the damage modeling approach. Within the MJC model + damage model framework, the influence of two fracture initiation strain model's (Autenrieth ε_{fi} model and Karp ε_{fi} model) capability to evaluate the transition from continuous chip to segmented chip is studied.

6. Results

The FE simulations were carried out using two different material models (JC and MJC). With the MJC+ damage model, two different parameter sets of fracture initiation strain models' ability to predict the continuous chip to segmented chip transition are evaluated. The FE simulations were run till a steady state was achieved, and the chip morphology was obtained. The chip morphology was plotted in the chip chart form. Further on, the continuous chip–segmented chip boundary is identified in the chip chart. The cutting forces predicted by the material models are also presented.

6.1. Validation of Material Model under Continuous Chip Formation Conditions

This section presents the MJC model validation in the simulation of the chip formation process. Experimental results with continuous chips are used to avoid the damage model influence. The rake angle of 10° was used with feeds of 0.05 mm.rev^{-1} to 0.6 mm.rev^{-1} and Figure 8 presents the cutting forces.

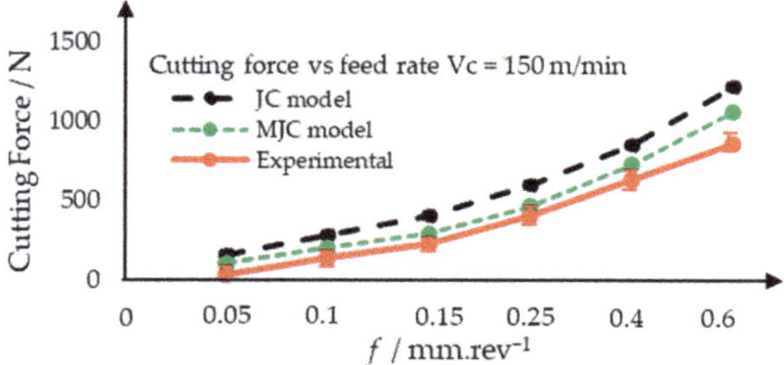

Figure 8. Cutting force predicted by the JC model and modified the JC model for a constant cutting speed of 150 m/min and rake angle of 10°.

The experimental result showing the increase in cutting force with feed is well established within the metal cutting with the influence of uncut chip thickness increase. The cutting forces predicted by the JC model and the MJC model have a similar trend. The MJC model improves the prediction for all of the cutting conditions quantitatively. This improvement in cutting force prediction as compared to the JC model is attributed to the material testing used and the ability of the MJC model to incorporate the material behavior with improved accuracy. The highest discrepancies in quantitative prediction for both JC and MJC models are at the maximum chip thickness conditions. The MJC model's cutting force prediction error can be attributed to the extrapolation that was carried out with the fitting of the MJC model at very high strain rates. Further improvement has to be carried out, among other things, by improving the friction models.

6.2. Experimental and FE Modeling of Continuous Chip–Segmented Chip Transition for Varying Rake Angle and Feed

The chips obtained from experimental investigation for varying rake angle and feed rates at a constant cutting speed of 150 m/min are plotted in the form of chip chart in to evaluate the ability of the different fracture initiation strain models in chip segmentation prediction. Figure 7 shows that, at constant cutting speed, continuous chips are produced for rake angles 10° and 20°. With a rake angle of 5°, the cutting process is in a transition zone. Continuous chips are produced for feed of 0.05 mm.rev^{-1} and 0.10 mm.rev^{-1} and segmented chips are produced for feed from 0.15 mm.rev^{-1} to 0.6 mm.rev^{-1}. For a rake angle of −5°, the produced chips are segmented for all feeds. At a constant feed of 0.40 mm.rev^{-1}, a change in the rake angle from 5° to 10° changes the chips from being continuous to segmented.

Similarly, at a constant rake angle of 5°, the change in feed rate from 0.10 mm.rev^{-1} to 0.15 mm.rev^{-1} leads to continuous to segmented chip. This leads to the rake angle of 5° being a transition zone. The experimental results clearly show that chip segmentation/continuous chip is stable under certain cutting conditions (−5°, 10°, and 20°) and in transition mode under certain conditions (5°). It is also seen that, in the transition zone, the feed also plays a role in chip segmentation. This states that minor changes in the cutting zone to be highly sensitive to the machining output. This would lead to challenges in the process capability in a production environment.

6.3. Prediction of Chip Segmentation Using MJC Model and Two Different Fracture Initiation Strain Models

Initial simulations were run with the JC flow stress model [37] and the JC fracture model [38] reported for AISI 1045 steel to predict chip segmentation. The continuous chip to segmented chip transition predicted by the JC flow stress-JC fracture model, as shown in Figure 9, with the JC fracture parameters that were obtained from the literature [38]. The JC flow stress–JC fracture model is unable to predict the influence of the rake angle and, at the same time, predicts lower chip segmentation intensity when compared to experimental investigation. With the inability of the JC flow stress–JC fracture model established, the flow stress behavior is modified while using the damage model to improve chip segmentation prediction. The chip morphology predicted by the MJC model with the damage model using the two fracture initiation strain models is presented here.

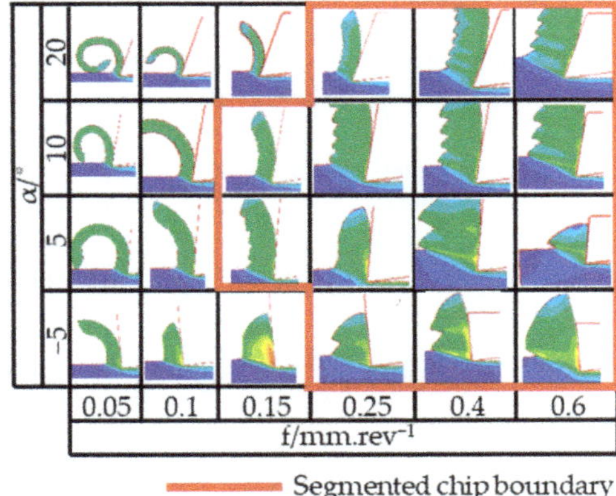

Figure 9. Continuous chip to segmented chip transition prediction predicted by the JC flow stress model in combination with the JC Fracture model. Note: Image magnification not to scale.

6.4. Autenrieth Fracture Initiation Strain Model Predicted Chip Segmentation Continuous Chip Transition

Figure 10 shows the transition of the chip morphology that was predicted by the Autenrieth fracture initiation strain model and MJC material model. The continuous chip to segmented chip transition is predicted with reasonable accuracy. At a constant feed of 0.6 mm.rev^{-1}, the model can predict the segmented chip for rake angles: −5° and 5° and the transition to continuous chip with a rake angle of 10°. The model can predict the transition from continuous chip to segmented chip as the feed is increased from 0.05 mm.rev^{-1} to 0.1 mm.rev^{-1}. At a feed of 0.05 mm.rev^{-1}, a continuous chip is predicted for both negative and positive rake angle. A qualitative comparison of chip segmentation between the experimental investigation and the Autenrieth fracture initiation strain model prediction shows that the level of chip segmentation intensity is lower when compared to the experimental results. This leads to a further quantitative evaluation in terms of chip segmentation intensity become counterproductive. Chip segmentation intensity has been predicted with improved accuracy in literature but only for a constant rake angle and varying cutting speeds [39]. The models in these studies have shown cutting speed's influence and not the rake angle's influence on chip segmentation.

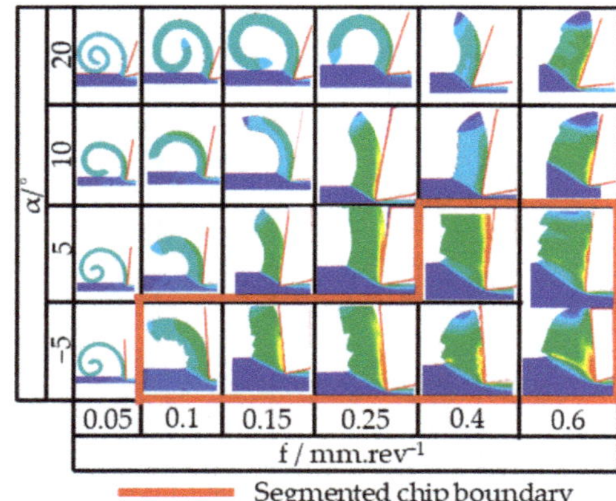

Figure 10. Continuous chip to segmented chip transition prediction predicted by the MJC flow stress model in combination with the Autenrieth Fracture initiation strain model. Note: Image magnification not to scale.

6.5. Fracture Strain Through Shear Compression Disk (SCD) Experiments Predicted Chip Segmentation Continuous Chip Transition

Figure 11 shows the transition of the chip morphology predicted by the fracture initiation strain that was obtained by Karp et al. [31] while using shear compression disk specimen. The model predicts segmented chips for all rake angles for feeds above and including 0.1 mm.rev^{-1}. For the feed of 0.05 mm.rev^{-1}, the segmented chip for a negative rake angle is not predicted correctly compared to experimental results. Qualitative evaluation of the chip segmentation intensity when compared to the experimental investigation is relatively low for a negative rake angle. On the other hand, the chip segmentation frequency is higher qualitatively compared to the Autenrieth fracture initiation strain model predicted chip segmentation frequency.

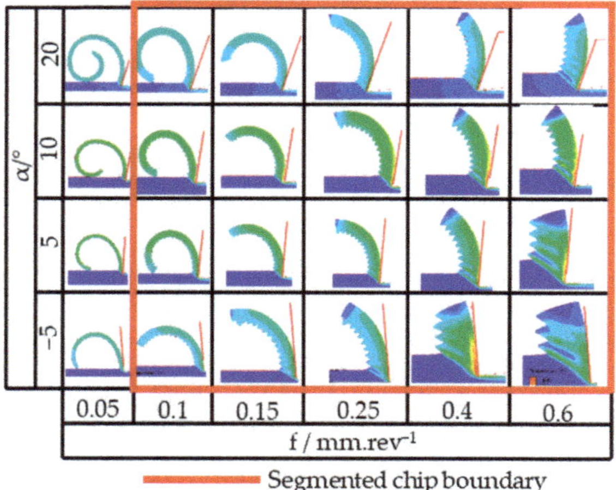

Figure 11. Continuous chip to segmented chip transition prediction by the fracture initiation strain obtained Shear compression specimen. Note: Image magnification not to scale. Note: Image magnification not to scale.

7. Discussion

The FE simulation of the machining process's ability to predict chip segmentation is dependent on the accuracy in which the material behavior is modeled to a large extent, as shown by Fernandez-Zeliaia et al. [40]. The other well-known influencing factors are friction behavior and thermal behavior input. Within flow stress behavior, the initial yield stress, the strain hardening behavior, the fracture initiation, and the material behavior after fracture are essential inputs. For the material under study, C45E steel, strain hardening behavior, and thermal softening behavior are more influential when compared to strain rate hardening [41]. The MJC model's ability to accurately predict three different strain hardening behavior as a function of temperature, therefore, systematically improves the accuracy of FE modeling of chip formation. The material behavior at the strain, strain rate, and temperature occurring in the chip formation conditions is vital in accurately predicting the chip morphology. In this study, the material flow stress is initially validated while using the flow stress that was obtained from compression testing through Gleeble tests (Figure 3). Although the temperatures at which the material behavior is validated through Gleeble tests are comparable, the strains and strain rates are lower by orders of magnitude. The material model validation through cutting process simulation is carried out by avoiding the cutting conditions where chip segmentation is not observed. The MJC model's DSA incorporation, which is active in temperatures occurring at the primary and secondary deformation zones, is an important outcome of this work. The ability to incorporate DSA keeping all other things constant has been shown to improve the cutting forces prediction accuracy, as shown in Figure 8. The fracture initiation models with the JC model have not been able to predict the chip segmentation boundary. This shows the influence of the MJC model on the improved prediction of chip segmentation boundary.

The damage behavior modeling involves fracture initiation strain and the flow stress after the fracture initiation strain. The results of Figure 9, Figure 10, and Figure 11 show that the continuous chip to segmented chip transition and chip segmentation intensity is sensitive to the fracture initiation strain input. The Autenrieth fracture initiation strain model better predicts the transition between the continuous chip and segmented chip when compared to the Karp fracture initiation strain model. Simulations run with the JC material model combined with the fracture initiation strain models have shown poor results in terms of prediction of the chip segmentation (Figure 9). This shows that the ability of the fracture initiation strain models to predict chip segmentation is dependent on the components of the flow stress behavior, i.e., initial yield stress and strain hardening behavior. The strain hardening behavior and the resulting temperature increase influences the fracture initiation strain. The strain rate difference and the initial yield stress variation between the two experimental conditions (for the Autenrieth fracture initiation strain model and Karp fracture initiation strain model) have been reported as being possible causes for the significant variation in the fracture initiation strain. The significant fracture initiation strain variation correspondingly leads to significant variation in the continuous chip to segmented chip transition and the chip segmentation intensity prediction in the finite element models that were developed in this study.

The experimental investigation (Figure 7) shows that the chip segmentation at a constant cutting speed is influenced by tool geometry and chip thickness. In this study, the normal rake angle is the tool geometry parameter modified, and its influence on the chip formation is studied. With the normal rake angle being modified from negative to positive rake angle under certain uncut chip thickness (feed) conditions, the chip formation process is transformed from a continuous chip to a segmented chip, as observed with the segmented chip boundary.

With the normal rake angle modification, the mechanical loading of the workpiece ahead of the tool is modified with varying levels of shear and compression. The mechanical loading is not uniform along the shear plane from the cutting edge to the free surface in the primary deformation zone. The stress that was observed by the workpiece material close to the cutting-edge is highly compressive when compared to the free end of the primary deformation zone, where the stresses are a combination of shear and compression. The workpiece material that is close to the cutting-edge rounding is mainly under the stress state defined by combined shear and compression defined by a large negative stress triaxiality. The workpiece material on the free end of the primary deformation zone has a relatively lower negative stress triaxiality; still, they are in the negative stress triaxiality conditions. Figure 6 shows that the fracture initiation strain in the negative stress triaxiality condition to be sensitive and varies based on testing conditions. This fracture initiation strain sensitivity is attributed to be the main reason for chip segmentation to be sensitive to minor changes in cutting conditions. At a constant chip thickness and cutting speed, the tool geometry influences this mechanical loading and it is observed to influence the onset of chip segmentation. Another critical factor that influences the material behavior is the material's history. From a machining perspective, the influence of the previous cut on the strain profile has been shown by Childs et al. [26]. The tool edge geometry controls the influence of the previous cut on the stress profile on the workpiece surface, tool wear, and tool vibration. The tool edge geometry could also influence the transition zone between the continuous chip to segmented chip. In this work, the influence of the previous cut is not taken into account and it could be a source of improvement in chip segmentation prediction in future work.

It is clearly understood that the uncut chip thickness, in addition to the rake angle, influences the transition from continuous to segmented chip at constant cutting speed. This is understood from the results, showing that, at a constant rake angle, the chip thickness influences chip segmentation. The chip thickness increase leads to the heat being concentrated in the primary shear zone. This concentration of heat leads to a segmented chip, as predicted in the presented FE simulations. On the other hand, chip thickness increase needs to be significant in this case at a rake angle of 5° up to 0.6 $mm.rev^{-1}$ to produce segmented chips.

8. Conclusions

In this study, a new modified JC model is developed based on the Voyiadjis–Abed–Rusinek [8] constitutive modeling approach. The modified JC model with the fracture modeling approach by Childs et al. [26] has been implemented for the first time in the FE framework for the chip formation simulation. The influence of the fracture initiation strain model on the continuous to segmented chip transition prediction has been successfully demonstrated. The following conclusions are made from the study.

1. The modified JC model can incorporate the DSA influence on flow stress curves of normalized C45E steel for varying temperatures and strain rates with reasonable accuracy.
2. The strain hardening behavior modeling of the modified JC model can predict the temperature's influence on strain hardening accurately when compared to the JC model.
3. The strong influence of DSA in the simulation of chip formation in machining of C45E steel is established with an improvement of cutting force prediction accuracy.
4. The transition from continuous chip to segmented chip is established to be a function of normal rake angle and feed at a constant cutting speed in orthogonal cutting.
5. The Autenrieth fracture initiation strain model can predict the chip segmentation boundary better as compared to the Karp fracture initiation strain model.

Author Contributions: Conceptualization, A.M.D.; Formal analysis, A.M.D.; Supervision, P.V.S., T.B. and M.E.; Validation, A.M.D. and P.V.S.; Writing–original draft, A.M.D.; Writing–review & editing, P.V.S., T.B. and M.E. All authors have read and agreed to the published version of the manuscript.

Funding: This research was funded by Sandvik Coromant AB and the Knowledge Foundation through the Industrial Research School SiCoMaP, Dnr 20110263, 20140130.

Acknowledgments: The authors would like to thank T.H.C. Childs for his support and guidance with the finite element implementation of the damage model and for invaluable discussions.

Conflicts of Interest: The authors declare no conflict of interest.

Nomenclature

Symbol	Description	Unit
A	Initial yield stress	MPa
A_0	Initial yield stress at room temperature and reference strain rate	MPa
A_Δ	Flow stress increase function in DSA range in modified JC model	MPa
a	Damage function parameter (Childs damage function)	-
B	Strain hardening coefficient	MPa
C	Strain rate hardening coefficient	-
D	Damage	-
D_+	Steady-state damage value	-
D_{Xi}	Fracture initiation parameter	-
D_3	Strain rate component coefficient of fracture initiation function	-
D_4	Temperature component coefficient of fracture initiation function	-
f	Feed rate	$mm.rev^{-1}$
T_l	Lower temperature bound function in DSA range (MJC Model)	°C
T_{ll}	DSA range's lower temperature bound at room temperature and reference strain rate (MJC Model)	°C
T_h	Upper-temperature bound function of DSA range	°C
T_{hl}	Upper temperature bound of DSA range at room temperature and reference strain rate (MJC Model)	°C
T	Temperature	°C
T_o	Reference temperature	°C
T_m	Melting temperature	°C
T_S	Lower bound healing temperature (Childs damage function)	°C
T_E	Upper bound healing temperature (Childs damage function)	°C
T^*	Homologous temperature	-
t	DSA regime fitting parameter of MJC Model	
m	Thermal softening coefficient	-
n	Strain hardening exponent	-
α	Rake angle	°
ξ	Temperature dimensioned parameter of the VAR model	-
$\Delta\sigma$	Flow stress increase of DSA range in MJC model at room temperature and reference strain rate	MPa
$\overline{\varepsilon}_{fi}$	Fracture initiation strain	-
$\overline{\varepsilon}$	Equivalent plastic strain	-
ε_{fi-X}	Fracture initiation strain	-
$\dot{\varepsilon}$	Strain rate	s^{-1}
$\dot{\varepsilon_0}$	Reference strain rate	s^{-1}
η	Stress triaxiality	-
$\overline{\theta}$	Lode angle parameter	-
μ_i	Internal Coulomb friction coefficient (Childs damage function)	-
$\overline{\sigma}$	Flow stress	MPa

Abbreviation

FE	Finite element
DSA	Dynamic strain aging
VAR	Voyiadjis–Abed–Rusinek model
JC	Johnson-Cook model
MJC	Modified Johnson-Cook model
SCD	Shear compression disk specimen

References

1. Schulz, H. High-speed machining. In *Manufacturing Technologies for Machines of the Future: 21st Century Technologies*; Dashchenko, A.I., Ed.; Springer: Berlin/Heidelberg, Germany, 2003; ISBN 978-3-642-62822-1.
2. Childs, T.; Maekawa, K.; Obikawa, T.; Yamane, Y. *Metal Machining–Theory and Applications*; Arnold: London, UK, 2000; ISBN 9780470392454.
3. Zahavi, E.; Torbilo, V. *Fatigue Design: Life Expectancy of Machine Parts*, 1st ed.; CRC Press: Boca Raton, FL, USA, 2019; ISBN 9780203756133.
4. Arrazola, P.J.; Ozel, T.; Umbrello, D.; Davies, M.; Jawahir, I.S. Recent advances in modelling of metal machining processes. *CIRP Ann.* **2013**, *62*, 695–718. [CrossRef]

5. Tasan, C.C.; Diehl, M.; Yan, D.; Bechtold, M.; Roters, F.; Schemmann, L.; Zheng, C.; Peranio, N.; Ponge, D.; Koyama, M.; et al. An Overview of Dual-Phase Steels: Advances in microstructure-oriented processing and micromechanically guided design. *Annu. Rev. Mater. Res.* **2015**, *45*, 391–431. [CrossRef]
6. El-Magd, E.; Treppman, C.; Korthäuer, M. Description of flow curves over wide ranges of strain rate and temperature. *Int. J. Mater. Res.* **2006**, *97*, 1453–1459. [CrossRef]
7. Hokka, M.; Rämö, J.; Mardoukhi, A.; Vuoristo, T.; Roth, A.; Kuokkala, V.-T. Effects of microstructure on the Dynamic Strain Aging in Ferritic-Pearlitic Steels. *J. Dyn. Behav. Mater.* **2018**, *4*, 452–463. [CrossRef]
8. Voyiadjis, G.Z.; Song, Y.; Rusinek, A.; Rusinek, Y.S.A. Constitutive model for metals with dynamic strain aging. *Mech. Mater.* **2019**, *129*, 352–360. [CrossRef]
9. Childs, T.H. Revisiting flow stress modelling for simulating chip formation of carbon and low alloy steels. *Procedia CIRP* **2019**, *82*, 26–31. [CrossRef]
10. Buchkremer, S.; Wu, B.; Lung, D.; Münstermann, S.; Klocke, F.; Bleck, W. FE-simulation of machining processes with a new material model. *J. Mater. Process. Technol.* **2014**, *214*, 599–611. [CrossRef]
11. Jaspers, S.; Dautzenberg, J. Material behaviour in metal cutting: Strains, strain rates and temperatures in chip formation. *J. Mater. Process. Technol.* **2002**, *121*, 123–135. [CrossRef]
12. Childs, T.H.C.; Otieno, A.W. Simulations and experiments on machining carbon and low alloy steels at rake face temperatures upto 1200 °C. *Mach. Sci. Technol.* **2012**, *16*, 96–110. [CrossRef]
13. Nemat-Nasser, S.; Guo, W.-G. Thermomechanical response of HSLA-65 steel plates: Experiments and modeling. *Mech. Mater.* **2005**, *37*, 379–405. [CrossRef]
14. Devotta, A.M.; Sivaprasad, P.; Beno, T.; Eynian, M.; Hurtig, K.; Magnevall, M.; Lundblad, M. A Modified Johnson-Cook Model for Ferritic-Pearlitic Steel in Dynamic Strain Aging Regime. *Metals* **2019**, *9*, 528. [CrossRef]
15. Calamaz, M.; Coupard, D.; Girot, F.; Mata, F.A.G. A new material model for 2D numerical simulation of serrated chip formation when machining titanium alloy Ti–6Al–4V. *Int. J. Mach. Tools Manuf.* **2008**, *48*, 275–288. [CrossRef]
16. Childs, T.H.C. Ductile shear failure damage modelling and predicting built-up edge in steel machining. *J. Mater. Process. Technol.* **2013**, *213*, 1954–1969. [CrossRef]
17. Liu, J.; Bai, Y.; Xu, C. Evaluation of Ductile fracture models in finite element simulation of metal cutting processes. *J. Manuf. Sci. Eng.* **2013**, *136*, 011010. [CrossRef]
18. Nasr, M.N.; Ammar, M.M. An evaluation of different damage models when simulating the cutting process using FEM. *Procedia CIRP* **2017**, *58*, 134–139. [CrossRef]
19. Vaziri, M.R.; Salimi, M.; Mashayekhi, M. Evaluation of chip formation simulation models for material separation in the presence of damage models. *Simul. Model. Pr. Theory* **2011**, *19*, 718–733. [CrossRef]
20. Wojciechowski, S.; Matuszak, M.; Powałka, B.; Madajewski, M.; Maruda, R.W.; Krolczyk, G.M. Prediction of cutting forces during micro end milling considering chip thickness accumulation. *Int. J. Mach. Tools Manuf.* **2019**, *147*, 103466. [CrossRef]
21. Sima, M.; Ozel, T. Modified material constitutive models for serrated chip formation simulations and experimental validation in machining of titanium alloy Ti–6Al–4V. *Int. J. Mach. Tools Manuf.* **2010**, *50*, 943–960. [CrossRef]
22. Atlati, S.; Haddag, B.; Nouari, M.; Zenasni, M. Analysis of a new Segmentation Intensity Ratio "SIR" to characterize the chip segmentation process in machining ductile metals. *Int. J. Mach. Tools Manuf.* **2011**, *51*, 687–700. [CrossRef]
23. Pan, H.; Liu, J.; Choi, Y.; Xu, C.; Bai, Y.; Atkins, A. Zones of material separation in simulations of cutting. *Int. J. Mech. Sci.* **2016**, *115*, 262–279. [CrossRef]
24. Wang, B.; Liu, Z. Serrated chip formation mechanism based on mixed mode of ductile fracture and adiabatic shear. *Proc. Inst. Mech. Eng. Part B: J. Eng. Manuf.* **2013**, *228*, 181–190. [CrossRef]
25. Johnson, G.R.; Cook, W.H. A constitutive model and data for metals subjected to large strains, high strain rates and high temperatures. In Proceedings of the 7th International Symposium on Ballistics, The Hague, The Netherlands, 19–21 April 1983; Volume 21, pp. 541–547.
26. Childs, T.H.; Arrazola, P.J.; Aristimuño, P.X.; Garay, A.; Sacristan, I. Ti6Al4V metal cutting chip formation experiments and modelling over a wide range of cutting speeds. *J. Mater. Process. Technol.* **2018**, *255*, 898–913. [CrossRef]

27. Zhang, W.; Cai, Y. *Continuum Damage Mechanics and Numerical Applications; Advanced Topics in Science and Technology in China*; Springer: Berlin/Heidelberg, Germany, 2010; ISBN 978-3-642-04707-7.
28. Bai, Y.; Teng, X.; Wierzbicki, T. On the application of stress triaxiality formula for plane strain fracture testing. *J. Eng. Mater. Technol.* **2009**, *131*, 021002. [CrossRef]
29. Bridgman, P.W. Effects of high hydrostatic pressure on the plastic properties of metals. *Rev. Mod. Phys.* **1945**, *17*, 3–14. [CrossRef]
30. Autenrieth, H.; Schulze, V.; Herzig, N.; Meyer, L.W. Ductile failure model for the description of AISI 1045 behavior under different loading conditions. *Mech. Time-Depend. Mater.* **2009**, *13*, 215–231. [CrossRef]
31. Karp, B.; Shapira, G.; Rittel, D. Experimental investigation of fracture under controlled stress triaxiality using shear-compression disk specimen. *Int. J. Fract.* **2017**, *209*, 171–185. [CrossRef]
32. Bai, Y.; Wierzbicki, T. A new model of metal plasticity and fracture with pressure and Lode dependence. *Int. J. Plast.* **2008**, *24*, 1071–1096. [CrossRef]
33. Johnson, G.R.; Cook, W. a Fracture characteristic of three metals subjected to various strains, strain rates, temperatures and pressures. *Eng. Fract. Mech.* **1985**, *21*, 31–48. [CrossRef]
34. Devotta, A.; Beno, T.; Löf, R.; Espes, E. Quantitative characterization of chip morphology using computed tomography in orthogonal turning process. *Procedia CIRP* **2015**, *33*, 299–304. [CrossRef]
35. Thirdwave Systems. Available online: www.thirdwavesys.com/advantedge/ (accessed on 1 March 2020).
36. Marusich, T.D.; Ortiz, M. Modelling and simulation of high-speed machining. *Int. J. Numer. Methods Eng.* **1995**, *38*, 3675–3694. [CrossRef]
37. Jaspers, S.; Dautzenberg, J. Material behaviour in conditions similar to metal cutting: Flow stress in the primary shear zone. *J. Mater. Process. Technol.* **2002**, *122*, 322–330. [CrossRef]
38. Abushawashi, Y.; Xiao, X.; Astakhov, V. A novel approach for determining material constitutive parameters for a wide range of triaxiality under plane strain loading conditions. *Int. J. Mech. Sci.* **2013**, *74*, 133–142. [CrossRef]
39. Kouadri, S.; Necib, K.; Atlati, S.; Haddag, B.; Nouari, M. Quantification of the chip segmentation in metal machining: Application to machining the aeronautical aluminium alloy AA2024-T351 with cemented carbide tools WC-Co. *Int. J. Mach. Tools Manuf.* **2013**, *64*, 102–113. [CrossRef]
40. Fernandez-Zelaia, P.; Melkote, S.N. Statistical calibration and uncertainty quantification of complex machining computer models. *Int. J. Mach. Tools Manuf.* **2019**, *136*, 45–61. [CrossRef]
41. Laakso, S.V.A. Heat matters when matter heats—The effect of temperature-dependent material properties on metal cutting simulations. *J. Manuf. Process.* **2017**, *27*, 261–275. [CrossRef]

Article

Assessment of Chip Breakability during Turning of Stainless Steels Based on Weight Distributions of Chips †

Hongying Du [1,*], Andrey Karasev [1], Thomas Björk [2], Simon Lövquist [3] and Pär G. Jönsson [1]

[1] Department of Materials Science and Engineering, KTH Royal Institue of Technology, Brinellvägen 23, 10044 Stockholm, Sweden; karasev@kth.se (A.K.); parj@kth.se (P.G.J.)

[2] Department of R&D, Ovako, 81382 Hofors, Sweden; thomas.bjork@ovako.com

[3] Department of R&D Verification of Component and Material, AB Sandvik Coromant, Mossvägen 10, 81181 Sandviken, Sweden; simon.lovquist@sandvik.com

* Correspondence: hongying@kth.se; Tel.: +46-8790-6151

† This paper is an extended version of paper published in the Conference SPS2018, Stockholm, Sweden, 16–18 May 2018.

Received: 15 April 2020; Accepted: 18 May 2020; Published: 21 May 2020

Abstract: Currently, the available evaluation methods for determining the chip breakability in the industry are mainly based on subjective visual assessment of the chip formation by an operator during machining or on chips that were collected after the tests. However, in many cases, these methods cannot give us accurate quantitative differences for evaluation of the chip breakability of similar steel grades and similar sets of machining parameters. Thus, more sensitive methods are required to obtain more detailed information. In this study, a new method for the objective assessment of chip breakability based on quantitative determination of the weight distribution of chips (WDC) was tested and applied during machining of stainless steels without Ca treatment (316L) and with Ca treatment (316L + Ca). The obtained results show great consistencies and the reliability of this method. By using the WDC method, significant quantitative differences were obtained by the evaluation of chips, which were collected during the machining process of these two similar grades of steel at various cutting parameters, while, visually, these chips look very similar. More specifically, it was found that the Ca treatment of steel can improve the chip breakability of 316L + Ca steel in 80% of cutting trials, since a fraction of small light chips (Type I) from this steel increased and a fraction of large heavy chips (Type III) decreased accordingly. Moreover, the WDCs that were obtained at different cutting parameters were determined and compared in this study. The obtained results can be used for the optimization of chip breakability of each steel at different cutting parameters. The positive effect of Ca treatment of stainless steel was discussed in this study based on consideration of the modification of different non-metallic inclusions and their effect on the chip breakability during machining.

Keywords: stainless steel; Ca treatment; machinability; turning; chip breakability; weight distribution; non-metallic inclusions

1. Introduction

During the last six decades, steelmakers have developed a considerable amount of new steel grades with low-level impurities, which significantly improved the mechanical properties of steel. However, the production industry also faces the increasing challenge during machining these high-quality steels, because of the increased tool wear, difficult-break chips, and the significantly reduced tool life, which increases the consumption of time, energy, and money in machining operations [1,2].

Non-metallic inclusions (NMIs) play a significant role with respect to the mechanical properties and machinability of steels. Non-metallic inclusions, especially large inclusions, are generally harmful to the property of the steel, by reducing mechanical properties, like toughness, fatigue life, and corrosive resistance, which may increase the failure probability of the final product [3,4]. For example, MnS will result in the anisotropy of mechanical property. It will also provide the origin for fatigue crack and corrosion. [1,5] From another side, because of different properties of NMIs, such as hardness and thermal expansions, as compared to the metal matrix, the NMIs will have different roles during the machining of materials. The effect of different inclusions on machinability characteristics (such as tool life, tool wear, chip breakability, etc.) was evaluated in many studies [1,6–11]. For instance, NMIs could act as a source of stress concentrations, which is beneficial for a crack formation making the chip easy to break. However, some hard inclusions will damage the tool surface and increase the tool wear [1,7,12]. It is reported that some NMIs (such as sulfides and Ca modified oxides) are beneficial for improved steel machinability [6–11]. These NMIs can help to improve the steel machinability in two ways: (1) as a source of stress concentrations, which decrease the cutting force and increase the chip breakability and (2) as a lubricant (constant tool protection layer) in the contact zone between the cutting tool and material, which reduces the abrasive and chemical wear of the tool and extends the tool life. Thus, it is necessary to modify NMIs in steels during steelmaking to obtain favorable machinability properties, while the mechanical properties are not affected in a negative manner.

Machinability is a complex concept that can be described by a wide range of parameters and factors. A large number of criteria, such as tool wear (TW), tool life (TL), cutting forces (CF), chip characteristics (CC), and surface roughness (SR), have been developed to evaluate the machinability of different workpiece materials [1]. Among them, chip breakability has an important impact on the easiness and operation of the machining process. If long composite chips, which combine distinct size and shape chip segments, are obtained in the machining, the production efficiency will be threatened when chips stuck in the rotating cutting tool or the chuck. Thus, chips with shapes, like short broken corkscrews or spirals, are preferred during the machining operation [13]. Chip breakage can take place in three different stages. Chip breakage for a material with a low ductility is preferred to take place during the chip formation phase. If the material has a low mechanical strength, chip breakage is more likely to happen during the chip curl phase. Additionally, when the formed chip is long and contacted with the workpiece surface or tool flank face, it is possible to break [14].

Today, there are several methods for evaluating the chip breakability. One common method that is applied in the industry is to use chip breaking curves made by visual inspections during the longitudinal turning operation. The operators defined the transition point between acceptable and unacceptable chip breakages, based on their subjective experience. Another test is the chip chart, which suggests appropriate cutting parameters to obtain a satisfactory chip breakage, which is supplied from the cutting tool companies [13]. Photos of chips that are collected from different cutting parameters are put together and compared. There are some other complex evaluation methods to determine the chip breakability, like a fuzzy rule-based system, to describe chip breakability performance [15]. However, when it comes to the problem of the sensitive quantitative comparison of similar steel grades (such as modified steels and reference steels) or during optimization of chip breakability of steel at different cutting parameters, the results from these methods often look quite similar. Therefore, it is difficult to conclude which of the two materials that has the better chip breakability or to decide the optimized machining parameters. In the previous study [16], a quantitative method of evaluation of chip breakability, which was based on measurement of weight distribution of chips (WDC) obtained during machining of two similar stainless steels, was tested. The obtained results showed a clear difference in chip breakability for investigated steels. However, the weight of chips during machining depends not only on the quality of steel, but also on the method and parameters of machining (such as the cutting speed, feed rate, and cutting depth). Therefore, only the measurement of WDC cannot be applied for the accurate evaluation and comparison of chip breakability in the cases when the machining is carried out at different parameters. Thus, more systematic investigations are necessary in order to improve the accurate quantitative assessment of chip breakability of steels during machining at different parameters.

The present study aims to develop and test a sensitive and quantitative evaluation method that is based on measurements of the chip weights to estimate the chip breakability of different steels and comparison of obtained results with the results obtained by using the standard chip charts. Here, the weight distributions of chips that were obtained at different cutting parameters were investigated and compared for industrial 316 stainless steels produced without and with Ca-treatment.

2. Materials and Methods

2.1. Workpiece Materials and Cutting Tool

This study investigated one Ca treated 316L continuous casting stainless steel (316L + Ca) and one reference stainless steel without Ca treatment (316L) from industrial trials. The only difference between the two trials is that there was a CaSi wire addition at the end of ladle treatment for the 316L + Ca steel. Their chemical compositions are given in Table 1. The results of mechanical tests and impact tests show that there is no significant difference in mechanical properties between the two materials (around +2.2% for Young's modulus and +10% for impact test in the transversal direction for 316L + Ca steel as compared to 316L steel).

Table 1. Composition of samples of the reference steel (316L) and the Ca-treated steel (316L + Ca).

Steel Grade	Ca-Treated	C	Si	Mn	Cr	Ni	S	O	Ca	Al
		(mass %)					(mass ppm)			
316L	No	0.02	0.38	1.60	16.82	11.18	70	20	-	40
316L + Ca	Yes	0.01	0.46	1.58	16.86	11.14	90	59	28	40

CNMG120408-MM 2025 cemented carbide inserts (Sandvik Coromant, Gimo, Sweden), with CVD coating (Ti (C, N) + Al_2O_3 + TiN), were used for the machining testings.

2.2. Conventional Evaluations of Chip Breakability

Conventional chip breakability evaluations were conducted during longitudinal turnings for 36 operations in two different cutting speeds (130 m/min and 180 m/min). A flood coolant (7–9% Hocut 4160 Emulsion, Barcelona, Spain) was applied during the turning test.

The first six operations were used for the visual inspection to make chip breaking curves with six different cutting depths (a_p = 0.5, 0.75, 1, 2, 3, and 4 mm). It was started with the first fixed depth of cut (0.5 mm), then the feed rate was slowly increased until chips changed from long chips to short chips, and then this feed rate was recorded. Subsequently, the depth of cut was increased to the next one and increased the feed slowly again. Repeat the previous operations for all six depths of cuts. Afterwards, a chip breaking curve was drawn with a_p versus f_n.

The next 30 operations were used for making chip charts. Each operation corresponded to one combination of the following machining parameters shown in Table 2.

Table 2. The cutting data used for chip breakability curves and chip charts.

Type of Test	Feed Rates (f_n) (mm/rev)	Cutting Depths (a_p) (mm)	Cutting Speed (m/min)
Chip breakability curves	0.15, 0.2, 0.3, 0.4, 0.5	0.5, 0.75, 1, 2, 3, 4	130, 180
Chip chart	0 → 0.5 (continuous)	0.5, 0.75, 1, 2, 3, 4	130, 180

Broken chips from each set of cutting parameters were collected by a collector close to the cutting operation during the adequate time for a representative amount separately. Subsequently, photographs for each set of chips were placed in a chart with f_n as the x-axis and a_p as the y-axis. In addition, based on visual subjective evaluation of the operator, all sets of short chips, which have one arc or two and more connected arcs, were marked as acceptable.

2.3. Chip Weight Measurement

The chips that were obtained from longitudinal turnings with similar sizes and shapes in the chip charts were compared by weight in this study. For each set of cutting parameters, more than 100 chips were randomly collected and weighted individually. An analytical balance with yield readability to four decimal places to the right of the decimal point (up to 0.0001 g) was applied for measurements. The chips were classified into three groups, according to the weights and shapes. Thereafter, the percentage of the number of each group were calculated and finally summarized.

2.4. Investigation of Non-metallic Inclusions

The non-metallic inclusions in the two steels were extracted using electrolytic extraction. During the electrolytic extraction process, the steel samples were partially dissolved, while inclusions did not dissolve. Accordingly, the inclusions were easily collected by a filtration process and investigated using a scanning electron microscope (SEM, S3700N-Hitachi, Hitachi High-Technologies Corporation, Tokyo, Japan) combined with an energy dispersive spectroscopy (EDS, Bruker, Karlsruhe, Germany).

In addition, the fracture sections of chips obtained from the previous cutting tests were also observed by using the SEM (S3700N-Hitachi, Hitachi High-Technologies Corporation, Tokyo, Japan).

3. Results and Discussion

3.1. Evaluations by Conventional Methods

Two different cutting speeds, 130 and 180 m/min, were applied during the longitudinal turnings to study the chip formation ability of the 316L and 316L + Ca steels. Figure 1 shows the chip breaking curves that were obtained at 130 m/min (a) and 180 m/min (b). The point in the graph is the parameter set that the operator thought was a watershed between acceptable chips and unacceptable chips. The right side of the curves in this figure represents good chip formation conditions. From the results of the chip breaking curves, the experimental grade 316L + Ca shows a slightly wider range of cutting parameter combinations (the spotted area) when compared to the reference 316L steel for both cutting speeds. However, for the same cutting parameters in the grey area, the obtained curves cannot show any difference between the two steel grades. In this case, it is difficult to evaluate the differences between the chip breakabilities for the reference and experimental steels at different cutting parameters in the selected area.

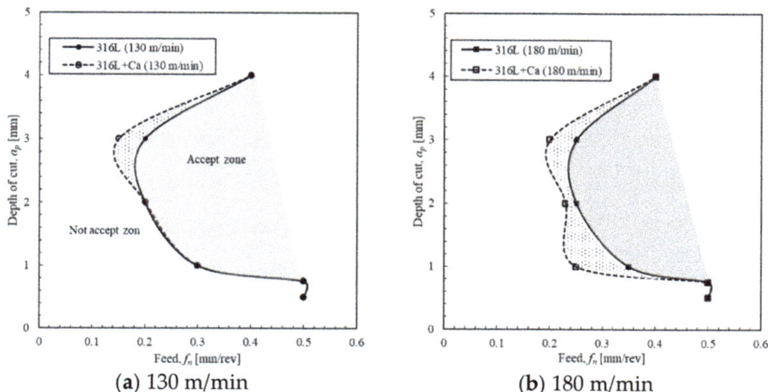

Figure 1. Chip breaking curve at cutting speeds of 130 m/min (**a**) and 180 m/min (**b**).

Figure 2 shows the results of another chip breakability evaluation method, the chip charts, obtained at a cutting speed of 130 m/min. Chips in the red polygons are acceptable chips. In general, the charts show more detailed information than the chip breaking curves. However, the polygons are visually quite similar, and it is hard to tell the difference between the chips of the reference sample 316L and the experimental 316L + Ca sample based on the same set of cutting parameters. These two steel grades share similar chip breakabilities, according to this chip chart method.

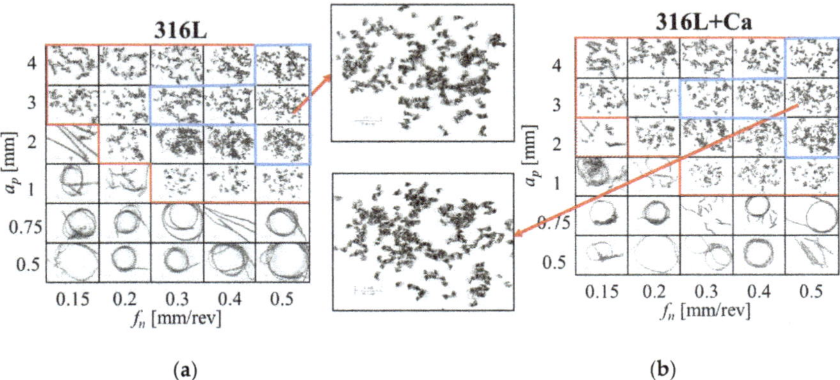

Figure 2. Chip charts of (**a**) the reference sample (316L) and (**b**) the experimental sample (316L + Ca) at a cutting speed of 130 m/min, reproduced from Ref. [16].

Chips from 5 sets of cutting parameters (blue zone in Figure 2, which were determined from both speeds as being acceptable based on ordinary means) were collected and weighted to better quantitatively evaluate and compare these two materials.

3.2. Classification of Chips

The chips from the accepted area (grey zone in Figure 1), which have good breakability, are mostly the chips with one arc or two and more arcs hard connected in one construction. In this study, a measured weight of each chip was used as a quantitative index for evaluating and comparing the chip breakability. It is apparent that the steel, which has the chips with a small number of arcs and correspondingly a lower weight, has the better chip breakability. Therefore, it can be expected that this steel can have better machinability and lower tool wear, which is beneficial from short broken chips at the selected cutting parameters. The chips could be divided into three groups according to the shapes and weights (as shown in Figure 3): Type I-light chips (<0.2 g) with one arc, Type II-middle chips (0.2–0.4 g) with two-three arcs, and Type III-heavy (>0.4 g) multi-arcs chips, as was classified in the previous study [16].

However, when the feed rate (f_n) and cutting depth (a_p) change, the thickness and width of chips will also change. As a result, the weight of chips and their corresponding gradation can significantly change. Accordingly, for a quantitative comparison of the chip breakability between different cutting conditions, a unified index should be used. Such an index can be the length of a chip (l) having the same cross-section area (A), which can be determined by using the following equation:

$$l = W_{chip}/(\rho_{me} \cdot A) = W_{chip}/(\rho_{me} \cdot w \cdot d) \tag{1}$$

where W_{chip} is the measured weight of the chip, ρ_{me} is the density of the metal chip, A is the undeformed cross-sectional area of the chip, and w and d are the width and depth of the measured chip. In this study, it was assumed that the ρ_{me} value is constant for all chips independently on the cutting parameters. The values of w and d cannot be precisely measured for each chip. However, they can approximately be estimated based on the feed rate f_n, cutting depth a_p and major cutting edge angle κ (the angle between the edge of cutting tools and the longitude direction of role material) using the following equations [13]:

$$w \approx \frac{a_p}{\sin \kappa}$$
$$d \approx fn \cdot \sin \kappa \tag{2}$$

Hence, the A value can be estimated by the following relationship:

$$A = w \cdot d \approx fn \cdot a_p \tag{3}$$

Since the cross-sectional area (A) of chips are proportional to the feed rate (f_n) and cutting depth (a_p), these cutting parameters were used in this study to evaluate the unified index of the chip weight (I_{Wchip}) and classification of chips that were obtained at different cutting conditions:

$$I_{Wchip} = W_{chip} / (a_p \cdot fn) \tag{4}$$

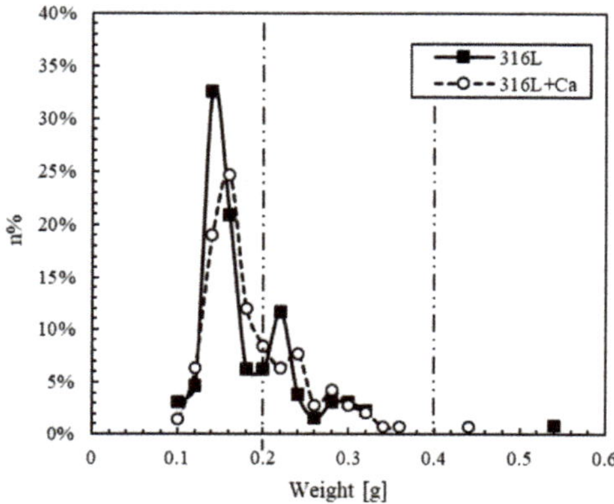

Figure 3. Distribution of chip weight with a step of 0.02 g at the following cutting parameters: v = 130 m/min, a_p = 3 mm, f_n = 0.5 mm/rev.

Table 3 shows the classification of chips depending on shape, weight, and I_{Wchip} index: Type I: one-arc chips with I_{Wchip} smaller than 0.13; Type II: chips having two or three arcs with I_{Wchip} 0.13–0.27; Type III: multi-arc chips (usually at least 3 curves) with I_{Wchip} values larger than 0.27.

Table 3. Classification of different chips.

Type of Chips	I	II	III
Photograph	5 mm	5 mm	5 mm
I_{Wchip}	<0.13	0.13–0.27	>0.27

3.3. *Chip Distributions*

Table 4 shows the results of the weight distribution of chips from 10 sets of cutting parameters corresponding to the blue zone in Figure 2. It was found that a clearer difference in chip breakability, as compared to conventional breakability test methods, could be detected for the 316L and 316L + Ca steels. The results show that compared to the 316L steel, the frequency of large Type III chips ($I_{Wchip} > 0.27$) is significantly smaller for the 316L + Ca steel. Especially in sets 1, 5, and 10, the frequency of Type III chips decreases by 11–32%. On the other hand, it is also evident that the number of small Type I chips ($I_{Wchip} < 0.13$) in the 316L + Ca steel increases for most cutting parameter sets. It is especially clear in sets 2, 4, 7, and 8, in which the frequency increases by 15–23%. The frequency of Type II chips is quite scatted. However, an obvious improvement of chip breakability was quantitatively detected for the 316L + Ca steel for the cutting parameters in set 2, 4, and 6–8 (due to significant increase of the number of small chips (Type I) and decrease of the number of middle size chips (Type II)) and in set 1, 5 and 10 (due to the significant decrease of the number of large chips (Type III)). No clear difference between the 316L and 316L + Ca materials is found between sets 3 and 9. It might be due to that, for cases where Type I chips are in the majority (more than 70%, in sets 3 and 9) in both steels, no clear difference could be detected. Thus, it can be concluded that a Ca treatment is beneficial for the chip breakability in most cases according to the obtained results (in 80% of trials in this study: sets 1, 2, 4–8, 10).

Table 4. Summary of results based on weighting measurements and calculated I_{Wchip}.

Set	1	2	3	4	5	6	7	8	9	10
Speed (m/min)	130	130	130	130	130	180	180	180	180	180
a_p (mm)	3	3	3	2	4	3	3	3	2	4
f_n (mm/rev)	0.3	0.4	0.5	0.5	0.5	0.3	0.4	0.5	0.5	0.5
Type I in 316L	-	6%	74%	62%	11%	25%	50%	38%	77%	0%
Type I in 316L + Ca	-	21%	72%	81%	13%	33%	69%	61%	72%	2%
Δ (Type I)	-	+15%	−2%	+19%	+2%	+8%	+19%	+23%	−5%	+2%
Type II in 316L	15%	90%	23%	35%	69%	68%	42%	62%	23%	36%
Type II in 316L + Ca	47%	75%	24%	18%	78%	63%	26%	39%	27%	58%
Δ (Type II)	+32%	−15%	+1%	−17%	+9%	−5%	−16%	−23%	+4%	+22%
Type III in 316L	85%	4%	3%	3%	20%	8%	8%	-	0%	64%
Type III in 316L + Ca	53%	4%	4%	1%	9%	5%	5%	-	1%	40%
Δ (Type III)	−32%	0%	+1%	−2%	−11%	−3%	−3%	-	+1%	−24%

The weight distributions of chips for 316L and 316L + Ca steels at different feeds with a fixed cutting depth (a_p = 3 mm) are shown in Figures 4 and 5 for cutting speeds of 130 and 180 m/min, respectively. The obtained results correspond to sets 1–3 and 6–8 in Table 4. It can be seen that, at the same cutting parameters, the experimental 316L + Ca steel led to the production of a larger fraction of small Type I chips in most cases (sets 6–8 at V_c = 180 m/min and set 2 at V_c = 130 m/min and f_n = 0.4 mm/rev). Moreover, the chips of a 316L + Ca steel from set 1 (f_n = 0.3 mm/rev) at the low cutting speed show an increase of Type II chips instead and a decrease of large Type III chips. However, no clear benefits of using a Ca-treatment steel was detected in the case of set 3 (V_c = 130 m/min, f_n = 0.5 mm/rev).

Although the chips obtained from different feed rates look visually quite similar in the chip charts, the percentage of the small Type I chips in both steels increases drastically with an increased feed rate (f_n) at both cutting speeds. This is true for all sets, except for set 8 (V_c = 180 m/min, f_n = 0.5 mm/rev). Moreover, it should be pointed out that an increase of the cutting speed from 130 up to 180 m/min promotes an increase of the fraction of small Type I chips for both steels, except for the cutting conditions in set 8. This can be connected with some increase of temperature in the local cutting zone with an increase of feed rate and cutting speed during cutting [17,18]. However, a too high temperature in the cutting zone during the higher cutting speed might significantly decrease the chip breakabilities, as was obtained for the set eight cutting parameters (V_c = 180 m/min, f_n = 0.5 mm/rev). This negative effect might be due to the increased plasticity of the steel matrix and inclusions, resulting in too few formations of void around inclusions in the cutting zone where the temperature is too high. This can decrease the possibility of chip breaking.

On the other hand, the fraction of the small Type I chips significantly decreases in most cases for both steels with an increased cutting depth (a_p) at the fixed feeding rate (f_n = 0.5 mm/rev). This is shown in Figures 6 and 7 at cutting speeds of 130 and 180 m/min, respectively, for the sets 3–5 and 8–10 in Table 4. At the same time, the fraction of large Type III chips tends to increase with increased cutting depths. It shows that, with a larger cutting depth, the breadth of chips is wider and the helical radius is larger, so the chips need a larger length to break. In this case, the Ca-treatment and increased cutting speed did not show clear benefits for the chip breaking with respect to the small Type I chips, except for in sets 4 and 8. However, the frequency of large Type III chips in the 316L + Ca steel is smaller than that in the 316L steel in sets 4 and 8.

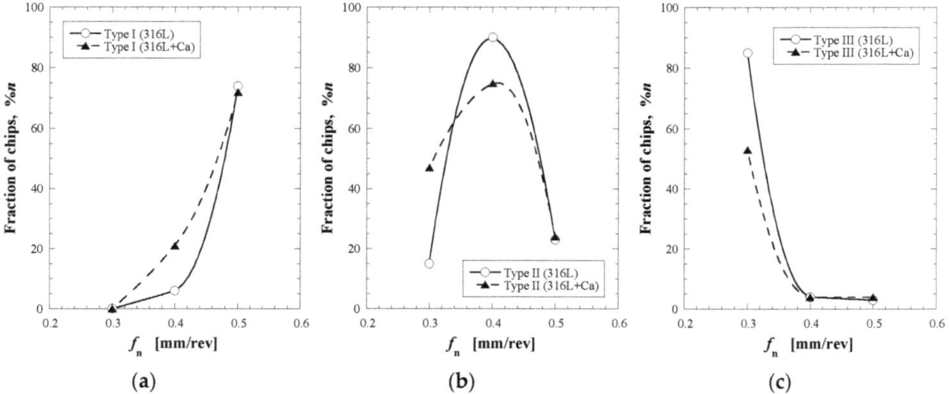

Figure 4. Chip weight distributions for Type I (**a**), Type II (**b**) and Type III (**c**) chips obtained at an a_p value of 3 mm and a cutting speed of 130 m/min.

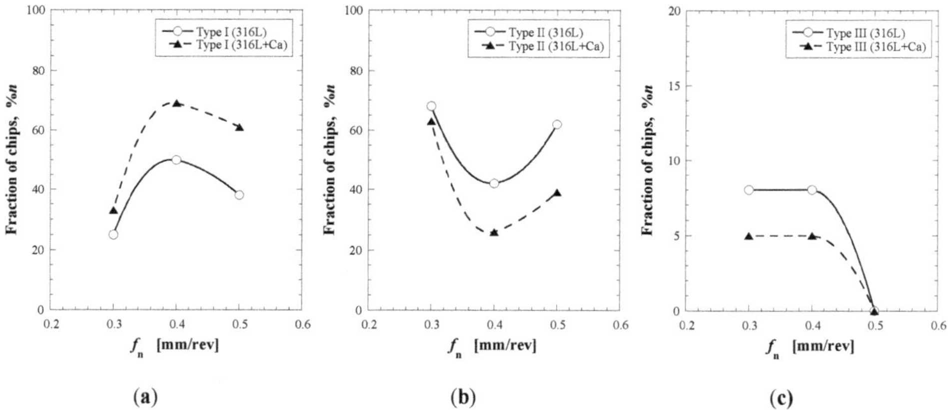

Figure 5. Chip weight distributions for Type I (**a**), Type II (**b**) and Type III (**c**) chips obtained at an a_p value of 3 mm and a cutting speed of 180 m/min.

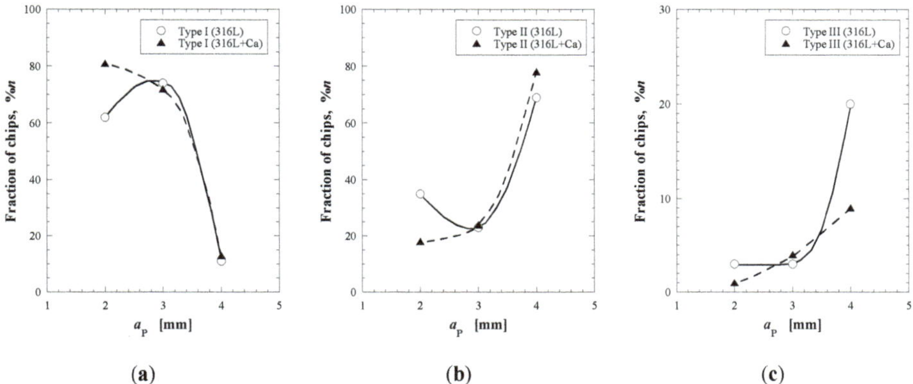

Figure 6. Chip weight distributions for Type I (**a**), Type II (**b**), and Type III (**c**) chips obtained at an f_n value of 0.5 mm/rev and a cutting speed of 130 m/min.

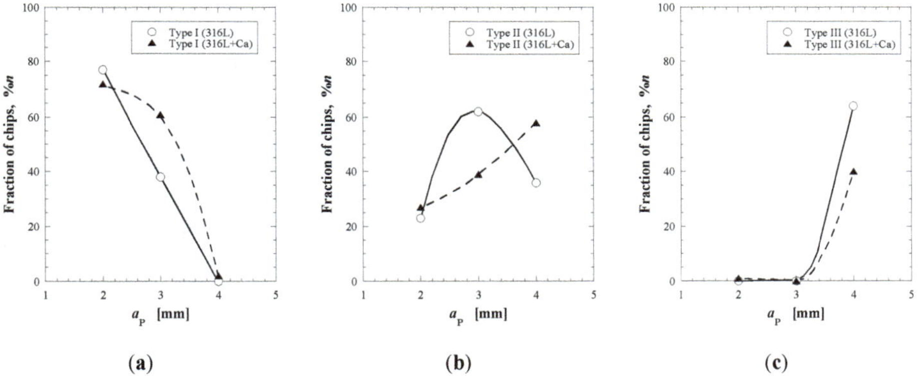

Figure 7. Chip weight distributions for Type I (**a**), Type II (**b**), and Type III (**c**) chips obtained at an f_n value of 0.5 mm/rev and a cutting speed of 180 m/min.

Overall, based on obtained results, it can be concluded that this weighting method can provide more detailed chip breakability differentiations of similar steel grades for different sets of cutting parameters.

3.4. The Difference of the Non-Metallic Inclusions in Two Steel Grades

The difference in chip breakability between these two steel grades 316L and 316L + Ca comes from the difference of the non-metallic inclusion content in the steels. The investigation results of non-metallic inclusions in the 316L + Ca sample show a significant increase of the amount (on ~40%) of two kinds (undeformed and elongated) of Al_2O_3-CaO-SiO_2-MgO oxide inclusions (as shown in Figure 8a,b), while the 316L sample contains mainly MnS inclusions and quite a small amount (4.4%) of Al_2O_3-MgO-MnO oxide inclusions (as shown in Figure 8c), according to a previous study [19]. More details of the evaluation of the non-metallic inclusions can be found in the previous paper [19]. The difference between oxide and sulfide inclusions with respect to plasticity at different temperatures lead to a different effect on the chip breakability for different sets of cutting parameters. For example, the hardness of pure MnS at 800 °C is only 20% of that at room temperature, whereas the hardness of silicate inclusions with a composition 34–40% SiO_2-29–40% Al_2O_3-28% MnO-3% TiO at 800 °C drops to only 10% of that at room temperature [20]. Many of undeformed Al_2O_3-CaO-SiO_2-MgO oxide inclusions were found in some big dimples of the fracture section of the broken chips of the 316L + Ca steel, which is supposed to contribute to the void nucleation and be beneficial for chip breaking for the current cutting conditions, as shown in Figure 9 (the arrows marked). A more systematic study of the mechanism of contributions of this kind inclusions on chip breakability will be considered in a separate paper.

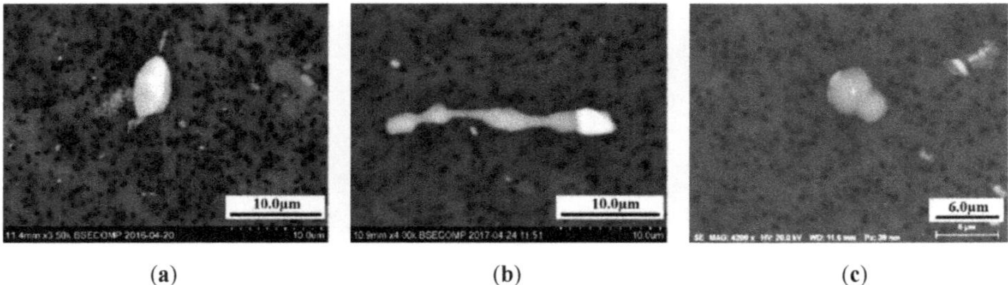

(**a**) (**b**) (**c**)

Figure 8. Typical non-metallic inclusions (NMI) investigated by using scanning electron microscope (SEM) after electrolytic extraction: undeformable oxides (**a**), elongated oxides (**b**) in 316L + Ca steel, and undeformed oxides (**c**) in 316L steel.

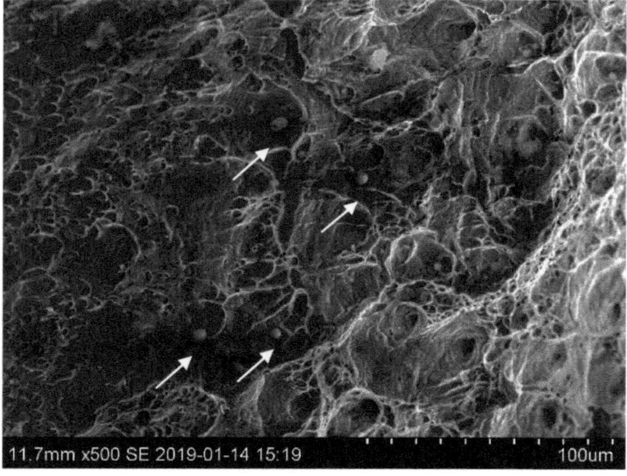

Figure 9. Oxide NMIs found in the large dimples on the fracture section of the chips obtained during machining of the 316L + Ca steel.

4. Conclusions

The focus of this study is to develop an alternative method for quantitative assessments of the chip breakability during machining of similar steels at acceptable cutting conditions, which are difficult to obtain while using conventional evaluation methods. Based on the results obtained in this study, the conclusions were summarized as follows:

1. It is hard to quantitatively differentiate the chip breakability of the 316L and 316L + Ca workpieces by using conventional visual subjective evaluation methods, because the chips obtained at different feed rates or different cutting depths often visually look very similar in the chip chart. However, the unified index of chip weight (I_{Wchip}) and chip distributions can be used for a differentiation and quantitative estimation of the chip breakability for these two steel grades.

2. In most cases, the Ca treated steel (316L + Ca) workpieces show a considerably improved chip breakability at the same cutting parameters when compared to the conventional reference 316L steel.

In these cases, the fraction of small chips (Type I) in the Ca-treated steel can be increased by 8–32%, while the fraction of large chips (Type III) can be decreased by 11–32%. However, if the fraction of the small chips is more than 70%, no clear difference could be detected between these two steel grades.

3. Although the chips obtained from different feed rates look visually similar in the chip charts, the fraction of the small chips (Type I) in both steel grades increases with an increased feed rate (f_n) and cutting speed. However, the fraction of small chips can significantly decrease at the higher cutting speed (180 m/min), larger feeding rate (0.5 mm/rev), and at an increased cutting depth (a_p).

4. The difference in chip breakability between the 316L and 316L + Ca steels derives from the different non-metallic inclusion contents in these two steel grades. More undeformed oxide NMIs, which could contribute to voids nucleation and be beneficial for chip breakability, were found in large dimples on the fracture section of the chips that were obtained from 316L + Ca steels.

Author Contributions: Conceptualization, A.K., P.G.J. and H.D.; methodology, T.B., H.D. and A.K.; formal analysis and material preparation, S.L.; investigation, H.D.; writing—original draft preparation, H.D.; writing—review and editing, A.K., P.G.J., T.B. and S.L.; supervision, A.K. and P.G.J. All authors have read and agreed to the published version of the manuscript.

Funding: This research received no external funding.

Acknowledgments: Hongying Du would like to thank Nils Stavlid for the assistance and contribution of discussion. Hongying Du also acknowledges the financial support and cooperation from VINNOVA, Jernkontoret, China Scholarship Council (CSC) and all other colleagues who participated in this study.

Conflicts of Interest: The authors declare no conflict of interest.

References

1. Ånmark, N.; Karasev, A.; Jönsson, P.G. The effect of different non-metallic inclusions on the machinability of steels. *Materials* **2015**, *8*, 754–783. [CrossRef] [PubMed]
2. Sousa, V.F.C.; Silva, F.J.G. Recent Advances in Turning Processes Using Coated Tools—A Comprehensive Review. *Metals* **2020**, *10*, 170. [CrossRef]
3. Schäfer, B.J.; Sonnweber-Ribic, P.; ul Hassan, H.; Hartmaier, A. Micromechanical Modeling of Fatigue Crack Nucleation around Non-Metallic Inclusions in Martensitic High-Strength Steels. *Metals* **2019**, *9*, 1258. [CrossRef]
4. Jamil, M.; Khan, A.M.; Hegab, H.; Sarfraz, S.; Sharma, N.; Mia, M.; Gupta, M.K.; Zhao, G.; Moustabchir, H.; Pruncu, C.I. Internal Cracks and Non-Metallic Inclusions as Root Causes of Casting Failure in Sugar Mill Roller Shafts. *Materials* **2019**, *12*, 2474. [CrossRef] [PubMed]
5. Temmel, C.; Ingesten, N.G.; Karlsson, B. Fatigue Anisotropy in Cross-rolled, Hardened Medium Carbon Steel Resulting from MnS Inclusions. *Metall. Mater. Trans. A* **2006**, *37*, 2995–3007. [CrossRef]
6. Akasawa, T.; Sakurai, H.; Nakamura, M.; Tanaka, T.; Takano, K. Effects of free-cutting additives on the machinability of austenitic stainless steels. *J. Mater. Process. Technol.* **2003**, *143–144*, 66–71. [CrossRef]
7. Fang, X.D.; Zhang, D. An investigation of adhering layer formation during tool wear progression in turning of free-cutting stainless steel. *Wear* **1996**, *197*, 169–178. [CrossRef]
8. Wang, Y.; Yang, J.; Bao, Y. Effects of non-metallic inclusions on machinability of free-cutting steels investigated by nano-indentation measurements. *Metall. Mater. Trans. A* **2015**, *46*, 281–292. [CrossRef]
9. Qi, H.S.; Mills, B. On the formation mechanism of adherent layers on a cutting tool. *Wear* **1996**, *198*, 192–196. [CrossRef]
10. Ånmark, N.; Björk, T.; Ganea, A.; Ölund, P.; Hogmark, S.; Karasev, A.; Jönsson, P.G. The effect of inclusion composition on tool wear in hard part turning using PCBN cutting tools. *Wear* **2015**, *334–335*, 13–22. [CrossRef]
11. Gutnichenko, O.; Bushlya, V.; Zhou, J.M.; Stahl, J.E. Tool Wear and Vibrations Generated When Turning High-chromium White Cast Iron with pcBN Tools. *Wear* **2017**, *390–391*, 253–269. [CrossRef]
12. Gouveia, R.M.; Silva, F.; Reis, P.; Baptista, A.M. Machining Duplex Stainless Steel: Comparative Study Regarding End Mill Coated Tools. *Coatings* **2016**, *6*, 51. [CrossRef]
13. Ståhl, J.-E. *Metal Cutting—Theories and Models*, 1st ed.; Division of Production and Materials Engineering, Lund University: Lund, Sweden, 2012.

14. Buchkremer, S.; Klocke, F.; Veselovac, D. 3D FEM simulation of chip breakage in metal cutting. *Int. J. Adv. Manuf. Tech.* **2016**, *82*, 645–661. [CrossRef]
15. Fang, X.D.; Fei, J.; Jawahir, I.S. A hybrid algorithm for predicting chip form/chip breakability in machining. *Int. J. Mach. Tool Manu.* **1996**, *36*, 1093–1107. [CrossRef]
16. Du, H.; Karasev, A.; Stavlid, N.; Björk, T.; Lövquist, S.; Jönsson, P.G. Using chip weight distribution as a method to define chip breakability during machining. *Procedia Manuf.* **2018**, *25*, 309–315. [CrossRef]
17. Kara, F.; Aslantaş, K.; Çiçek, A. Prediction of cutting temperature in orthogonal machining of AISI 316L using artificial neural network. *Appl. Soft Comput.* **2016**, *38*, 64–74. [CrossRef]
18. Monkova, K.; Monka, P.P.; Sekerakova, A.; Hruzik, L.; Burecek, A.; Urban, M. Comparative Study of Chip Formation in Orthogonal and Oblique Slow-Rate Machining of EN 16MnCr5 Steel. *Metals* **2019**, *9*, 698. [CrossRef]
19. Du, H.; Karasev, A.; Sundqvist, O.; Jönsson, P.G. Modification of Non-Metallic Inclusions in Stainless Steel by Addition of CaSi. *Metals* **2019**, *9*, 74. [CrossRef]
20. Gove, K.B.; Charles, J.A. The high-temperature hardness of various phases in steel. *Met. Tech.* **1974**, *1*, 279–283. [CrossRef]

MDPI
St. Alban-Anlage 66
4052 Basel
Switzerland
Tel. +41 61 683 77 34
Fax +41 61 302 89 18
www.mdpi.com

Metals Editorial Office
E-mail: metals@mdpi.com
www.mdpi.com/journal/metals

www.ingramcontent.com/pod-product-compliance
Lightning Source LLC
LaVergne TN
LVHW070147170726
843515LV00007B/1867

* 9 7 8 3 0 3 6 5 2 4 9 4 8 *